Praise for *Prof*

"Thomas Povey's enthusiasm for his subject is infectious, and any student who reads the introduction and initial chapter will fully understand why this book has the potential to change their life. It will keep an aspiring mathematician, physicist, or engineer (and their teachers) happy for months."

—Mike Nicholson, Director of Student Recruitment and Admissions, University of Bath, Director of Undergraduate Admissions and Outreach, University of Oxford (2006-14)

"This book is about the excitement of mathematics and physics at the school-university boundary. It contains a remarkable collection of intriguing intellectual challenges, all with complete solutions. You can read it to increase your ability in maths and physics; you can read it to develop your problem-solving skills; but most of all you can read it for fun. Professor Povey has an eye for entertaining puzzles and a gift for making even complex concepts easy to understand."

—Professor Richard Prager, Head of Cambridge University School of Technology

"This instructive book will allow aspiring young scientists to test and expand their knowledge of physics and mathematics, and gain greater confidence in their abilities. It will also entertain and inspire them."

—Professor Sir Chris Llewellyn Smith FRS, Director General of CERN (1994-98)

"By challenging students with unfamiliar but engaging problems, I hope Professor Povey inspires many more budding physicists and mathematicians to realise their potential."

—Sir Peter Lampl, Founder and Chairman of the Sutton Trust

"A real tour de force of scientific problems. Thomas Povey is clearly a master at this kind of conundrum ... Fascinating."

—John Farndon, author of *Do You Think You're Clever?*

"Albert Einstein is rumoured to have said: 'Everything should be made as simple as possible, but not simpler.' Thomas Povey's lucid and imaginative *Professor Povey's Perplexing Problems* shows you how."

—Andrew Robinson, Author of *Einstein: A Hundred Years of Relativity*

"This book will amuse those who like puzzles. The topics vary from the practical to the quirky. Readers with a basic knowledge of maths and physics will be entertained and encouraged by the interesting and challenging questions they are able to solve."

—Professor Dame Ann Dowling DBE FRS FREng, President of the Royal Academy of Engineering

Professor Povey's Perplexing Problems

Pre-University Physics and Maths Puzzles with Solutions

Thomas Povey

A Oneworld Book

First published in North America, Great Britain and Australia by

Oneworld Publications, 2015

int t ti in 2015

A CIP record for this title is available from the British Library

ISBN 978-1-78074-775-0

Printed and bound in Great Britain by Clays Ltd, St Ives plc

Oneworld Publications

10 Bloomsbury Street, London WC1B 3SR, England

To Mike Leask and Terry Jones—brilliant teachers, insatiable adventurers.
With thanks to Tet Amaya, for substantial contributions throughout.

Contents

Introduction

This is a personal book which celebrates a passion I have always had for playful problems in physics and maths. Questions that rely on material no harder than that studied in high school, but which encourage original thinking, and stretch us in new and interesting directions.

In practically every country that has an established higher education system, the most competitive universities operate some kind of admissions test or interview. These generally have only one object: to distinguish from many brilliant applicants those with the greatest potential. To test this potential, questions are often designed to test an applicant's ability to think creatively. In many countries, these pre-university tests are set by university professors. Whilst the questions might reference a particular high-school syllabus, they would not normally be constrained by it. The questions are often deliberately off-beat. This is to see if applicants can apply what they have already learnt in new and more challenging situations. The questions test the applicants' ability to pick an unusual problem apart, reduce it to its essential components, and use standard tools to solve it.

This book is a collection of some of my favourite pre-university problems in physics and maths. These questions are devised to encourage curiosity and playfulness, and many are of the standard expected in some university entrance tests. These questions are perplexing, puzzling, but—most of all—fun. You should regard them like *toys.* Pick up the one that most appeals, and play with it. When you have exhausted it, you can entertain a friend with it. It might seem impossibly nerdy, but for me almost nothing is as enjoyable as being baffled by an apparently simple problem in an area of classical physics I *thought* I understood. This book is a way of sharing the pleasure I have taken in some of the physics and maths puzzles I have most enjoyed. The questions will appeal to those high-school students who have mastered the basics, and feel they have room to play a little with something more unusual. Teachers can also use the problems to stretch their pupils with something more challenging or unconventional. Here I am grateful to those teachers who allowed their classes to pilot these problems, and to the students who bravely attempted them.

I invented many of the questions myself. Others were suggested or inspired by friends and colleagues with an interest in this book. Still more are well-known classics. Even working outside a syllabus however, it is hard to be entirely original with questions in physics and maths. Although these questions can be asked in essentially unlimited ways, the number of fundamental concepts underpinning them is relatively small. Even the ones I am most proud of will no doubt have been asked in numerous other forms over the years.

There was one problem I was unable to solve. Many of these questions—especially the harder ones—work best as *discussion problems*. They are not meant to be asked and answered straight. Instead they should be discussed as part of a tutorial-style conversation, with an experienced tutor helping the student along, giving hints, correcting mistakes, providing encouragement, and so on. There is no real substitute for this kind of help. A question which might seem impenetrable can suddenly yield when a tutor provides just the right hint. I would encourage you to try the problems with friends, or subject tutors, or anyone with a sound knowledge of physics and maths.

I should say a few words about the level of difficulty of the problems. Every single one of them is challenging in some way. The publisher asked me to use stars to indicate their relative difficulty. Although I think this is quite subjective, I had to concede that it would be a shame for someone to happen to try all the most difficult problems first and feel defeated. So I have indicated how hard *I* found the problems the first time I tried them. If you think I have rated some of them too generously, it simply means your intuition is better than mine in that particular area. In general, those questions with one or two stars should yield to a student who is on top of the pre-university high-school syllabus[1] in physics and maths. The following descriptive gradings should give you some idea of what to expect.

> ★ Not too difficult. Requires thought, and some insight, but should be soluble without hints.
>
> ★★ Challenging. Requires considerable thought, and some insight, and may require simple hints.
>
> ★★★ Hard. Requires significant thought, and considerable insight, and more substantial hints. In some cases, questions with three stars are only suitable as discussion problems.

I have also included a handful of exceptionally hard questions, to give the most confident students an opportunity for a real challenge.

> ★★★★ Exceptionally hard. Suitable for discussion. May be too difficult for most students to solve without substantial help. Complex and unusual, these questions allow for different and non-standard approaches.

[1]Here I mean a student who has completed the first year of A-level, International Baccalaureate, Scottish Highers, SAT, etc., and is about a year away from entering higher education.

This book has been what I call my "Saturday project", and there are, I increasingly discover, always fewer Saturdays than I hope there will be. I have not had the luxury of someone to work with, so there will certainly be mistakes, and errors of logic. They are all my fault. But I hope they are few enough that this book brings more enlightenment than confusion. The most challenging aspect of putting it together was striking the right balance with the difficulty of questions. The questions range from being quite straightforward to really quite hard—they are as unpredictable as many university entrance tests. Here I am indebted to the many teachers, students and colleagues who were kind enough to provide detailed comments on the difficulty of the questions. I would be delighted to receive further constructive notes and any genuinely unusual questions which could improve a future edition of this book, and thank in advance anyone willing to provide this feedback.[2] I hope you will find friends you can share the fun of these questions with, and argue the case for improved or more elegant solutions (many of which exist). To share your solutions with a wider audience, and to learn from the ideas of others, you can contribute to the Puzzle Forum at

www.PerplexingProblems.com

If you are feeling brave, you can also play The Deadly Game of Puzzle Points, which is explained in the final chapter of this book. I've provided scoresheets and ticklists so you can keep track of your progress.

Finally, good luck, and have fun.

T.P., Oxford, 2015

---

[2] You can contact me by emailing tom@tompovey.com

Acknowledgements

I am indebted to all the teachers, headteachers, students, parents of students, and colleagues who provided many useful comments, particularly on the difficulty and range of questions that were eventually included in this book. Among those who were kind enough to provide feedback were: Oscar Van Nooijen, Brian Smith, Terry Jones, Frances Burge, Julie Summers, Robin Parmiter, Elaine Cook, Georgina Allan and Glenn Black. With thanks to Ben Sumner for copyediting the manuscript.

For errata, I would like to thank: Alex Bellos, Robin Chapman, Jonathan Gray, Ezra Webb, Marco Cheng, Gregory Lazarakos.

T.P., Dec. 2015

An odd journey

There was no early evidence that I was destined to be a physicist or engineer. I can't even say retrospectively whether my interest in playful physics problems—celebrated in this book—stems from judicious intervention by parents or a divinity to save a wayward child, or is simply the result of a series of accidents. I can explain some of it, at least, by the things I enjoyed and the people I met. Here is how it happened.

Early days

Though untrained, my mother was an engineer of some calibre, and would have taken a degree in the subject were it not for the sexism of the 1970s, which made the climate insufferable for her. That, and her interest in education, must certainly have had some influence on me, but as a child I was neither a particularly gifted nor hard-working. I was, in modern parlance, a late developer.

It is said that from an early age I possessed some practical ingenuity, however. Some of the first manifestations of this were my attempts to storm the barricades of the large wooden playpen my parents had installed in our living room. Each morning they would erect the playpen around my brother, who is a year older than me. It was there not to prevent him escaping, however, but to stop me attacking him. He spent most of his early years in that playpen, reading maths books voraciously, performing calculations on an abacus, and placidly ignoring my ever more sophisticated attempts to lay siege to his fortress. Even at the age of three, my brother was both gifted and hard-working.

Tree-climbing and fireworks

When I was between the ages of about five and twelve we lived on a small tropical island off the coast of Venezuela. It was here that I had my first experiences as an engineer and a chemist. During those years, pretty much all my pocket money was spent on pulleys, potassium nitrate, and electrical wire.

My brother and I used the pulleys to ascend ever larger trees, with complex rope systems that we used to haul ourselves to the top of the rainforest canopy. When it was getting dark, our father would come out into the jungle to find us, following the sounds of voices drifting down from the treetops as much as a hundred feet above him. I would descend on a little wooden seat I had made, letting out three hundred feet of rope hand over hand through a three-way pulley system. Then I would haul the seat all the way back up for my brother to descend. It is remarkable that we survived this home-made contraption.

When we had exhausted all the tree-climbing possibilities, I developed an interest in fireworks. I started with innocuous things made from match heads and tin foil, which made a satisfying pop that delighted dinner guests of a certain persuasion. Later I progressed to small, but ear-splitting devices made from recycled umbrella tubes. It is probably best not to say too much about the design of these devices, but they were tremendously effective, and could be heard exploding from over a mile away. There was no perfectly safe way of constructing them, but even at the age of seven I had worked out that meticulous preparation, cleaning and shock avoidance were the only things that prevented accidents, such as fingers being blown off—or worse. I transported them to the detonation site in a box lined with cotton wool. The world was a different place back then, and we got away with things that nowadays are highly illegal in most countries. It's also a miracle that we did not sustain serious injuries.

When I started building larger fireworks, the process of shaving off enough match heads became rather time consuming. I also realised that I was in serious danger of losing entire limbs if I accidentally dropped one of them. It was safer, I realised, to move to mixtures that could not so easily be ignited by percussion. And so it was that a deep interest in chemistry was born.

Fortunately my mother was teaching A-level chemistry at the time, so I had a ready supply of reference books. By the age of eight I was an expert on every common oxidizing agent and easily oxidizable substance. Soon I was impressing my mother with technical words like *brisance*, *deflagration*, and *detonation*.[1] Fortunately, my parents took a fairly enlightened view of my education, but my burgeoning interest in science must have brought them as much concern as it did relief. At least, they must have thought, there was a possibility that my enthusiasm might translate into an interest in studying.

Somehow I managed to persuade a local pharmacist to import half-pound bags of potassium nitrate, to be used, I said, for preserving corned-beef. I bought sulphur from another pharmacist, who never thought to ask what it was for. With the larger fireworks came the problem of safe detonation. I did this using electrical initiation from a safe distance. Six months' pocket money bought me several hundred feet of electrical cable—quite an investment. In no time at all, I saw to it that our land was heavily cratered, and strewn with Roman candle tubes and casings. My interest in fireworks was entirely theatrical

[1] *Brisance* is a measure of the detonation pressure of an explosive. *Deflagration* describes the relatively slow, subsonic combustion process that occurs through heat transfer in low explosive mixtures with slow decomposition rates, whereby successive layers of material are heated and ignited by adjacent material. *Detonation* describes the process of very rapid ignition of high explosive mixtures by the propagation of a shock wave at supersonic speeds through the mixture, driven by a rapid exothermic reaction.

in nature, but I expect that even in such a remote location the police would have taken an active interest in my pyrotechnic activities, had the political backdrop not given them something rather more serious to deal with. Only a year earlier Grenada had been heavily bombed during a US invasion. This followed the bloody overthrow of the government by Marxists in an internal coup. For quite some time after the invasion the island was littered with burnt out tanks and helicopter gunships. I remember spending several days on one of our favourite beaches freediving with my brother into the remains of a helicopter. We brought up unspent anti-tank shells which we handed over to US marines. In return they gave us chilled root beer, which at the time seemed very exotic. It is amazing what entertainment children can consider to be normal.

I think it is probably fortunate that I gave up firework-making before I had a chance to try anything more ambitious. My firework career ended fairly abruptly when I set off a small experimental device with unexpected consequences. I sometimes liked to entertain dinner-party guests with small table-top explosions that provided a satisfying diversion towards the end of the night. On one such occasion I proudly brought out a little masterpiece, mounted on a chopping board. The display was supposed to be a miniature choreography of fire, noise and smoke. The fuse, which was made of match-heads ground in water then dried, fizzed along the chopping board perfectly, and at the correct moment a charge ignited, sending a powerful cloud of smoke into the air. Next, a series of coloured fireworks were set off, followed by the finale: a bang. However, I'd miscalculated the bang somewhat. It was so loud it knocked all the spectators backwards away from the table, into a heap of chairs and cutlery on the floor. One particularly sensitive guest ran screaming from the house. I am told she never visited us again.

Puzzle-boxes, computers and kites

We moved back to England when I was in my first year of secondary school. I quickly discovered that rain and confined gardens provided less scope for outdoor interests, so I took to puzzle making, while my brother turned his attention to computers.

When most people think of puzzles, they have in mind jigsaw puzzles, or cheap newsstand magazines. I was interested in mechanical puzzles. I think my initial inspiration was a half-remembered story my father told me, about a wooden replica book-case his aunt had owned. This carved object, about the size of a small loaf of bread, was apparently impenetrable. A small opening could be found by manipulating one of the carved books. This revealed a key. Another sliding panel revealed a keyhole. Eventually, a larger secret compartment was opened. Or so the story went.

I set about learning carpentry, and became reasonably proficient with a fairly modest set of hand tools. It was still before the days of the internet, and puzzle making was a sufficiently niche and fragmented activity that, despite my best efforts in the local library, I turned up almost nothing on the history of the subject. In retrospect, this was a huge advantage, because it allowed me to devise entirely unique designs free of the influence of traditions that had developed around the world. I would later learn that the genre that had captivated me was called the *secret box puzzle*,[2] a sub-classification of the *opening puzzle*.[3]

Over the next few years I spent thousands of hours in the garage, perfecting designs for ever more elaborate puzzle boxes. Their design and manufacture required considerable care. Once closed, the only way in was to use the perfectly operating internal mechanism, or a sledgehammer. My most complex puzzle box took the form of a footstool. Two sides were decorated with gilded resin-cast dragons, their open mouths leading into the box. The other two sides had four wooden drawers, carved with hieroglyphic reliefs. Once decoded, the hieroglyphs gave instructions for moving the drawers in a particular sequence. The final move caused two metal balls to drop into the mouths of the dragons, activating a second mechanism. And so on, with five mechanisms in total. One false move set the entire puzzle back to the beginning, frustrating the would-be puzzlist.

Years later, when I was seventeen, I happened across a puzzle box in the antique market on Portobello Road. I couldn't afford to buy it, but I came away with something much more valuable, the phone number of the most prolific collector of mechanical puzzles in human history. I called him and invited myself to visit his collection of mechanical puzzles—just over 150,000 of them. I am pleased to say that my puzzles passed muster even under his critical eye. From then on I had a standing invitation to an annual party which brought together leading puzzle collectors, puzzle makers and recreational mathematicians from around the world.

Whilst I was making puzzles, my brother was learning BASIC[4] programming. It was the early days of home computing, and our mother had decided to become a programmer. She had bought an Amstrad CPC6128[5] computer, with 128 KB of RAM, and a built-in BASIC interpreter. This almost certainly influenced my brother in his career path—he is now a programmer and a researcher in the field of speech recognition. Between the ages of about thirteen and sixteen, he spent most of his time writing computer games. When he wasn't programming he would be making electronic gadgets. He was a solitary worker, and I was rarely allowed to interfere with his inventions. Once I

[2] James Dalgety, a friend, a world authority on the mechanical puzzle, and the most prolific collector of puzzles in human history, uses the following eleven classifications to organise his puzzles: interlocking puzzles; jigsaw puzzles; assembly puzzles (non-interlocking); pattern puzzles; opening puzzles; disentanglement puzzles; route-finding puzzles; sequential movement puzzles; folding and hinged puzzles; dexterity puzzles; and jugs and vessels.

[3] Much later I found that opening puzzles had been made in Hakone in Japan, and Sorrento in Italy, since the early 19th century. Interestingly, the design of those puzzles hadn't changed much in two hundred years.

[4] BASIC, or Beginner's All-purpose Symbolic Instruction Code, is a programming language that was popular on home computers in the 1980s.

[5] This was the era of the first home machines—CPC stood for Colour Personal Computer.

went into his electronics lab, a chaotic tangle at the back of the shed, to find him crouched over a series of three-inch nails he had hammered right through the wooden counter top. Touching each nail in turn, he would say "ouch", and withdraw his hand with a grimace. Written in marker pen under each nail were numbers between 0 and 110. He said that he was setting up an improved power supply so he didn't have to switch between transformers. The nails offered direct connections to anything up to 110 volts, driven from transformers hidden underneath. The only problem, he explained, were the shocks he got when connecting wires. Fortunately he was never seriously hurt by the apparatus.

Just before my brother left for university, we developed a common interest in kites and rockets. The rocketry was a fairly benign activity involving kits bought from a local model shop. But it probably sparked my later—and rather more serious—interest in jet and rocket propulsion systems. The kite flying was a little more experimental—we became interested in ram-air single-line kites, and parafoil kites. We came up with all manner of curious designs. My brother became particularly obsessed with designs that resembled oblate hemispheres, with a ram intake at the front and stabilising fins at the sides. The ideal material would have been polyurethane-coated nylon, due to its low porosity and light weight. We were unable to source this, however, so made our own material by dipping lightweight silk into wax. We then cut out panels according to our paper designs, and used an electric sewing machine to assemble them. For several months the house was awash with kite-making equipment.

It is fair to say that our most innovative designs were outright failures. The fabric was so heavy that many of the kites stood little chance of flying in anything less than a gale. Neither of us was old enough to drive, but we persuaded our friend Arthur to assist in field trials on a disused Second World War airfield. First, we laid the kite out carefully along the runway. Then, with my brother holding the kite aloft, me at the controls and sitting in the open boot of the car facing backwards, and Arthur at the wheel, we would slowly drive off down the runway until we were going as fast as my brother could run. Then I would give the controls a gentle tug, and Arthur would floor the accelerator pedal in the hope of getting the kite airborne. If we were lucky the kites would fly, but only just.

Later, we switched to parafoil kite stacks. These were more reliable, and far more exciting to fly because they had terrific lift. On one occasion the wind picked up quickly, and became so strong that Arthur insisted I tie him to the tow-bar of the car to prevent him taking off. I soon had him safely secured with a length of hemp rope, looped around his waist. It turned out that Arthur was stronger than the rope, and in a sudden gust the rope parted and he was lifted into the heavens. When his feet were about six feet off the ground he let go of the kite and came to earth with a bump. He dislocated his left shoulder, and had a small but impressively bloody gash above his eye. I packed up the kites and we set off for Scarborough General Hospital with Arthur's left arm in a sling. It was an hour away, and by the end of the journey we were a well-practised team: Arthur operating the pedals and steering, and me on the gears and handbrake. In the casualty room, the ward sister met us with great amusement, saying she would have to find an anaesthetist before she could stitch up Arthur's forehead. The source of her amusement was Arthur, who at that time was the consultant anaesthetist at Scarborough General Hospital.

An interview

On a crisp and sunny morning in December 1996, I was standing in a rather echoey stairwell in St Catherine's College, Oxford. On the duck-egg-blue door, in inch-high letters, was the inscription "Dr M.J.M. Leask". I had come for a physics interview. In a few moments, the door would open, allowing me for the first time into that strange, wonderful, privileged and hugely demanding world of intellectuals and scientists. But for the time being I stood there in my jeans and a bright Christmas jumper, an old one I had inherited from my grandmother, staring down at my battered green Nikes and the most highly polished linoleum I had ever seen. In those days my hair was below my shoulders. I must have been a slightly odd sight.

I can't remember exactly what I was thinking, but I must have been wondering what on Earth I was doing there. It wasn't that I was nervous. In a strange way I was enjoying myself. The thing was that I felt I had no realistic chance of "getting in", so had determined to enjoy my time in Oxford. After all, I reasoned, I might never visit this peculiar place again. The last thing on my mind was that I might be offered a place. And I had not even considered that I might not only take a degree in physics at Oxford, but then a doctorate in engineering, before becoming a tutorial fellow, or *tutor*.[6] Very few people from my school (a perfectly respectable comprehensive in the North of England) applied to Oxford, and I suspect this had coloured my judgement. Another thing affecting my chances at that precise moment was the fact that the day before I had completely bombed the admissions test. I had found it simply impossible. Almost every question on the paper had the strange variable "i" in it. Or so I thought. In fact "i" was simply the unit imaginary number. Any student who had studied AS Further Mathematics, or who had completed most standard high-school mathematics courses, would have covered some topics that use *complex notation*. At the time I had not.

[6] At Oxford and Cambridge, the tutorial fellows (those who lecture in the departments and take tutorials in the colleges) are most commonly referred to simply as *tutors*, as indeed is anyone with a similar teaching role. The terminology is clearly confusing, however. My mother calls me a *don* (I think with an element of humour), a word which went out of circulation in about the 1960s, and which some now regard affectionately but others find obnoxious. That strange word was apparently particular to Oxford, Cambridge and the Catholic Priesthood. According to the Oxford English Dictionary, which ought to have an authoritative word or two to say on the subject, "In the colloquial language of the English universities: A [don is a] head, fellow or tutor of a college." Not to be confused with the second entry for the same word: "(A respectful name for) a high-ranking or powerful member of the Mafia." Gulp!

So when the door opened, I entered with intrigue rather than expectation.

Inside was a sun-filled room lined with bookcases and whiteboards, and full of scientific instruments and curiosities. A small and obviously energetic man with a shock of grey hair was sitting legs akimbo in an egg-chair, holding—to my horror—my test paper. The very first thing he said to me, indeed the very first thing any tutor said to me, was, "My dear boy, you have done disastrously in the test. What on earth happened?" He emphasised the word "disastrously", but addressed the question to me as though it was some fascinating but abstract matter that we might ponder together until we solved it. There was nothing for it but complete honesty. I explained that I had been absolutely at sea during the test, and had only realised after finishing that the letter "*i*" represented an imaginary number.

"How infinitely[7] amusing," said the tutor, tipping his head back and waving his legs about in delight. He clearly *was* genuinely amused. Not by my lack of knowledge, I should stress, but at the humorous predicament of sitting a paper primarily on complex algebra without ever having studied the topic. The amusement didn't last long. In a gymnastic manoeuvre he was on his feet and at the whiteboard, brandishing a pen, and quickly sketching out an optics problem. An eye appeared, followed by a circle, which he explained represented a glass sphere, then finally a dot at the centre of the circle—a speck of dust. Then he turned around, eyes twinkling in evident delight at the thought of discussing the problem he had just set. "We'd better see if you know any physics," he said triumphantly. The energy in the room was palpable. It was impossible not to like this man, and I determined not to let him down.

Over the course of the next forty-five minutes I stood at the whiteboard trying to solve some of the most peculiar and difficult problems I had ever encountered. At one point the tutor leapt to the other side of the room and rummaged in a drawer, reappearing with a curious mirrored bowl that had a curved mirrored lid with a hole in it. A sort of clamshell. He dusted the mirrors with a handkerchief, assembled the clamshell on a side table, and dropped a tiny plastic pig, about the size of the last joint of your little finger and vivid pink, into the hole. A floating pig appeared, hovering just above the top of the instrument and looking out at us with tiny black eyes. Quite an impressive illusion. Had he not been kneeling on the floor at this point, I felt certain that the tutor would have been hopping with delight. He wanted me to explain how it worked. Then he asked a confusing question about high-voltage transmission lines, an elegant little question on the thermal expansion of solids

[7]Our year group became very fond indeed of Dr Leask, and enjoyed the fact that he spoke as he thought, always in mathematical language. The quote we most like to remember comes from the only trip I took with him outside Oxford. We were rock climbing at a place called Symonds Yat, in the Wye Valley. Dr Leask was a very long way up a vertical cliff precariously perched on his tiptoes with nothing securing him to the rock. A fall would certainly have been fatal. I was really quite worried and could not decide whether it was better to alert him to my concern, or to say nothing for fear of distracting him. I decided to say something. "Should you not put some protection in?" I asked. To which he immediately replied, "Don't distract me, my dear boy, I'm standing on the square root of nothing." The imagery was apt.

(illustrated by dropping a ball through a hypothetical plate that was heated or cooled), and finally a question about target shooting, in which the rifle was rotated about the horizontal axis.

I must have said many things that were wrong in that interview, and the tutor would ask me if I was sure, or gently contradict me with a "but that would mean that...", to see if I could work out why what I had said was incorrect. On the rare occasions I got something right, he would say "brilliant, absolutely brilliant!" And before I had time to catch my breath we would be on to the next question. Then, in a flash, the forty-five minutes were up. "Time is our enemy, my boy," he said, "time is our enemy." And with that, he shook my hand warmly, and I was back in the lobby, staring at my green Nikes on the polished linoleum and the words "Dr M.J.M. Leask" inscribed on the door.

It felt a little bit like the contents of my brain had been decanted, rearranged, and put back in. Rather stunned, I stood there trying to work out whether the encounter had really happened. I had expected an *Oxford experience*, but nothing like this.

I walked slowly to the train station to catch a train north. As I started to recall the interview blow by blow I realised that this tutor was without doubt the most enthusiastic, charming and engaging teacher I had ever met. If there was one person in the world I wanted to be taught by, it was him.[8] The train left Oxford station, and I was downhearted at the thought that in all probability our paths would never cross again.

Shortly after my interview I received the offer of a deferred place to read physics—I had decided to go travelling for a year before university. I assumed the offer was a clerical error, but wrote back immediately to confirm my acceptance, half-expecting to receive a reply explaining the mix-up. To my surprise I got a warm, handwritten letter by return of post, telling me that the tutors were looking forward to welcoming me in a year's time.

Engineering a treehouse

That year I went wherever job opportunities took me, making speculative applications by letter around the world. I spent six months in Connecticut in the US, working first as a lumberjack, then designing and building a treehouse for an artist. I got the treehouse commission based on my climbing skills, and—I presume—my enthusiasm for a project nobody else wanted to take on. Fortunately I was never asked to produce any engineering qualifications.

The treehouse was a significant undertaking. Sixteen feet square and suspended thirty feet above the ground, it required complex metalwork, which I subcontracted to a local blacksmith. We hoisted four sixteen-foot beams into place over a period of several days, using lifting equipment borrowed from

[8]Dr Leask taught me for three years, and was every bit as brilliant as I believed he would be. He had an ability to bring subjects to life with an insatiable curiosity and infectious enthusiasm I have never seen matched by anyone else.

the local crane-hire company. I had pre-rigged the hoists and installed the bracketry with the help of a tree surgeon. We needed a lot of people for the lift, but the treehouse project had caused something of a stir locally, so man-power proved easy to come by. We commanded the ground troops from the platform—no doubt with an imperious air—as each beam was positioned. I allow myself a measure of pride in the design of the treehouse, which could cope with the relative movement of the five trunks of the tree.[9] We achieved this by using a pivot and polished skid at opposing ends of each beam, which allowed the whole structure to deform without internal stress in high winds. Likewise, the joists and top deck were pinned only at one end, and retained at the other in sliding grooves.

I am pleased to say that my first engineering consultancy project was completed in time and on budget. To celebrate, we had dinner thirty feet in the air. I understand that the treehouse was later extended to include walls, a roof and a wood-burning stove, and served as guest quarters for quite a number of years.

Physics

Shortly after I came back to England I took the train down to Oxford to start my physics degree. There were eight people in my year group in college.[10] We had very different backgrounds, but did have one thing in common—we all really liked physics. Feeling slightly ashamed at having done so badly in the college entrance test, I had resolved to correct my shortcomings. During my year out, I had studied maths religiously after work. The strategy paid off—I was no longer an embarrassment, and just about managed to keep pace with my peers. We all found the first term quite hard. But the long evenings in the library were mitigated by camaraderie, and we—almost—began to enjoy them. Mike Leask was as inspirational as I had expected him to be. Tutorials were unscripted affairs, in which we were encouraged to ask questions on any topic we wanted. He clearly wanted to inspire in us a love of physics, and not just get us through exams. It worked.

Some of our most enjoyable evenings during those three years were our informal puzzle parties. The doyen of these occasions was our friend Tet, who was—and still is—an inexhaustible source of mathematical puzzles. The format was simple. We would turn up with a puzzle or two to share, and stay up till four or five in the morning arguing for alternative or improved solutions to them. The fun wasn't limited to mathematics however. Physics puzzles, mechanical puzzles and perception puzzles were equally acceptable, especially if they involved an experiment. I spent more than one evening tied up firmly with string whilst Tet convinced me I wasn't topologically entangled. Another evening was spent navigating neutrally buoyant helium balloons around the

[9] I had estimated the sway based on measurements I took on a windy day, using pieces of string which had been tied with slip knots.

[10] At Oxford, many individual colleges each take a small number of students.

room with just the power of static electricity. We cut a huge Möbius strip[11] down the middle, yielding a strip twice as long with two full twists. A whole evening was spent arguing about the fair division of cake. We realised that bagels could be cut in half as interconnected rings. We were all quite normal.

The parties were happenings. Tet and I were invited to talk about cake sharing on the BBC Radio 4 programme Home Truths,[12] bringing the topic to almost two million people. I went on to answer physics puzzles as a panellist on another BBC programme, Material World, and to present other Christmas radio specials answering listeners' questions on physics:

> You have just poured a cup of tea, and the doorbell rings. You like your tea as hot as possible. Is it best to put the milk in the tea before or after you return from the door?[13]
>
> When looking from the window of a moving car, why do the wheels of neighbouring cars sometimes appear stationary?[14]

Researching, scripting and presenting the shows was great fun, and led to a number of other radio appearances and puzzle specials in newspapers. Tet's puzzle parties had a lot to answer for. I suspect they played no small part in inspiring this book.

Jet engines and rockets

When I finished my degree I decided I wanted to do something in applied physics. I started a doctorate in propulsion. I was supervised by a physicist, Terry Jones, and a mathematician, Martin Oldfield. Having both defected to jet-engine research via hypersonics, they decided that as a fellow defector I was worth a punt. Together we designed and built huge experiments to test the performance of turbines for next-generation planes. Looking at the results for the first time was always very exciting. Following a particular breakthrough

[11] A one-sided surface which can be created by taking a long strip of material, introducing a rotation of one of the ends about the long-axis by 180 degrees, and joining the ends together. It was discovered by August Ferdinand Möbius in 1858, and has fascinating mathematical properties.

[12] A programme presented by the late, great, John Peel, which explored the eccentricities of domestic life. Peel was one of the most influential radio DJs in the UK between the early 1960s and his death in 2004.

[13] By putting the milk in before going to the door, the mean temperature difference driving heat out of the tea is reduced, and the tea is hotter (at the same time) than if you put the milk in when you return from the door.

[14] At night, this is generally due to the stroboscopic effect, and is caused by street-lights flickering on and off at a well-defined frequency (usually twice the mains frequency, equalling 100 Hz in the UK). During the day, an equivalent stroboscopic effect has been reported due to humming. In 1967, William Rushton (Rushton, W.A.H., 1967, "Effect of Humming on Vision," Nature 216, pp. 1173–1175) of the Cambridge Physiological Laboratory observed that "humming causes the eye to vibrate and this can produce a strobo-scopic effect when a rotating black and white strobe disk is viewed in non-fluctuating light." Some other authors failed to replicate his results, but reported similar effects due to "throat clearing" vibration. Under continuous daytime illumination, I have observed that I can get small parts of a car wheel to appear stationary by flicking my eyes from left to right, and vice versa—I included this personal experience in my Radio 4 explanation.

thanks to an experiment nobody else believed would work, Terry turned to me with a wink and summed it up. "It's criminal that we are paid to have this much fun." During those three years not a single moment seemed like *work*.

Ten years later I have my own research group. With brilliant students and collaborators I look at rocket-engine systems and optimising jet-engine components, and dabble with things that can be used by the oil and gas industries, and by ordinary consumers. Although we are in the lab late almost every night, it still doesn't feel like real work. And to see a plane take off for the first time with a new engine, or to watch a rocket blast into orbit carrying a critical system I've designed, is still heart-stoppingly exciting. The only thing that salves the conscience when a job is this much fun is keeping a stiff upper lip and reminding oneself that someone has to do it.

Your journey

You may read this book in the hope that it will prepare you better—either for university entrance tests, or for life at university when you arrive. I hope it will do both. Not by more thoroughly covering a syllabus, but by encouraging you to try harder, more unusual, and more open-ended problems, in a spirit of fun and adventure. The skills you will gain by doing so will serve you well for high-school exams, and university entrance tests, but more importantly for higher study, by inspiring a more independent approach to learning. However arbitrary the collection of problems in this book, they should test your ability to grapple with the unfamiliar. You will learn to tease new problems apart, and apply things you already know in ways you had never considered. You have all the tools you need, but you should see what amazing things you can do with them.

But usefulness was not the primary motivation for writing this book. First and foremost it is a celebration of all that is wonderful and fascinating about some of my favourite puzzles in physics and maths. If I have succeeded, it will be in giving you something you want to play with—puzzles designed for pleasure, not as a means to some lesser end. But I hope you will also discover that behind each of these puzzles lies something honest and purposeful, designed to allow you to explore the important concepts in more detail. You will see that the puzzles *need* to be off-beat so that we can challenge our understanding. To see if we come up trumps or fall flat on our face, but—more importantly—to take pleasure in both, because both have caused us to *think*. How, after all, can we test our understanding with a problem of a type we have already seen?

Whether you are a fourteen-year-old with an insatiable appetite for curious science, a seventeen-year-old preparing for university, or a physicist long-defected to another field, I hope you see the *fun* in these questions. I hope you will enjoy them for being curious, eccentric and peculiar, but I also hope you see the serious purpose behind each of them. I have been baffled by them. I have had the pleasure of sharing them with friends. And I have had the satisfaction of solving them. These puzzles now deserve a new home. They are

1.1 Shortest walk ★

AN ANT STARTS at one vertex of a solid cube with side of unity length. Calculate the distance of the shortest route the ant can take to the furthest vertex from the starting point. There is more than one solution to this problem.

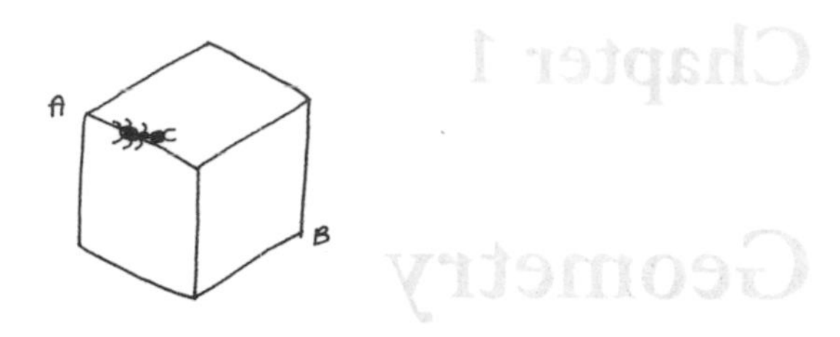

Answer

This is a well-known question, though not one with very much scope for discussion.

If we consider unwrapping any pair of sides that connect the starting vertex A to the furthest vertex from the starting vertex, B, it is clear that the unwrapped surface forms a rectangle with sides of length 1 and 2. The shortest distance is a straight line with length $\sqrt{5}$.

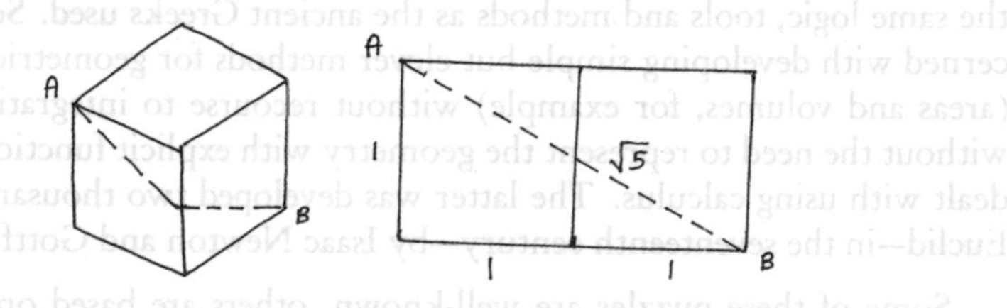

This solution is not unique. There are, in fact, six solutions of the same minimum length, which are shown in this diagram of the unwrapped cube.

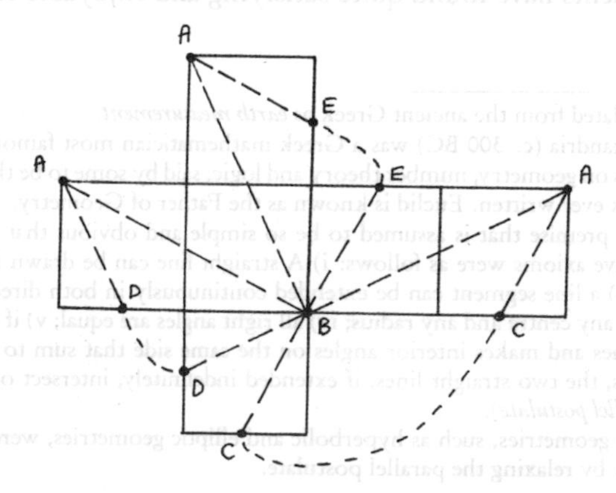

1.2 Intercontinental telephone cable ★

This rather neat little question is simple enough that it does not warrant much introduction. I have heard of a number of variants over the years.

A TELEPHONE COMPANY has run a very long telephone cable all the way round the middle of the Earth. Assuming the Earth to be a sphere, and without recourse to pen and paper, estimate how much additional cable would be required to raise the telephone cable to the top of 10 m tall telephone poles.

Answer

The circumference of the Earth is given by $C = 2\pi R_E$, where R_E is the radius of the Earth. The new circumference at the top of 10 m tall telephone poles is given by $C' = 2\pi\,(R_E + 10)$. The difference in the circumferences is $C' - C = 20\pi$. We should know that $\pi \approx 3.14$, so we should be able to work out in our heads that the additional length of cable is approximately equal to 62.8 m. The question is designed to be slightly tricky in that we might be expecting a big number. If we approach it rationally we see that questions don't come much easier than this!

There are, of course, intercontinental telephone cables, and the history of submarine cable-laying started not long after the invention of the telegraph. The first commercial telegraph was developed by William Cooke and Charles Wheatstone in 1838, and as early as 1840 Samuel Morse was working on the concept of submerged cables for transatlantic operation. Experiments under the waters of New York Harbour in 1842 confirmed the feasibility of such a project, and in 1850 the first telephone cable was laid under the English Channel (to France), by the Anglo-French Telegraph Company. Numerous other submerged telephone cables were laid across the English Channel, the Irish Sea and the North Sea in the years that followed, and the first attempt at a transatlantic cable was made in 1858. The first successful transatlantic project was in 1865, when undersea cables were laid by the SS Great Eastern, a huge iron steamship designed by the Bristolian Isambard Kingdom Brunel. Now there are numerous fibre-optic cables snaking across the Atlantic seabed.

1.3 Chessboard and hoop ★★

I invented this puzzle a couple of years ago. It is one of my favourites because students seem to enjoy it, and can generally make good progress without much help, although sometimes they need a hint to get started.

A THIN HOOP of diameter d is thrown on to an infinitely large chessboard with squares of side L. What is the chance of the hoop enclosing two colours?

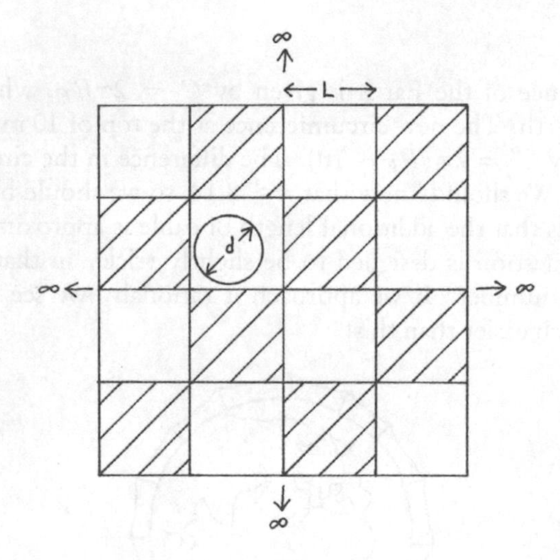

Answer

This question can be answered by applying very simple geometry, and should be accessible to anyone with a good grasp of GCSE-level maths. The first realisation is that if the hoop crosses any tile boundary it will enclose two colours. It is easier to first pose the reverse of the question asked—that is, "what is the chance of not enclosing two colours?" We then subtract this chance from the chance of *any* outcome, giving the answer to the question as posed. We will use notation familiar to students of *probability*, although only the most elementary understanding of that subject is required here. We use the notation $P(A)$ to denote the probability of the hoop enclosing two colours. By definition $0 \leq P(A) \leq 1$; that is, the probability lies somewhere between 0 (will not happen) and 1 (will definitely happen). We use $P(\bar{A})$ to denote the probability of the hoop *not* enclosing two colours. There are only two possible outcomes, so $P(\bar{A}) = 1 - P(A)$.

We are dealing with two-dimensional space, so we will probably need to define the landing *areas* which lead to one outcome or the other. If we were dealing with three-dimensional space we would need to define *volumes*. In one-dimensional space, we would need to define *distances*, and so on. We can define the location of the hoop unambiguously by referring to the position of the centre of the hoop. We could equally use the location of any identifiable point on the rim of the hoop, but then we would also need to express the angular orientation of the hoop. This introduces more complexity with no benefit.

We will now consider three cases: $d > L$; the special case $d = L$; and $d < L$. Let's consider each in turn.

- THE CASE OF $d > L$. In this case, in which the diameter is larger than the side of a tile, it is perfectly obvious that the hoop must enclose two colours. We see that $P(A) = 1$.

- THE SPECIAL CASE OF $d = L$. For the special case that $d = L$, it is possible for the hoop to land entirely within a single tile, but the chance of this happening is $1/\infty$. In this *limiting case* the circle touches the tile at four points. That is, a single solution out of infinitely many possible solutions satisfies the condition that only one colour is enclosed. We write $P(\bar{A}) = 0$, and therefore $P(A) = 1$.

- THE CASE OF $d < L$. Consider all the circles that can be drawn entirely within a single tile, and which touch the tile boundary either at one point (along an edge of the tile) or at two points (in the corner of the tile). For now we observe that the circles that touch the edges of the tile at either one or two points have centres which lie on a smaller square exactly in the middle of the larger tile. The *locus* of points is shown in the diagram. For a circle of diameter d, the shortest distance between the locus of points and the tile is $d/2$ everywhere. The side of the smaller square is therefore $L - d$.

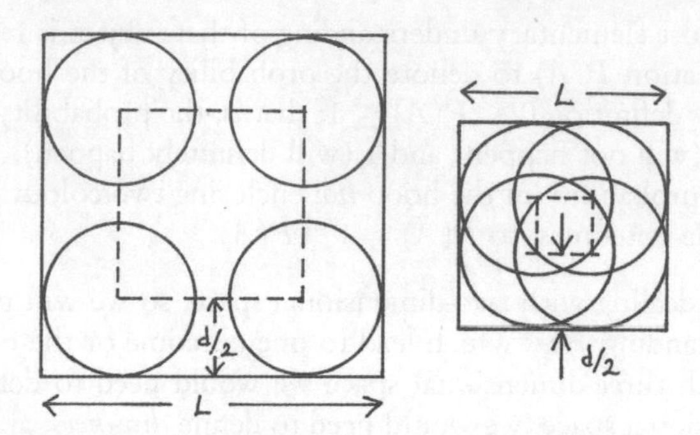

We now observe that any circle with a centre within the smaller square does not touch the tile boundaries. The locus of all points at which the hoop centre can be without touching the border of the tile is defined by the smaller square. The probability of the hoop centre landing in the smaller square is given by the ratio of the area of the smaller square to that of the larger square.

$$P(\bar{A}) = \frac{(L-d)^2}{L^2} \text{ for } d \leq L$$

So

$$P(A) = 1 - \frac{(L-d)^2}{L^2} \text{ for } d \leq L$$

1.4 Hexagonal tiles and hoop ★★

AN INFINITELY LARGE floor is tiled with regular hexagonal tiles of side L. Different colours of tiles are used so that no two tiles of the same colour touch. A hoop of diameter d is thrown onto the tiles. What is the chance of the hoop enclosing more than one colour?

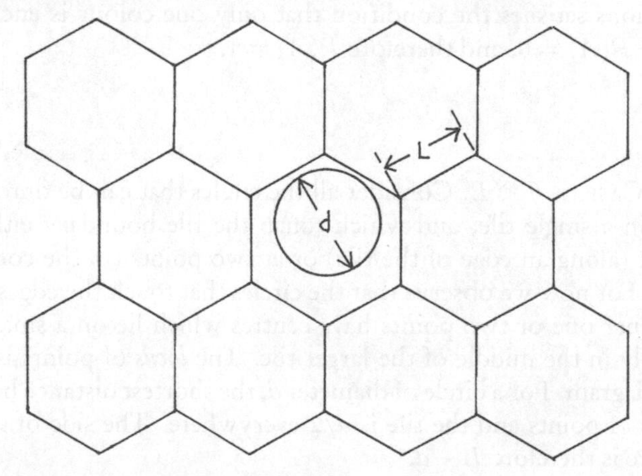

Answer

We can answer this question using a similar method to that used for the chessboard and hoop, but the geometry is a little more complex. We start by defining the locus of hoop centres that allows a hoop entirely enclosed within a tile to touch the edge of the tile at one position (edges) or two positions (corners). The locus of such points is defined by the line $CDEFGH$, which describes a smaller hexagon centred on the tile centre. The edges of the smaller hexagon are $d/2$ away from the edges of the large hexagon. That is, the perpendicular distance between AB and CD, for example, is $d/2$. The locus of all hoop centres for which the hoop does not intersect the edge of the tile is defined by the *area* $CDEFGH$.

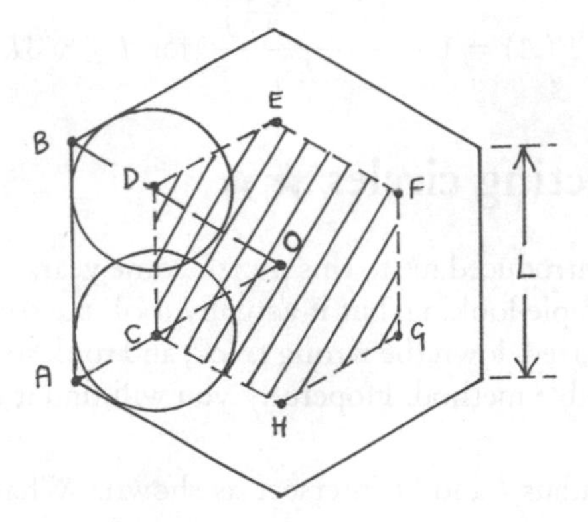

To calculate the area of the smaller hexagon we need the length of the side CD, which is denoted by $|CD|$. From the geometry of the regular hexagons we can see that $|CD| = |AB| - 2x = L - 2x$, where $x = (d/2)\tan 30° = d\left(\sqrt{3}/6\right)$. We have $|CD| = L - d\left(\sqrt{3}/3\right)$.

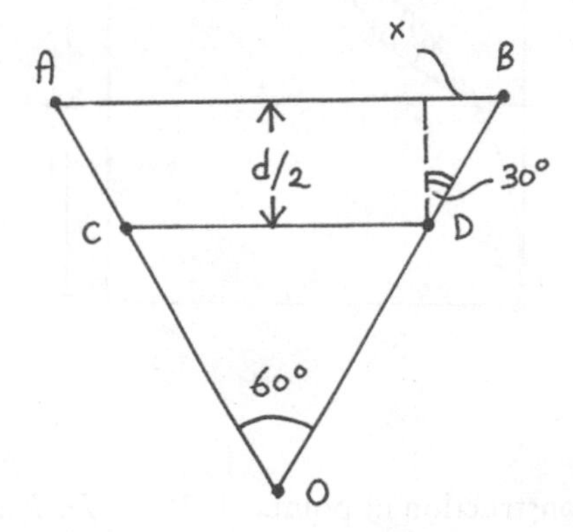

The chance of landing entirely within a tile, $P(\bar{A})$, is given by the ratio of the areas of the smaller tile and the larger tile. We do not need to calculate the areas explicitly, because we know they are of similar shape, so the area is proportional to the square of any *characteristic length*. For the characteristic length we could take the side length. The relevant lengths become $|CD|$ and $|AB|$. This gives

$$P(\bar{A}) = \frac{|CD|^2}{|AB|^2} = \frac{\left(L - \frac{d\sqrt{3}}{3}\right)^2}{L^2} \text{ for } d \leq \sqrt{3}L$$

The probability of enclosing more than one colour, $P(A)$, is given as $P(A) = 1 - P(\bar{A})$, or

$$P(A) = 1 - \frac{\left(L - \frac{d\sqrt{3}}{3}\right)^2}{L^2} \text{ for } d \leq \sqrt{3}L$$

1.5 Intersecting circles ★★

A friend of mine introduced me to this puzzle some years ago. I like it because it is deceptively simple-looking, but it actually took me some time to solve it. I think I simply started down the wrong track, and took some time to find my way back to a sensible method. Hopefully you will find it easier than I did.

Two circles of radius l and $2l$ intersect as shown. What is the area of the shaded region?

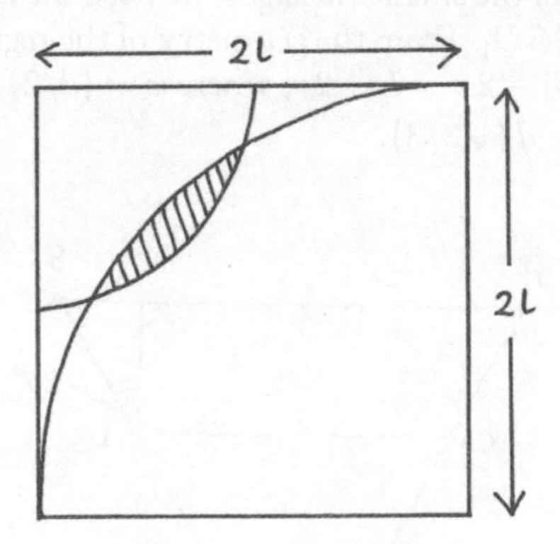

Answer

Consider first the construction of points A, B, C, D, E and F in the sketch below. We note first that the distance between A and C is equal to l, because AC forms a radius of the smaller circle. We write $|AC| = |AD| = l$. A similar statement leads to $|BC| = |BD| = 2l$.

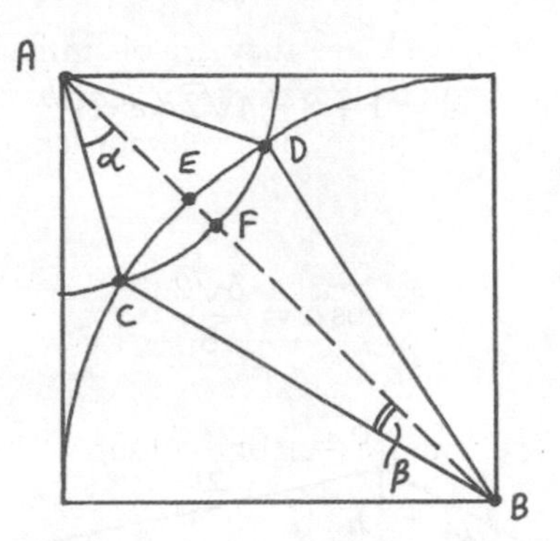

The area of the shaded element $CEDFC$ is equal to the area of sector $ADFCA$ plus the area of sector $BCEDB$ minus the area of the quadrilateral $BCADB$.

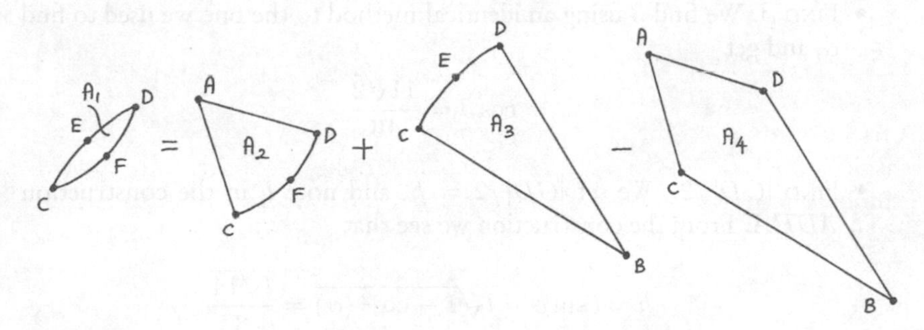

We represent these areas by A_1 to A_4 respectively, and write

$$A_1 = A_2 + A_3 - A_4$$

Expressing this explicitly we have

$$A_1 = l^2\alpha + 4l^2\beta - |AB| \frac{|CD|}{2}$$

To simplify this we require expressions for $|AB|$, $|CD|/2$, α and β. We now find each of these in turn.

- FIND $|AB|$. We see that AB is the longest diagonal of the square, and has length $|AB| = 2\sqrt{2}l$.
- FIND α. Using the cosine rule[5] on the triangle $ADBA$ we see that

$$(2l)^2 = l^2 + \left(2\sqrt{2}l\right)^2 - 2\,(l)\left(2\sqrt{2}l\right)\cos\alpha$$

Simplifying, we get

[5]Usually written in the form $c^2 = a^2 + b^2 - 2ab\cos C$, where a, b and c are the lengths of the sides of a triangle, and C is the angle between the sides of length a and b.

$$4 = 1 + 8 - 4\sqrt{2}\cos\alpha$$

That is

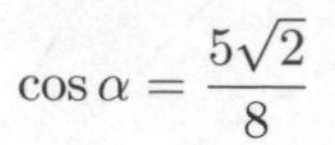

$$\cos\alpha = \frac{5\sqrt{2}}{8}$$

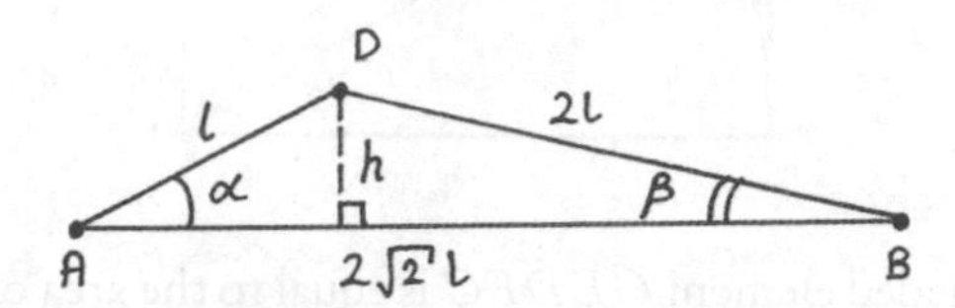

- FIND β. We find β using an identical method to the one we used to find α, and get

$$\cos\beta = \frac{11\sqrt{2}}{16}$$

- FIND $|CD|/2$. We set $|CD|/2 = h$, and note h in the construction $ADBA$. From the construction we see that

$$h = l\sin\alpha = l\sqrt{1 - \cos^2(\alpha)} = \frac{l\sqrt{14}}{8}$$

Now that we have expressions for $|AB|$, $|CD|/2$, α and β, we can evaluate A_1. We write

$$A_1 = l^2\left[\cos^{-1}\left(\frac{5\sqrt{2}}{8}\right) + 4\cos^{-1}\left(\frac{11\sqrt{2}}{16}\right) - \frac{\sqrt{7}}{2}\right]$$

This gives the area of the shaded region as

$$A_1 \approx 0.108\, l^2$$

My original method was considerably longer even than this! I like this problem because it is apparently trivial, but it takes some thought to solve with relatively compact algebra.

1.6 Cube within sphere ★

WHAT IS THE volume of the largest cube that fits entirely within a sphere of unity volume?

Answer

By symmetry, all eight vertices of the cube must touch the sphere. Again, by symmetry, the length of the longest diagonals of the cube, for example $|ab|$, must be equal to the diameter of the sphere D, because the line ab passes through the centre of the sphere.

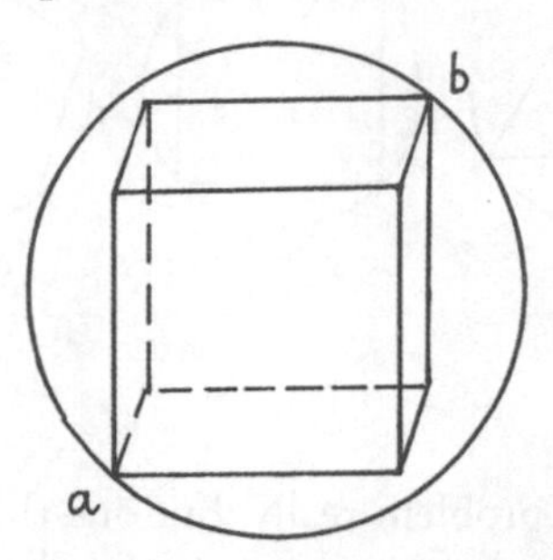

From the geometry of the cube, using Pythagoras' theorem

$$|ab|^2 = \left(\sqrt{2}L\right)^2 + L^2$$

where L is the length of the side of the cube. Thus

$$|ab| = \sqrt{3}L = D$$

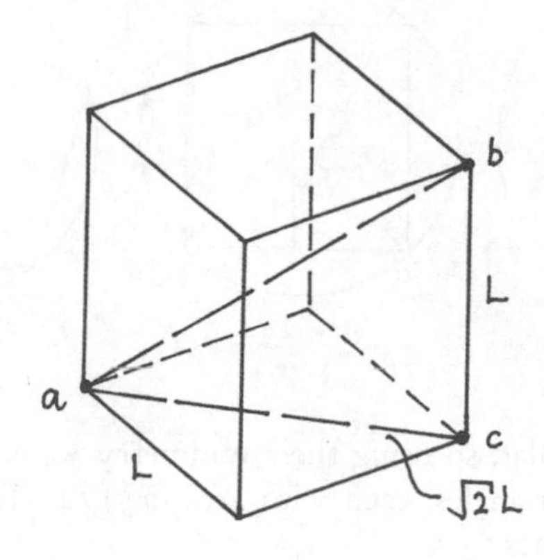

The volume of the cube is thus given by

$$V_C = L^3 = \frac{D^3}{3^{3/2}}$$

All that remains is to substitute for D^3 using the equation for the volume of the sphere, $V_S = (4/3)\,\pi r^3 = \pi D^3/6 = 1$. Rearranging $D^3 = 6/\pi$. The volume of the cube within the sphere is given by

$$V_C = \frac{2}{\pi\sqrt{3}} \approx 0.368$$

1.7 Polygon inscribed within circle ★

WHAT IS THE area of an n-sided regular polygon inscribed within a circle of radius r?

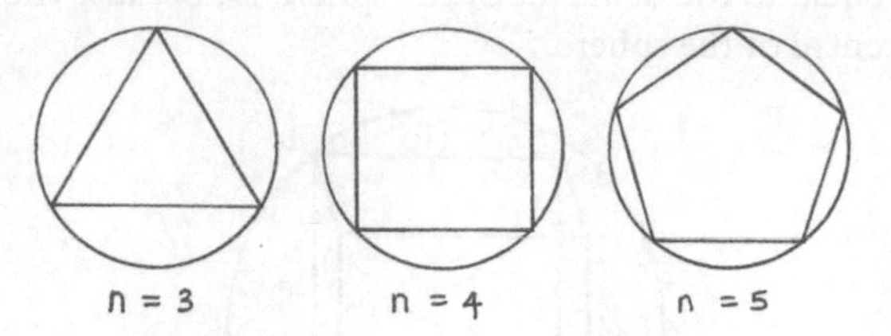

Answer

This is a straightforward problem really, but one that is a good test of a student's ability to derive simple formulae for unfamiliar situations. There is relatively little scope for practising this skill in most high-school courses, so I've included a number of problems of this type to develop this aspect of problem-solving.

Consider the constructions for $n = 3, 4$ and 5-sided polygons.

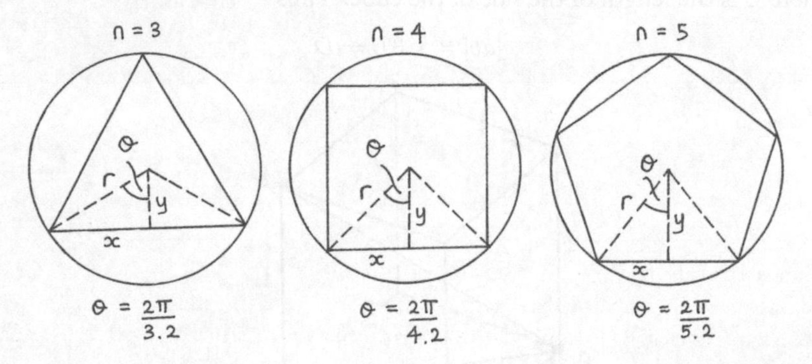

The polygons are regular, so using their symmetry we can decompose them into $2n$ right-angled triangles, each with area $(xy)/2$. The total area of the n-sided polygon becomes

$$A_n = 2n\frac{xy}{2} = nxy$$

Noting the angle θ in each of the constructions, we see that $\sin\theta = x/r$ and $\cos\theta = y/r$, where $\theta = (2\pi)/(2n) = \pi/n$. The area of the polygon now becomes

$$A_n = nr^2 \sin\frac{\pi}{n} \cos\frac{\pi}{n}$$

We can simplify this using the trigonometric identity $2\sin A\cos B = \sin(A+B) + \sin(A-B)$. We get

$$A_n = \frac{nr^2}{2} \sin \frac{2\pi}{n}$$

As a final note, we see that, for very large n, we can make the approximation $\sin(2\pi/n) \approx 2\pi/n$. The expression for area becomes

$$A_n \approx \pi r^2 \text{ (for large } n\text{)}$$

This is the equation for the area of a circle, as expected.

1.8 Circle inscribed within polygon ★★★

FOR A CIRCLE inscribed inside a regular n-sided polygon, what is the minimum n so that $A_{shaded}/A_{circle} \leq 1/1000$?

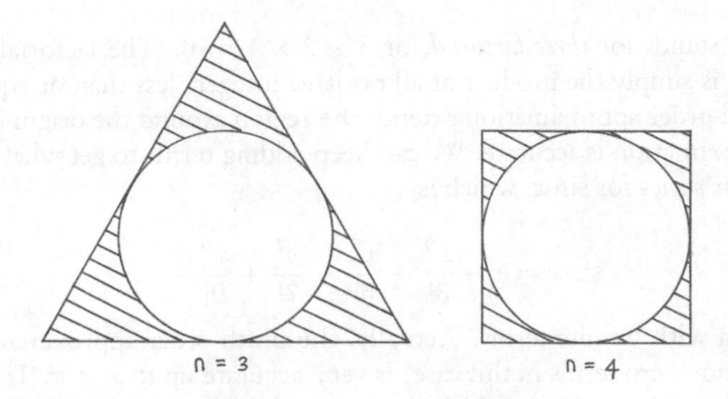

This is a variant on the previous question, but has the added problem that the final expression cannot be solved by simply rearranging the equation. You can use a number of techniques familiar to high-school students, however. And, unusually, the suggested methods are now given as hints.

One method is to use a graphical solution—that is, to find two expressions so that the intersection defines the solution. A second method is an iterative numerical technique—essentially trial and error of a thoughtful sort, in which you see if you are getting closer or further from the solution and correct your guess up or down accordingly until you *converge* on a solution. A third technique, and certainly the most elegant, is to use a *Taylor polynomial approximation* to the function. *Taylor series* are no longer taught in all high-school syllabuses but are certainly taught in some extension papers, and are covered in the first year of all university courses. Because the Taylor approximation is such an elegant technique we will say a few words about it here.

The first thing to say is that the concept of a Taylor series is very simple, so there is no need to be intimidated by it. It says we can represent a function about a particular point as an infinite polynomial series. If, for example, we wish to represent $\sin x$ around zero, we start with a linear (or *first-order*) approximation that only applies at very small angles, namely

$$\sin x \approx x \quad \text{for very small } x$$

This is an approximation that all high-school students should be familiar with,[6] and will use routinely. As we move away from zero the approximation starts to fail because, as we well know, the graph of $y = \sin x$ has curvature as we move away from the origin. To correct for the curvature, we need to add a higher-order term which itself is approximately zero when very close to the origin, so as not to harm our first-order approximation, but which gets larger (either positive or negative in sign) at a greater rate than x as we move away from the origin. We skip the second-order term because in this series it is zero, so the third-order approximation[7] is

$$\sin x \approx x - \frac{x^3}{3!}$$

where 3! stands for *three factorial*, or $3 \times 2 \times 1 = 6$. The factorial[8] of any integer n is simply the product of all positive integers less than or equal to n. The third-order approximation extends the region around the origin in which the approximation is accurate. We can keep adding terms to get what is called the Taylor series for $\sin x$, which is

$$\sin x = x - \frac{x^3}{3!} + \frac{x^5}{5!} - \frac{x^7}{7!} + \frac{x^9}{9!} - \ldots$$

and so on with infinite terms. Actually, the ninth-order approximation (the first five non-zero terms in this case) is very accurate up to $x = \pi$. Try it.

The general process by which the series approximation terms are derived for a particular function is due to Brook Taylor, a mathematician who worked on what later became known as *differential calculus*. He published this, his most famous result, in 1712.

---

[6]Here's a simple geometric proof. If we consider a thin sector of a circle of angle θ and chord length l, we see that (*in radians*) we have $l = r\theta$. The perpendicular distance h is given by $h = r \sin\theta$. We see from the geometry of the sketch that for very small θ, $h \approx l$, and therefore $\sin\theta \approx \theta$.

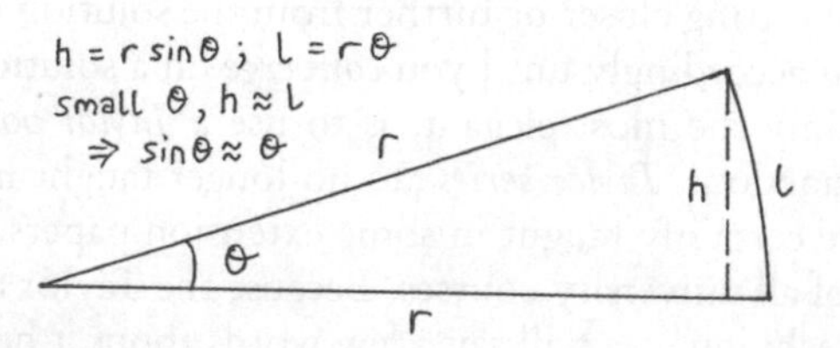

[7]By n-th order we mean including terms up to x^n.

[8]Zero factorial is equal to unity. The proof is as follows. Noting that $(n+1)! = (n+1) \times n \times \ldots \times 2 \times 1 = (n+1) \times n!$, and setting $n = 0$, gives $1! = 1 \times 0!$, so $0! = 1! = 1$.

The Taylor series for $\cos x$ and $\tan x$ are as follows

$$\begin{aligned}\cos x &= 1-\frac{x^2}{2!}+\frac{x^4}{4!}-\frac{x^6}{6!}+\frac{x^8}{8!}-...\\ \tan x &= x+\frac{1}{3}x^3+\frac{2}{15}x^5+\frac{17}{315}x^7+\frac{62}{2835}x^9-...\end{aligned}$$

Answer

Consider the constructions for $n = 3$ and 4-sided polygons.

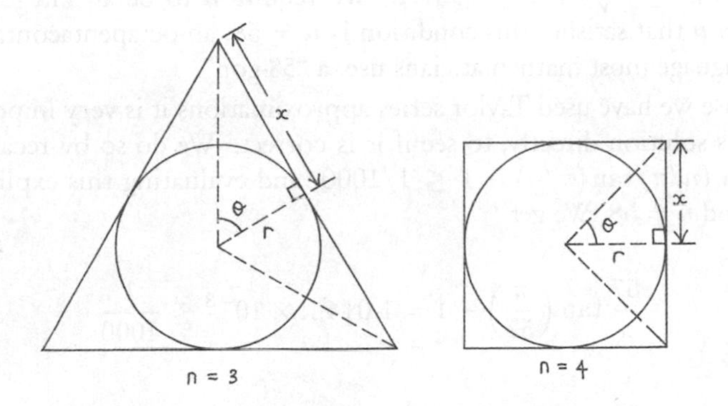

The polygons are regular, so using their symmetry we can decompose them into $2n$ right-angled triangles, each with area $(xr)/2$, where r is the radius of the circle, and x is the half-length of the side of the polygon, given by $x = r\tan\theta$, where $\theta = \pi/n$. The total area of the n-sided polygon becomes

$$A_n = nxr = nr^2 \tan\frac{\pi}{n}$$

Consider that

$$\frac{A_{shaded}}{A_{circle}} = \frac{A_n}{A_{circle}} - 1 = \frac{n}{\pi}\tan\frac{\pi}{n} - 1$$

So we require the minimum n that satisfies

$$\frac{n}{\pi}\tan\frac{\pi}{n} - 1 \leq \frac{1}{1000}$$

Here we must choose either a graphical, numerical (iterative) or series solution approach. Let us use the series solution, which is the most elegant, although trying the problem in other ways is also instructive. We need to approximate $\tan(\pi/n)$ using the Taylor series for $\tan x$. Taking a third-order approximation we have

$$\tan x \approx x + \frac{1}{3}x^3$$

Our inequality reduces to

$$\frac{n}{\pi}\left[\frac{\pi}{n}+\frac{1}{3}\left(\frac{\pi}{n}\right)^3\right]-1\leq\frac{1}{1000}$$

That is

$$\frac{1}{3}\left(\frac{\pi}{n}\right)^2\leq\frac{1}{1000}$$

This gives $n \geq \pi\sqrt{1000/3} = 57.4$. We require n to be an integer, so the minimum n that satisfies this condition is $n = 58$, an octapentacontagon, or, in the language most mathematicians use, a "58-gon".

Because we have used Taylor series approximations it is very important to check this solution directly, to see if it is correct. We do so by recalling the condition $(n/\pi)\tan(\pi/n) - 1 \leq 1/1000$, and evaluating this explicitly for $n = 57$ and $n = 58$. We get

$$\frac{57}{\pi}\tan\left(\frac{\pi}{57}\right)-1 = 1.014... \times 10^{-3} \not\leq \frac{1}{1000}$$

$$\frac{58}{\pi}\tan\left(\frac{\pi}{58}\right)-1 = 9.791... \times 10^{-4} \leq \frac{1}{1000}$$

We see that this agrees with our result.

1.9 Triangle inscribed within semicircle ★

This little problem is an exercise in using vectors.

GIVE A VECTOR *proof* that for a triangle inscribed within a semicircle, the included angle is always $\pi/2$.

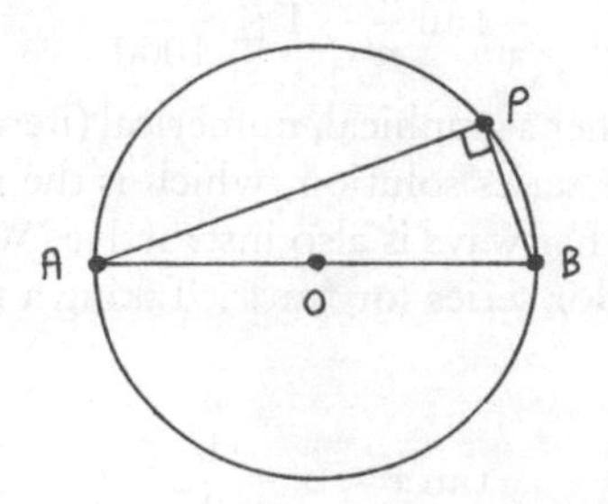

Answer

Consider the semicircle APB, formed by the intersection of a line AOB with a circle. From the centre of the circle, if B is represented by the vector[9] $\boldsymbol{a}$, then A is represented by the vector $-\boldsymbol{a}$. Let point P be represented by the vector $\boldsymbol{p}$.

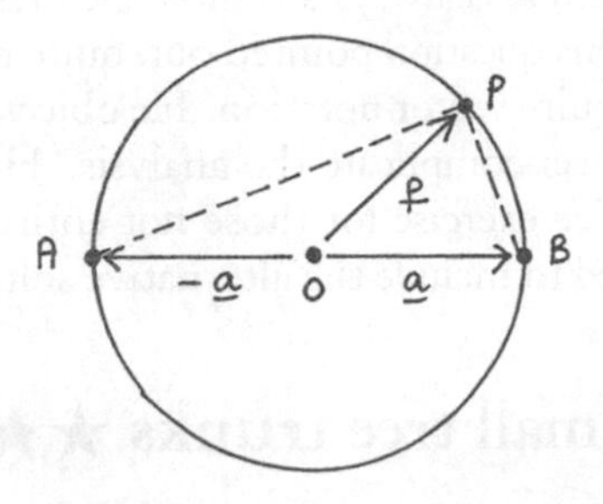

The line AP is represented by the vector $\boldsymbol{p}+\boldsymbol{a}$, and the line BP is represented by the vector $\boldsymbol{p} - \boldsymbol{a}$. For two general vectors $\boldsymbol{A}$ and $\boldsymbol{B}$, the dot product is defined by $\boldsymbol{A}.\boldsymbol{B} = |\boldsymbol{A}|\,|\boldsymbol{B}|\cos\theta$, where θ is the angle between the vectors. Thus, for $\theta = \pi/2$, $\boldsymbol{A}.\boldsymbol{B} = 0$. We recall two properties of the dot product. It is *commutative* (that is, $\boldsymbol{A}.\boldsymbol{B} = \boldsymbol{B}.\boldsymbol{A}$) and *distributive* (that is, $\boldsymbol{A}.(\boldsymbol{B}+\boldsymbol{C}) = \boldsymbol{A}.\boldsymbol{B} + \boldsymbol{B}.\boldsymbol{C}$).

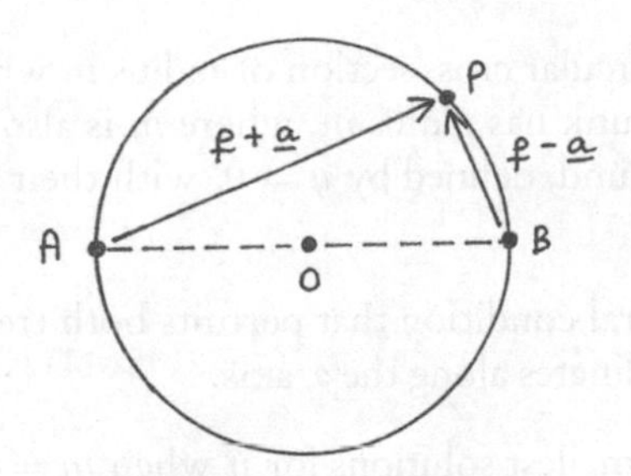

Consider the dot product of the two vectors defining AP and BP. Expanding using the distributive property we have

$$(\boldsymbol{p}+\boldsymbol{a})\,.\,(\boldsymbol{p}-\boldsymbol{a}) = \boldsymbol{p}.\boldsymbol{p} - \boldsymbol{p}.\boldsymbol{a} + \boldsymbol{a}.\boldsymbol{p} - \boldsymbol{a}.\boldsymbol{a}$$

Simplifying using the definition of the dot product and the commutative property, we have

$$(\boldsymbol{p}+\boldsymbol{a})\,.\,(\boldsymbol{p}-\boldsymbol{a}) = |\boldsymbol{p}|^2 - |\boldsymbol{a}|^2$$

We note that both $\boldsymbol{a}$ and $\boldsymbol{p}$ are radii of the circle, so

[9]In this book we will use bold symbols to distinguish a vector $\boldsymbol{a}$ from its scalar magnitude $|\boldsymbol{a}|$, or a. In handwriting (the hand-drawn diagrams in this book, for example) vectors are denoted with underscores (such as $\underline{a}$).

$$|\boldsymbol{a}| = |\boldsymbol{p}|$$

We see that the dot product $(\boldsymbol{p}+\boldsymbol{a}).(\boldsymbol{p}-\boldsymbol{a}) = 0$. However, $|\boldsymbol{p}+\boldsymbol{a}| \neq 0$ and $|\boldsymbol{p}-\boldsymbol{a}| \neq 0$. So we conclude from the definition of the dot product that for $(\boldsymbol{p}+\boldsymbol{a}).(\boldsymbol{p}-\boldsymbol{a}) = 0$ to be satisfied we must have $\theta = \pi/2$. Thus, a triangle inscribed within a semicircle is always a right-angled triangle.

A friend who read this question pointed out, quite rightly, that easier solutions exist that don't require vector notation. He objected to the use of vectors on the basis that they over-complicate the analysis. He was right, of course, but I still think it's a nice exercise for those not entirely familiar with using vectors. I did feel obliged to include the alternative solution,[10] however.

1.10 Big and small tree trunks ★★ or ★★★★

Quite a number of maths questions are designed to test a student's ability to set up equations for unfamiliar problems and then solve them. This is an extremely useful skill for a numerate person. The question here is one of several similar problems I came up with on a long-haul flight to the US a couple of years ago. The geometry is trivial, but I think most high-school students would struggle with it, simply because the ability to approach completely unfamiliar situations is valued too little in most curricula.

A TREE TRUNK has a circular cross section of radius n, where n is an integer. A second circular tree trunk has radius m, where m is also an integer. Both tree trunks rest on the ground, defined by $y = 0$, with their axes aligned with the z-axis.

1. Derive the general condition that permits both tree trunk centres to lie at integer co-ordinates along the x-axis.

2. Give the three smallest solutions for n when $m = 1$.

[10]Noting the construction $APBOA$, we see that $\phi = \pi - 2\alpha$ and that $\theta = \pi - 2\beta$. Adding these expressions we get $\phi + \theta = 2\pi - 2(\alpha + \beta)$. But $\phi + \theta = \pi$. Combining these expressions we get $\pi = 2\pi - 2(\alpha + \beta)$. Simplifying, we get $\alpha + \beta = \pi/2$, as required. This solution is undeniably simpler.

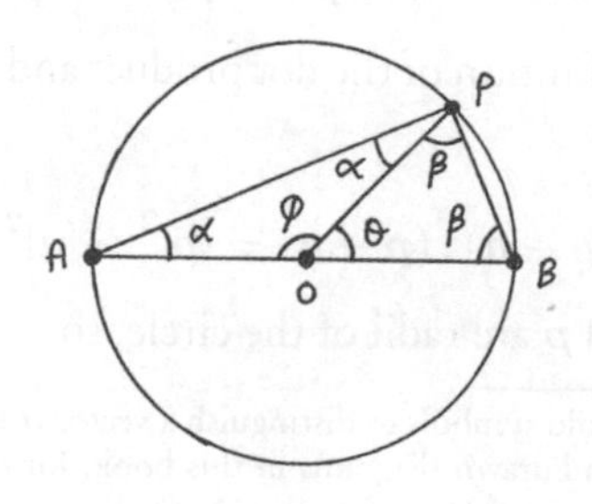

3. Give a general solution for n and m (extension for ★★★★).

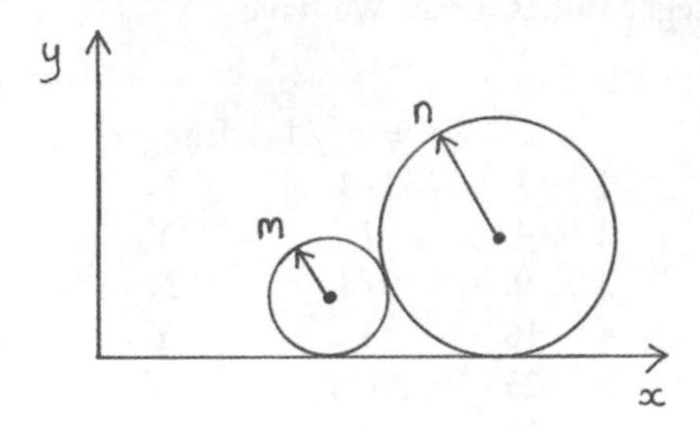

Answer

This is a simple geometry problem that I think a lot of students would struggle to unpack. Let's consider each of the three parts in turn.

1) DERIVE THE GENERAL CONDITION THAT PERMITS BOTH TREE TRUNK CENTRES TO LIE AT INTEGER CO-ORDINATES ALONG THE x-AXIS

We are told that n and m are integers, so the difference in the y co-ordinates of their centres is also an integer, $n - m$. We require that the difference in the x co-ordinate also be an integer. We denote the difference simply as x. We note that the tree trunks touch at a mutual point of tangency, so the distance between their centres is $n + m$.

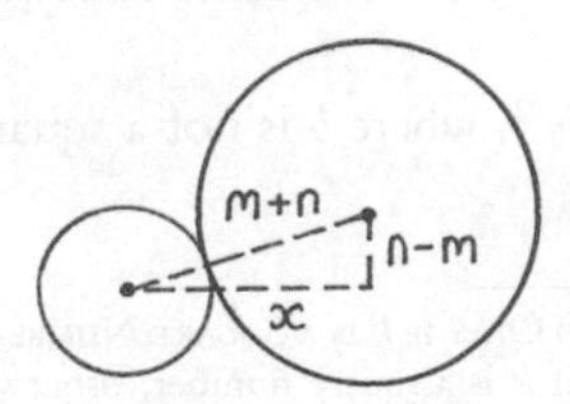

From Pythagoras' theorem we have

$$(n+m)^2 - (n-m)^2 = x^2$$

Simplifying, we get

$$4nm = x^2$$

This is the general condition that must be met.

2) GIVE THE THREE SMALLEST SOLUTIONS FOR n WHEN $m = 1$.

A mathematician would hardly regard it as elegant, but to obtain the first three solutions for n when $m = 1$ we can solve for $m = 1$ simply by substituting this into the expression and trying values for n.

$$\text{for } m = 1,\ n = \frac{x^2}{4}$$

We can probably see the first three solutions to this. If we can't, we can simply try the first few integer values for x. We have

x	x^2	$n = x^2/4$	Integer?
1	1	1/4	N
2	4	1	Y
3	9	9/4	N
4	16	4	Y
5	25	25/4	N
6	36	9	Y

The first three solutions are $n = 1$, 4 and 9.

3) GIVE A GENERAL SOLUTION FOR n AND m (EXTENSION FOR ★★★★).

Consider the general condition that we are required to meet, namely $4nm = x^2$, or $2\sqrt{nm} = x$. We know that x is an integer. We write this as $x \in \mathbb{Z}$. Here $\in$ means *is a member of*, and $\mathbb{Z}$ represents the set of all integers. By extension $2\sqrt{nm} \in \mathbb{Z}$. It is possible to prove[11] that $\sqrt{k} \in \mathbb{Z}$ if and only if k is a square number.[12] From this it follows that there are two possibles cases:

- n or m is equal to unity and the other is a square number.
- $m = a^2b$ and $n = bc^2$, where b is not a square number and where a, b and c are integers.[13]

---

[11] PROOF THAT $\sqrt{k} \in \mathbb{Z}$ IF AND ONLY IF k IS A SQUARE NUMBER.

THEOREM: $\sqrt{k}$ is an integer if k is a square number, otherwise it is irrational (it cannot be expressed as the ratio of two integers). Here $k \in \mathbb{Z}^+$; that is, it is a member of the set of all positive integers.

PROOF: a) If k is a square number we write $k = l^2 \Rightarrow \sqrt{k} = l \in \mathbb{Z}$; b) We assume that $\sqrt{k}$ is a rational number, that is, $\sqrt{k} \in \mathbb{Q}$, and prove that this is impossible by contradiction. Consider two sets of numbers $\{a_i\}$ and $\{b_i\}$ in which members are positive integers: $a_i \in \mathbb{Z}^+$ and $b_i \in \mathbb{Z}^+$. We assume that we can write $\sqrt{k} = a_i/b_i$, where (a_i, b_i) is an ordered pair of positive integers. There must be a unique least element $a \in \{a_i\}$. Similarly, unique least element $b \in \{b_i\}$ exists. The least element of $\{a_i\}$ must pair up with the least element of $\{b_i\}$. Hence $\sqrt{k} = a/b$, so $k = a^2/b^2$, so $b^2k = a^2$. Because $a, b, k \in \mathbb{Z}^+$, and because the left-hand side is divisible by k, the right-hand side must also be divisible by k, so we can write $a = kc$ (provided that k is not a square number, otherwise we could write $a = \sqrt{k}c$), where $c \in Z^+$. We write $b^2k = (kc)^2 \Rightarrow \sqrt{k} = b/c$. But for $k \in \mathbb{Z}^+$ and $k \neq 1$, $\sqrt{k} > 1$. Therefore, from $\sqrt{k} = a/b$, we see that $a > b$. Similarly, from $\sqrt{k} = b/c$, we see that $b > c$. So we have found $\sqrt{k} = b/c$ with $c < b < a$. However, a and b were the least elements of $\{a_i\}$ and $\{b_i\}$ respectively, which leads to a contradiction. $\{a_i\}$ and $\{b_i\}$ must both be empty, so $\sqrt{k}$ cannot be written as a rational number. We have proved that $\sqrt{k} \notin \mathbb{Q}$.

[12] A *square number* or *perfect square* is an integer that is the square of an integer.

[13] TO SHOW THAT n AND m SATISFY THIS CONDITION. If $m = a^2b$ and $n = bc^2$, then $mn = a^2b^2c^2 = (abc)^2$. Where $a, b, c \in \mathbb{Z}$, then $abc \in \mathbb{Z}$, so $mn = (abc)^2$ is a square number.

1.11 Professor Fuddlethumbs' stamp ★★

PROFESSOR FUDDLETHUMBS IS walking across the Great Lawn in College holding a small green stamp, which he was delighted to receive by mail that morning from a collector. He is going from arch A to arch B, a path which defines a straight line across the middle of the lawn, dividing it into two equal rectangles. The lawn is square, and measures exactly 80 yards on a side. Absent-minded as ever, the professor takes a random walk across the lawn. When he gets to B he realises in horror that he has lost his stamp. "Thank heavens," he says, "that I always count the number of paces it takes me to cross the Great Lawn." He stands and surveys the vast expanse of grass. "According to my calculations," he muses, "I walked exactly 100 yards. I wonder how far I could have strayed off the straight-line path between the arches?" To help the professor, calculate the furthest distance he could have wandered from the straight line connecting A and B.

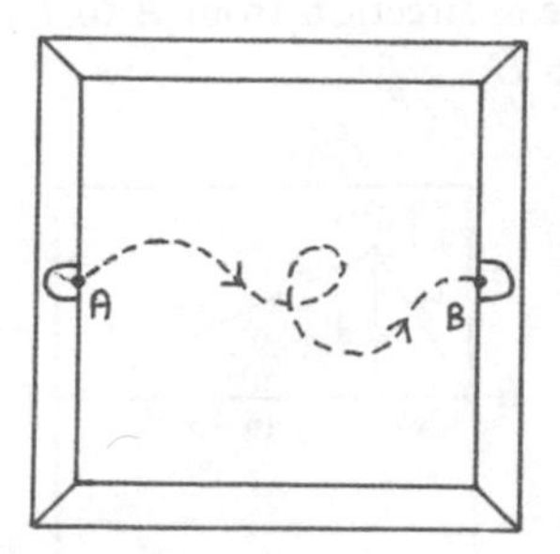

Answer

There are many ways of approaching this question that are more or less *formal* in the way one approaches the proof. If we could see the answer it would probably be relatively simple to convince ourselves that it is correct. I will now outline one method which is slightly more formal, however, and does not require a flash of insight. It is more mechanical in the way it approaches the proof. Mechanical approaches to questions lack elegance, but are perhaps more reliable as ways of solving problems. There are certainly proofs that we could consider as even more formal—you could start by deriving the equation of the ellipse, for example. But I leave those proofs as an exercise you can try for yourself.

Postulate first that for any point P there exists a curved route (shown by the dotted line) which is the shortest path APB. We see without needing a more rigorous proof that the straight lines AP and PB will always lead to a shorter route, no matter what the shape of the curve APB. Thus we see that the shortest route between A and B via a point P is always the two straight lines AP and PB.

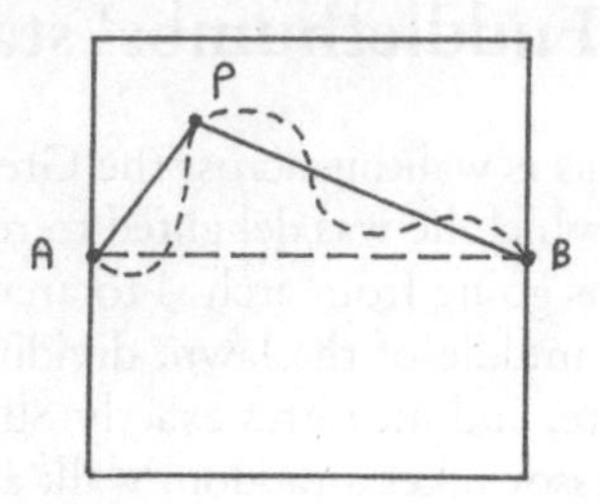

Consider now for a general point P the shortest distance APB. We denote this distance $l_1 + l_2 = L$. We see that if P is a given fixed perpendicular distance d from the line AB, L is given by

$$L = \left(x^2 + d^2\right)^{1/2} + \left[(80 - x)^2 + d^2\right]^{1/2}$$

where x is distance in the x-direction from A to P, and $80 - x$ is distance in the x-direction from P to B.

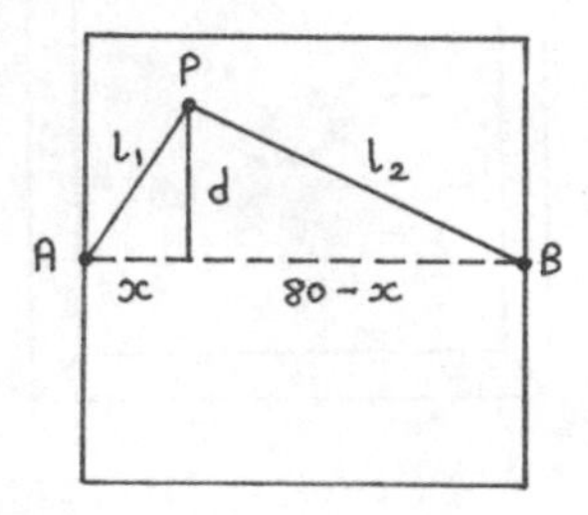

We could ask what value of x gives rise to the maximum d for constant L. This is found at the point $dd/dx = 0$. Differentiating both sides with respect to x we get

$$\frac{dL}{dx} = 0 = \frac{x + d\,(dd/dx)}{\left(x^2 + d^2\right)^{1/2}} + \frac{d\,(dd/dx) - 80 + x}{\left[(80 - x)^2 + d^2\right]^{1/2}}$$

Setting $dd/dx = 0$ and rearranging we have

$$x\left[(80 - x)^2 + d^2\right]^{1/2} = (80 - x)\left(x^2 + d^2\right)^{1/2}$$

We see that the solution of this is $(80 - x) = x$, or $x = 40$. This is exactly halfway between A and B, as expected.

We can find the maximum value of d by analysing the geometry of the problem. For $L = 100$ and $x = 40$, by symmetry we see that $l_1 = l_2 = 50$, and thus $d = \sqrt{50^2 - 40^2} = 30$.

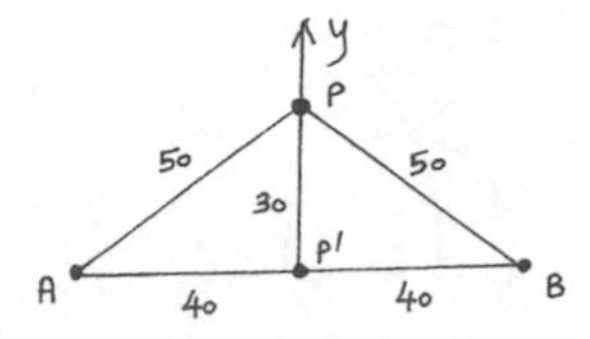

The furthest Professor Fuddlethumbs can stray from the path is 30 yards. It is going to be a long search for his stamp.

1.12 Captain Fistfulls' treasure ★

CAPTAIN FISTFULLS LANDS on what he thinks is Treasure Island and rubs his hands with glee. He has the famous Lost Map of Fiddle-de-dum, which shows where the treasure is buried. The other pirates unload the shovels and treasure chest whilst Captain Fistfulls consults the map. The map is a schematic of two trees. It reads very plainly, "Walk ye fifty paces from one tree and one hundred paces from the other. There lies the treasure." All Captain Fistfulls has to do now is work out where to dig. Assuming he can find the correct two trees, in how many places will he have to dig? How can he easily check to see whether a pair of trees is definitely the wrong pair?

Answer

We draw a circle with a radius of 50 paces from one tree (say, A), and a circle with a radius of 100 paces from the other tree (say, B). If the trees are more than 50 paces from each other, and within 150 paces of each other, the circles (solid lines) will intersect at two points X_1 and X_4. Repeating the process the other way round (dotted lines) we get points X_2 and X_3. Captain Fistfulls will have to dig in 4 locations to try and find the treasure. There are two limiting cases, however. When the separation of the trees, d, is equal to the sum of the given distances ($d = 150$) or the difference between the given distances ($d = 50$), Captain Fistfulls would have to dig in only two locations.

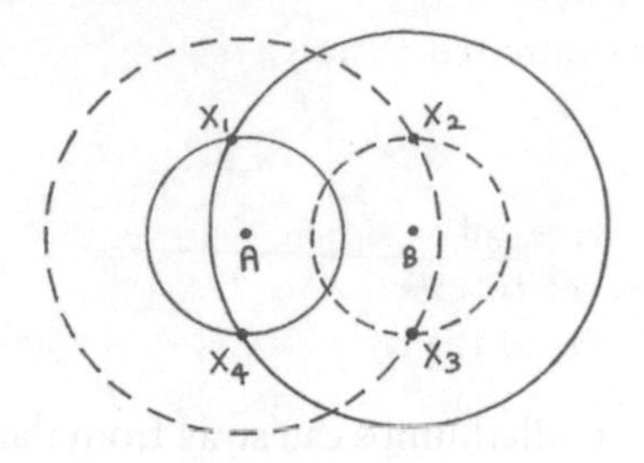

Captain Fistfulls can check whether a pair of trees is the wrong pair by simply checking whether the separation d lies in the range

$$r_1 - r_2 < d < r_1 + r_2$$

where r_1 and r_2 are the distances of the treasure from the trees. If d for a pair of trees does not lie in this range, he has the wrong pair of trees. For the case of $r_1 = 50$ and $r_2 = 100$, this reduces to $50 < d < 150$. If this requirement is not obvious, a diagram will quickly convince you.

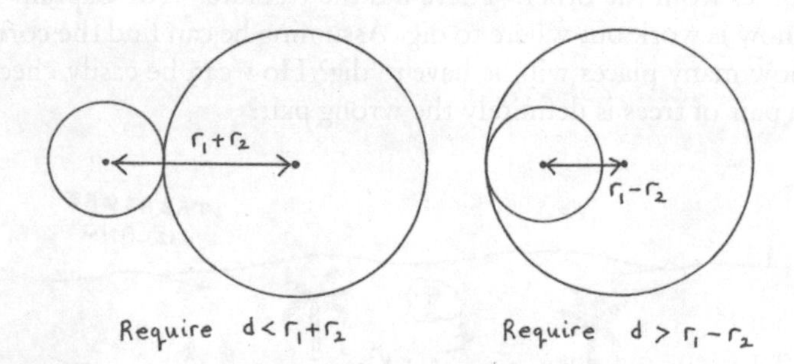

1.13 Captain Fistfulls' treasure II ★★

Captain Fistfulls is on his second treasure-hunting voyage. This time he is heading to Malpassant Island. His last trip failed due to a faulty sextant, that sent him to the wrong latitude, leading to a wasted trip across the tropics. Now in the French West Indies, he is sure he is onto some treasure. He is in possession of the Lost Map of Mauvaise Blague, the most famous of all French pirates. The map is a schematic of two trees. The instructions simply read "Walk ye fifty paces from one tree and fifty paces from the other. There lies the treasure." The island is flat and barren indeed. After a long walk, the party crest a faint ridge and look down into Desolate Valley, right in the centre of the island. It is the only place where there is enough water for trees to grow. "Blistering barnacles!" thunders the Captain, "what unnatural evolutionist allowed these trees to breed?" Sure enough, there are not two trees but sixteen. Captain Fistfulls paces the distance between the furthest trees. "Rum luck!"

he says. "Less than one hundred paces. Yer going to have to put your backs into it boys!"

ASSUME THERE ARE 16 trees, all less than 100 paces from each other, and that treasure is buried 50 paces from one tree and 50 paces from another. If we do not know the locations of the trees, in the worst case scenario what is the *maximum* number of places that it is necessary to dig to find the treasure?

Answer

Let r_1 and r_2 be the distances the treasure is from each of two trees. If the distances from the trees are *not* equal, $r_1 \neq r_2$, the condition that must be met for each pair of trees to define *four* possible locations is $r_1 - r_2 < d < r_1 + r_2$, where d is the separation of the trees. In the limiting cases $d = r_1 - r_2$ and $d = r_1 + r_2$, a pair of trees defines only *two* locations.

In the special case of $r_1 = r_2 = r$, a pair of trees defines two possible locations for $0 < d < 2r$. In the limiting special case $r_1 = r_2$ and $d = r_1 + r_2$, a pair of trees defines a single location. The other special limiting case $r_1 = r_2$ and $d = r_1 - r_2$ is impossible: this implies that $d = 0$, and the trees have to be at the same location.

In the present case we have $r_1 = r_2 = 50$, giving the range $0 < d < 100$ for pairs of trees which define *two* possible locations. We are told that $d_{\text{max}} < 100$, so every possible pair of trees will define two points. So the total number of locations is two times the number of combinations of trees.

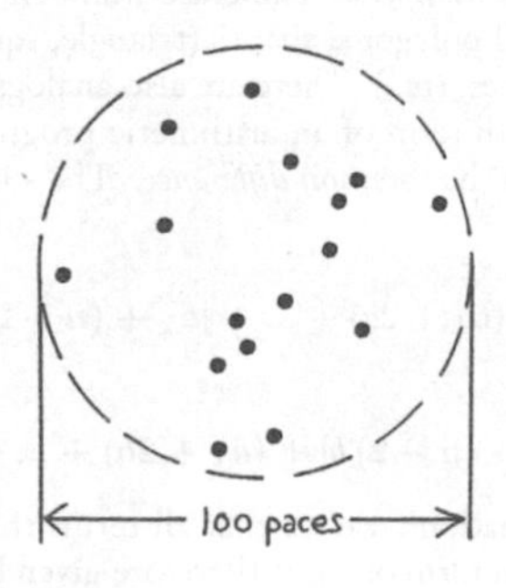

Let's look at the number of combinations of tree pairs we can pick. The possible unique pairs are

Tree	1	with trees	2, 3, 4...16	(15 pairs)
Tree	2	with trees	3, 4, 5...16	(14 pairs)
⋮	⋮	⋮	⋮	⋮
Tree	14	with trees	15, 16	(2 pairs)
Tree	15	with tree	16	(1 pair)

The sum of the numbers of pairs form what is called a *triangular number*,[14] a number which is the sum of elements that can be arranged in a triangular form. For example, the fourth triangular number T_4, represented by

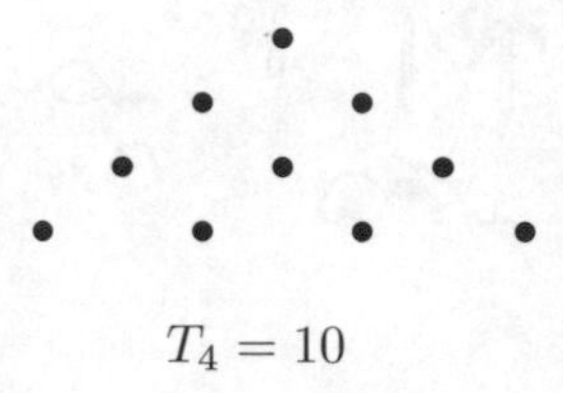

$$T_4 = 10$$

Here we require T_{15}, which is given by

$$T_{15} = \sum_{n=1}^{15} n = 1 + 2 + 3 + ... + 15$$

Of course, we could do this easily by adding the numbers together, but it is perhaps a little crude. What if we wanted T_{1000}, for example? We recognise that the sum is also a very simple case of an *arithmetic progression*,[15] for which the sum to n is given by

$$S_n = \frac{n}{2}(a_1 + a_n)$$

We see that T_{15} is given by

$$T_{15} = \frac{15}{2}(1 + 15) = 120$$

[14] A triangular number is one type of what are called *figurate numbers,* a class that includes *polygonal numbers* and *polyhedral numbers*. These are numbers that can be represented by objects arranged in two-dimensional polygonal shapes (triangle, square, etc..) or three-dimensional polyhedral shapes (pyramids, cubes, etc..). There are also analogs in higher dimensions.

[15] The general form for the n-th term of an arithmetic progression is $a_n = a_1 + (n-1)\,b$. Here a_1 is the first term and b is the *common difference*. The sum to n, S_n, can be expressed in two different ways

$$S_n = a_1 + (a_1 + b) + (a_1 + 2b) + ... + [a_1 + (n-2)\,b] + [a_1 + (n-1)\,b]$$

or

$$S_n = [a_n - (n-1)\,b] + [a_n - (n-2)\,b] + (a_1 + 2b) + ... + (a_n - 2b) + (a_n - b) + a_n$$

If we add both sides of these equations, we see that all terms that include b sum to zero, giving $2S_n = n\,(a_1 + a_n)$. The general form of S_n is therefore given by $S_n = \frac{n}{2}(a_1 + a_n)$.

The *maximum* total number of locations the captain must dig is given by $2T_{15} = 240$. It seems the pirates are going to be busy for some time.

Alternative solution

We might have solved this problem using language more familiar to a student of probability and combinatorics. In that language, finding the number of different pairs of trees is a problem of picking 2 objects from 16 unlike objects. The general formula for the number of combinations of r objects in n unlike objects is

$$^{n}C_{r} = \frac{n!}{r!\,(n-r)!}$$

For our particular case we have $^{16}C_2 = 16!/\left[2!\,(16-2)!\right] = (16 \times 15)/2 = 120$. For 120 combinations of trees we have 240 possible locations for the treasure, as before.

1.14 Captain Fistfulls' treasure III ★★★

I recommend that you attempt the easier variants of this question before trying this trickier version.

THERE ARE 19 TREES, all fewer than 80 paces from each other. Treasure is buried 49 paces from one of the trees and 31 paces from another. The particular arrangement of trees dictates the number of places it is necessary to dig to guarantee finding the treasure. Prove that at least one arrangement of trees exists that gives a number of possible places that is equal to the maximum number of possible places that could be achieved with 19 trees. That is, where the arrangement of trees and the two distances of the treasure from the trees is chosen to maximise the number of possible places.

Answer

This is essentially a *packing problem*. We will return to this concept in a moment. Let us first calculate the maximum number of places it is necessary to dig, and the required tree spacing to achieve this maximum number.

The maximum number of places we might need to dig is given by four times the 18th triangular number, T_{18}. That is

$$4\,[T_{18}] = 4\left[\frac{18}{2}\,(1+18)\right] = 684$$

Or, by the alternative method

$$4 \times {}^{19}C_2 = 4 \times \frac{19!}{2!\,(19-2)!} = 684$$

where the formulae were discussed in the previous questions. We also showed that to achieve the maximum number of possible places, each independently identifiable pair of trees must define four possible locations. So the pairs must meet the condition $r_1 - r_2 < d < r_1 + r_2$, where d is the separation of the trees and r_1 and r_2 are the distances from the two trees. In this case we have $r_1 = 49$ and $r_2 = 31$, so the condition for an arrangement of trees that can give the maximum possible number of locations is

$$18 < d < 80$$

That is, all trees must be within 80 paces of all other trees, and no tree must be within 18 paces of a second tree. The problem now is one of packing circles. We represent each tree as a dot with a circular "keep-out zone" of diameter 18 paces around it. We need to pack the circles so that their centres do not go outside a larger circle of diameter 80 paces. We can think of two very simple packing schemes: regular square packing, and regular hexagonal packing.

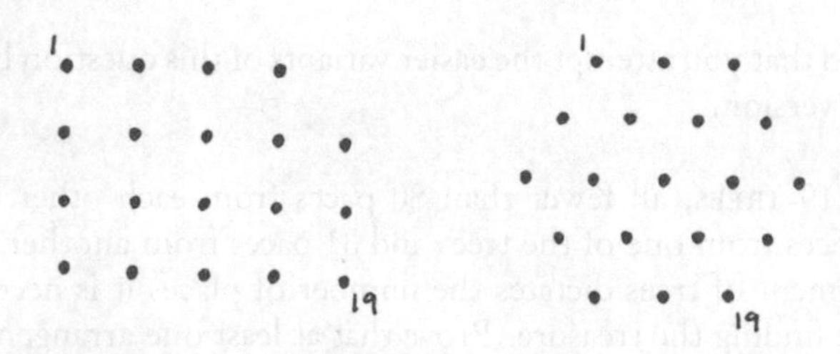

It is immediately apparent that in most situations with a *large* number of objects something closer to regular hexagonal packing will give a better result. If we consider the most compact regular hexagonal packing of 19 objects, we get an arrangement with three objects on each side of the hexagon. There are five objects across the longest diagonals of the hexagon. From the geometry of the packing arrangement, the maximum separation between objects, d_{max}, is four times the minimum separation between trees, d_{min}. We write $d_{max} = 4d_{min}$. We have two constraints

$$18 < d < 80 \text{ and } d_{max} = 4d_{min}$$

and we can easily see that there are infinitely many solutions, such as

$$d_{min} = 19 \text{ and } d_{max} = 76$$

The geometry of this solution is shown here.

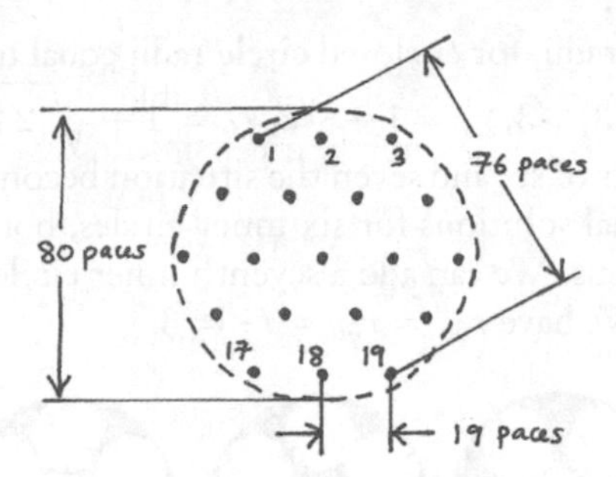

So there are infinitely many solutions that satisfy the constraints and allow the maximum number of possible locations for the treasure.

Packing is a very interesting topic in general. As we have noted, we can cast problems of the sort shown above as variants on the classic *circle packing in a circle* problem. The problem of packing circles of many diameters within another circle is very complex indeed. Even the problem of packing circles of single diameter within a large circle is harder than you might imagine. It is, in fact, still a serious research topic. Take, for example, the paper by Graham et al. entitled *Dense packings of congruent circles in a circle.*[16] The authors write

> Problems of packing congruent circles in different geometrical shapes in the plane were raised in the 1960s, and many results—mainly for small packings—were obtained...The development of new, effective optimization algorithms for packing problems and the ever increasing performance of computing systems have recently brought these problems into focus again; computer-aided methods can now be used to construct good large packings. We consider the problem of packing congruent circles inside a larger circle...Given n, we want to place n congruent circles without overlaps inside the larger circle in such a way that their common radius is as large as possible...This circle packing problem has another equivalent presentation, where n points (rather than circles) are placed inside a circle...The goal is to maximize the minimum pairwise distance between the points.

Optimum solutions for circle numbers from two to five are trivial.

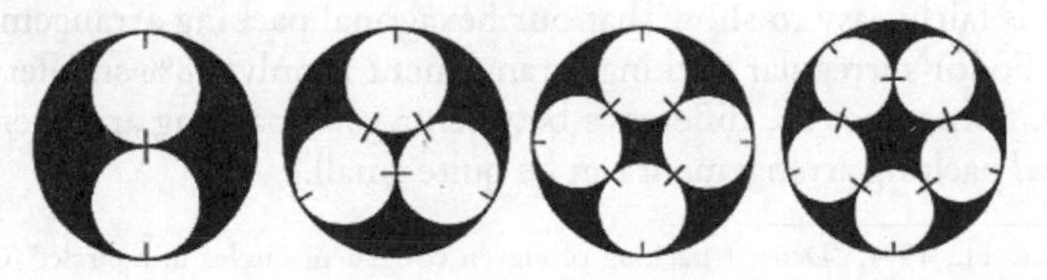

---

[16]Graham, R.L., Lubachevsky, B.D., Nurmela, K.J., Ostergard, P.R.J., 1998, "Dense packings of congruent circles in a circle," Journal of Discrete Mathematics, 181, pp. 139–154.

The enclosing circle radii, for enclosed circle radii equal to unity, are given by $r_2 = 2$, $r_3 = 1 + (2/3)\sqrt{3}$, $r_4 = 1 + \sqrt{2}$, $r_5 = 1 + \sqrt{[2(1 + 1/\sqrt{5})]}$. When we get to circle counts of six and seven the situation becomes more interesting. There are two optimal solutions for six inner circles, both of which have the same outer circle radius. We can add a seventh inner circle with no increase in outer circle radius. We have $r_{6a} = r_{6b} = r_7 = 3$.

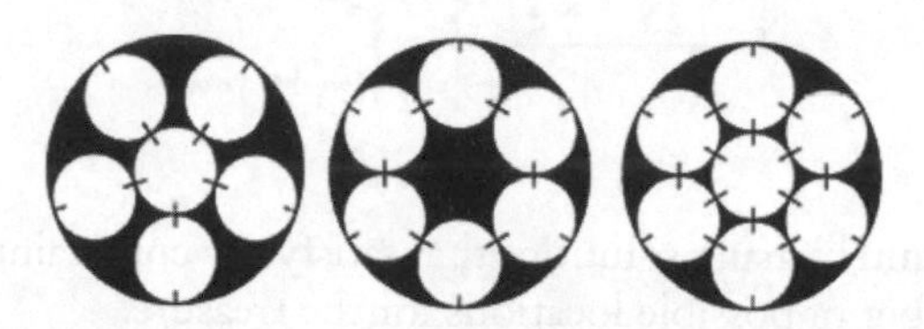

Optimum solution proofs for 8, 9 and 10 inner circles were demonstrated in 1969, but it was not until 1994 that the optimal packing configuration for 11 inner circles was proven.[17] The result was $r_{11} = 1 + 1/\sin(\pi/9) \approx 3.92$. There is an entire paper devoted to the topic, showing the considerable complexity of this apparently simple-looking problem.

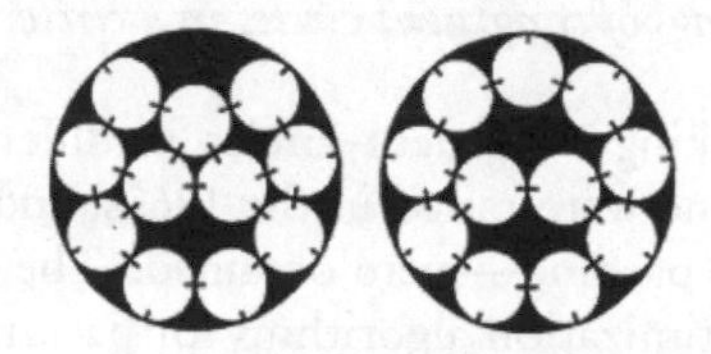

Optimum solutions for 12 and 13 inner circles were demonstrated as recently as 2000 and 2003, respectively. What's fascinating is that no optimal solution has yet been proven for $n = 14$ to $n = 18$, but in 1999 Fodor[18] proved the optimal condition for $n = 19$, with $r_{19} = 1 + \sqrt{2} + \sqrt{6} \approx 4.86$.

Let's compare this with the regular hexagonal packing arrangement presented above. It is fairly easy to show that our hexagonal packing arrangement gives $r_{19} = 5$. Fodor's irregular packing arrangement is only 2.8% smaller in terms of radius. Sometimes the difference between a *good* packing arrangement and an *optimal* packing arrangement can be quite small.

[17] Melissen, H., 1994, "Densest packing of eleven congruent circles in a circle," Geometriae Dedicata 50 (1994), pp. 15–25.

[18] Fodor, F., 1999, "The densest packing of 19 congruent circles in a circle," Geometriae Dedicata 74, pp. 139–145.

This is one of a number of packing problems that are still of great interest to mathematicians. Another well-known problem is the *kissing number problem*. The question is as follows. In $(n+1)$-dimensional Euclidean space, what is the maximum number of n-dimensional *hyperspheres* of the same radius that can be touched (or *kissed*) simultaneously by a single such hypersphere? For the case of a line (one-dimensional space), or the plane (two-dimensional space), the solutions are trivial: $k_1 = 2$ and $k_2 = 6$ respectively, the latter corresponding to the hexagonal packing of circles. For three-dimensional space, the proof for $k_3 = 12$ was not derived until the end of the nineteenth century, though Sir Isaac Newton (1642–1727) originated the problem. The only other exact solutions that have been obtained are for 4, 8 and 24-dimensional space.

1.15 The geometry of Koch Island ★★★★

The *coastline paradox* or *Richardson effect* is the observation that the measured length of an irregular border (a coastline, for example) depends on the length scale being used. Where the border has *fractal* geometry,[19] which is approximated by many natural systems over a wide range of scales, the measured length of the border increases indefinitely as the length scale becomes smaller. The concept originates with Lewis Fry Richardson,[20] who was researching the likelihood of two countries going to war based on the length of their borders! He found that some borders were reported as having wildly different lengths by the countries on either side. The border between Spain and Portugal, for example, was reported to be either 987 km long or 1,214 km long.

Richardson later published a study of the lengths of parts of the coastlines of Australia, South Africa, Britain and Portugal, measured using a number of different polygon length scales, and mapping line segments of equal length so that their vertices were on the real curve of the coastline.[21] The resulting trend with length scale depended on the characteristic shape of the supposed fractal landscape being considered. The research was picked up much later by a community of scientists interested in the study of fractals from a purely mathematical standpoint, most famously Benoit Mandelbrot, who wrote a 1967 paper entitled *How long is the coast of Britain?*[22]

One of the best known fractal shapes is the Koch Snowflake or Koch Island, devised in 1904 by Helge von Koch. It is built iteratively using an equilateral triangle as its base shape, $n = 0$. At each step, $n = 1, 2$ etc., the middle third of each side is removed and replaced with an equilateral triangle to increase the

---

[19] Fractals are objects/quantities that have self-similarity, meaning that they look characteristically similar whatever scale they are viewed on.

[20] An English mathematician, physicist, psychologist and pacifist (1881–1953).

[21] Richardson, L.F., 1961, "The problem of continuity," General Systems Yearbook, Vol. 6, pp. 139–187.

[22] Mandelbrot, B., 1967, "How long is the coast of Britain?" Statistical Self-Similarity and Fractional Dimension Science, New Series, Vol. 156, No. 3,775, pp. 636–638.

area of the previous shape. The shapes corresponding to $n = 0, 1$ and 2 are shown. It is obvious by inspection that the total perimeter L_n and enclosed area A_n increase with each iteration.

For a base island $n = 0$, with a side of unity length, derive expressions for the perimeter L_n and area A_n for a modified island of order n.

Answer

First consider the number of sides N_n. We start with $N_0 = 3$, and at each step transform each side into four new sides, effectively multiplying the number of sides by four. We have

$$N_n = 3\,(4)^n$$

Now consider the length of the smallest unit, l_n. We start with $l_0 = 1$, and at each step reduce the length of the smallest unit by a factor of $1/3$. We have

$$l_n = \left(\frac{1}{3}\right)^n$$

The total length of the perimeter, L_n, is equal to the number of sides, N_n, multiplied by the side length. We have

$$L_n = N_n l_n = 3\left(\frac{4}{3}\right)^n$$

We see immediately that this number explodes. The coastline of Koch Island has no finite limit. We write $L_{n\to\infty} \to \infty$. On Koch Island the coastline paradox certainly holds true.

What about the area A_n? We see that in a step leading to N_n sides, we create $N_n/4$ new triangles with side length l_n. The area created in this step is therefore given by

$$\Delta A_n = \frac{N_n}{4}\left(\frac{l_n}{l_0}\right)^2 A_0$$

where the individual small triangles created in the n-th step each have area $(l_n/l_0)^2 A_0$, and A_0 is the area of the original base triangle. By substituting for N_n and l_n, and noting that $l_0 = 1$, we have

$$\Delta A_n = \frac{3}{4}(4)^n\left(\frac{1}{3}\right)^{2n} \times A_0 = \frac{3}{4}\left(\frac{4}{9}\right)^n A_0$$

The progression follows immediately

$$A_n = A_{n-1} + \Delta A_n$$

Or, explicitly

$$A_1 = A_0 + \frac{3}{4}\left(\frac{4}{9}\right)^1 A_0$$

$$A_2 = A_1 + \frac{3}{4}\left(\frac{4}{9}\right)^2 A_0$$

$$A_n = A_{n-1} + \frac{3}{4}\left(\frac{4}{9}\right)^n A_0$$

We can represent this as a sum

$$A_n = \left[1 + \frac{3}{4}\sum_{k=1}^{n}\left(\frac{4}{9}\right)^k\right] A_0$$

In the limit $n \to \infty$ we see that

$$A_{n\to\infty} = \left[1 + \frac{3}{4}\sum_{k=1}^{\infty}\left(\frac{4}{9}\right)^k\right] A_0$$

We recall that the sum to n and sum to infinity for a geometrical series[23] with initial term a and ratio r are $S_n = a\left(1 - r^n\right)/\left(1 - r\right)$ and $S_\infty = a/\left(1 - r\right)$, for $|r| < 1$. Applying these general results to the current problem we have

$$A_n = \left[1 + \frac{1}{3}\frac{(1-(4/9)^n)}{(1-4/9)}\right] A_0 = \left[1 + \frac{3}{5}\left(1 - \left(\frac{4}{9}\right)^n\right)\right] A_0$$

$$A_{n\to\infty} = \left[1 + \frac{(3/4)}{1-(4/9)} - \frac{3}{4}\right] A_0 = \frac{8}{5} A_0$$

We see that as $n \to \infty$ the area $A_{n\to\infty}$ approaches a finite value determined by the limit of a converging series. This is in contrast to the length of the

[23] For a geometric series with the form $S_n = a\left(1 + r + r^2 + ... + r^{n-1}\right)$, by noting that $rS_n = a\left(r + r^2 + ... + r^n\right)$, we see that $S_n\left(1 - r\right) = a\left(1 - r^n\right)$, so $S_n = a\left(1 - r^n\right)/\left(1 - r\right)$. For $|r| < 1$, we have $S_\infty = a/\left(1 - r\right)$.

fractal, which explodes. The area of an equilateral triangle of side unity is $A_0 = \sqrt{3}/4$, so we have, explicitly

$$A_{n\to\infty} = \frac{2\sqrt{3}}{5}$$

1.16 An easyish fencing problem ★

This *fencing problem* is a well-known and popular puzzle, which I have seen in a number of forms over the years. It is about the simplest example one can give of an *optimisation* problem. In this sort of question we take a given physical situation, parametrise it by writing equations linking some object of interest to the variables, and then solve for a minimum or maximum subject to a given constraint.

To understand the general process, consider first the *box and can optimisation problems*, also popular questions. In these problems we calculate the *shape*[24] of a cylindrical can or cuboid box that maximizes the volume (the object of interest) for a given surface area (the *constraint* on the problem). The variables in the case of the cylinder optimisation problem would be the height and diameter (or radius) of the cylinder. The problem could be rephrased as calculating the minimum surface area for a given volume. The parameter that we are assuming is constant may *look* like a variable—that is, we may not know its value. Again, this does not change the problem.

Now we have seen the general principle, we turn our attention to the classic fencing problem.

A FARMER HAS a very large field bounded on one side by a long straight cliff. He has a fixed length of fence with which he wishes to create the largest possible rectangular fenced-in area he can. What ratio x/y should he choose for the lengths of the sides, x and y, of the fenced-in area?

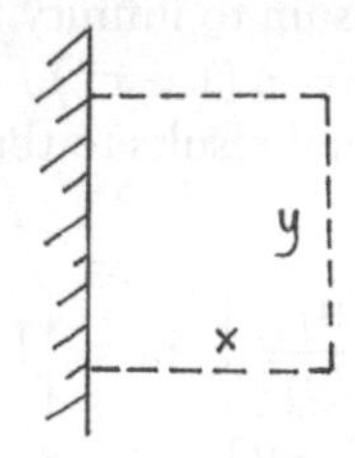

Answer

We answer this simple question with two methods, one using calculus, and one where we use the symmetry of the problem to find the maximum.

[24]Sometimes we refer to the shape as the *aspect ratio*, or the ratio of the height to the diameter, for example.

- **METHOD 1: USING CALCULUS.**

  As I remarked in the introductory notes, the aim is to parametrise the problem by expressing the object of interest, in this case the area A, in terms of the lengths x and y, and the constraint L, the overall length of fence available.

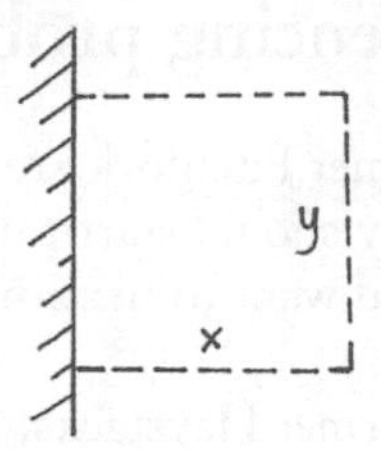

It should be clear from the construction that the constraint gives us $y = L - 2x$. The area is given by $A = xy = x(L - 2x)$. We are interested in the maximum area for a given x, for example. We differentiate and set the result equal to zero to determine the maximum. We write

$$\frac{dA}{dx} = L - 4x = 0$$

The solution to this is $x = L/4$, giving $y = L/2$. So the ratio of the lengths is $x/y = 1/2$. The field with the largest area for a given length of fence is a rectangle with an aspect ratio of $2:1$.

- **METHOD 2: SYMMETRY ARGUMENT**

  Placing a symmetry line at the centre of the fence, we divide the fence into regions of side a and b, and area ab. Noting that by doing so we divide the total length of fence L into four sections, we express a as $(L/4) + k$, and b as $(L/4) - k$. These satisfy $2a + 2b = L$, which is the constraint on the length of the fence.

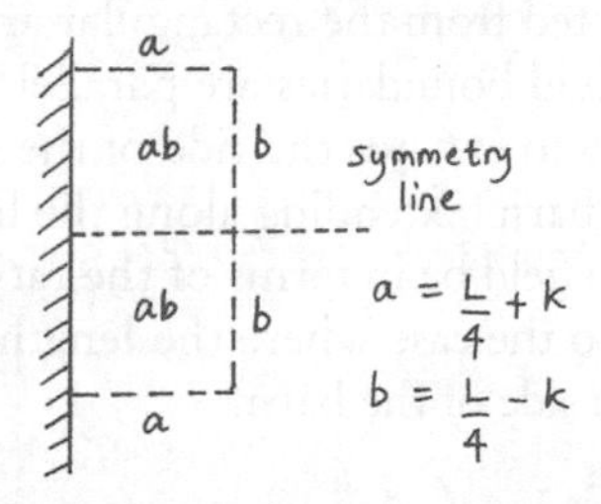

The enclosed area is given by

$$A = 2ab = 2\left(\frac{L}{4} + k\right)\left(\frac{L}{4} - k\right) = 2\left[\left(\frac{L}{4}\right)^2 - k^2\right]$$

which is clearly maximised when $k = 0$, giving $a = L/4$ and $b = L/4$. That is, $a = b$. This agrees with the previous solution—the field with the largest area for a given length of fence is a rectangle with an aspect ratio of $2 : 1$.

1.17 A hardish fencing problem ★★★

Farmer Haystacks and Farmer Pitchfork are standing in the noonday sun trying to work out where they should start putting up a new fence. They need to pen some animals in, and want to maximise the area they can get with the fence they have available.

"The problem," says Farmer Haystacks, "is that there barn." He points to a large square barn sitting in the middle of open farmland.

"That barn's no problem," says Farmer Pitchfork, "that's free fencing that is. Just lay the fence along the line of one wall and use the wall to extend the fence. Simple."

"Pitchfork, you ain't as sharp as you used to be," says Farmer Haystacks. "I can see that there barn is free fencing, but I can't work out whether to line the fence up with one wall, or with the longest diagonal."

"You need to get out of the sun, Haystacks. Lining up with the longest diagonal will mean a big triangular chunk out of your field."

"Pitchfork, you just don't get it, do you?" says Farmer Haystacks. "We need to extend that fence as much as possible and that diagonal is longer."

The farmers continue to argue in the heat but reach an impasse. Farmer Haystacks wants to line the fence up with the longest diagonal, and Farmer Pitchfork wants to line it up with one wall of the barn. This is the sort of problem that can best be solved by mathematics.

A FARMER HAS a very large field in the middle of which is a square barn. He has a fixed length of fence with which he wishes to construct the largest possible *rectangular* fenced-in area he can (accounting for protrusions due to the barn, which must be subtracted from the rectangular area). Is it better to construct the fence so that the field boundaries are parallel with the sides of the square barn (so that one of them extends the side of the barn), or so that they are at 45° to the sides of the barn (extending along the line of the longest diagonal)? What shape should the field be in terms of the ratio of the width x and height y? Restrict solutions to the case where the length of the fence is at least $3\sqrt{2}$ times the length of the side of the barn.

Answer

Although the mathematics is relatively straightforward, this is a pretty difficult problem because more than one optimum solution exists. The optimum solution depends on the ratio of the length of fence L to the length of the side of the square barn l. We will refer to configuration A as that where the fence is aligned with the wall of the barn, and configuration B as that where the fence is aligned with the longest diagonal. We need to consider both possible configurations for every ratio L/l. Consider the possible situations in turn.

- CONFIGURATION A: FENCE LONG ENOUGH TO USE FULL WALL OF BARN OF LENGTH l.

  In this case we have $L = 2x + 2y - l$. The area is given by

$$A_A = xy = \frac{(L+l)}{2}x - x^2$$

  Differentiating with respect to x, and setting $dA_A/dx = 0$, we obtain the solution $x = (L+l)/4$, which gives $y = (L+l)/4$. This is a square field with area

$$A_A = \frac{1}{16}(L+l)^2$$

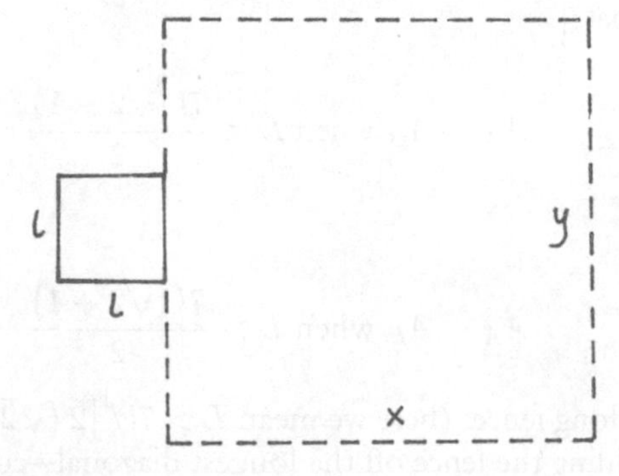

- CONFIGURATION B: FENCE LONG ENOUGH TO INCLUDE LONGEST DIAGONAL OF LENGTH $\sqrt{2}l$.

  In this case we have $L = 2x + 2y - \sqrt{2}l$. The area is given by

$$A_B = xy - \frac{l^2}{2} = \frac{(L+\sqrt{2}l)}{2}x - x^2 - \frac{l^2}{2}$$

  Differentiating with respect to x, and setting $dA_B/dx = 0$, we obtain the solution $x = (L+\sqrt{2}l)/4$, which gives $y = (L+\sqrt{2}l)/4$. This is a square field with a notch cut out, with area

$$A_B = \frac{1}{16}\left(L+\sqrt{2}l\right)^2$$

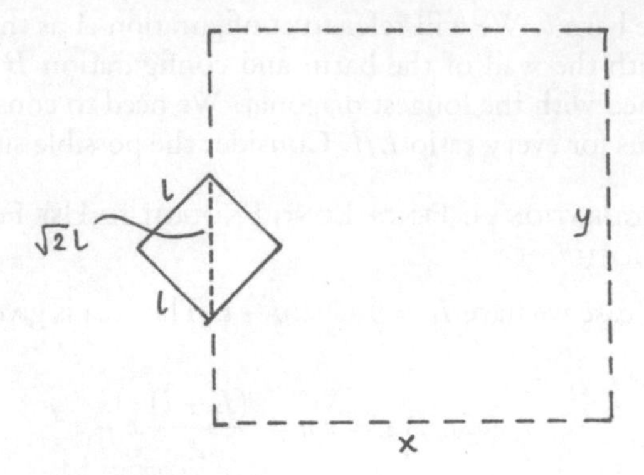

We now ask ourselves which of A_A and A_B is larger. Subtracting one from the other we obtain

$$A_A - A_B = \frac{l}{8}\left[L\left(1-\sqrt{2}\right)+\frac{7}{2}l\right]$$

Thus we find that

$$A_A > A_B \text{ when } L < \frac{7l\left(\sqrt{2}+1\right)}{2}$$

and

$$A_A < A_B \text{ when } L > \frac{7l\left(\sqrt{2}+1\right)}{2}$$

We see that for long fences (here we mean $L > 7l/\left[2\left(\sqrt{2}-1\right)\right]$) we optimise the area by building the fence off the longest diagonal—configuration B. The additional increase in area we achieve by extending the fence by $\sqrt{2}l$, instead of by only l in configuration A, is greater than the reduction in area due to the triangular protrusion caused by the barn. For shorter fences it is better to line the fence up with the wall of the barn, as in configuration A. There is, of course, the special case that occurs for the critical length $L_c = 7l/\left[2\left(\sqrt{2}-1\right)\right]$, for which the areas are equal: $A_A = A_B$. Evaluating the lengths x and y in the critical case for configurations A and B we get

$$x_A = y_A = \frac{9+7\sqrt{2}}{8}l \approx 2.362\,l$$

$$x_B = y_B = \frac{7+9\sqrt{2}}{8}l \approx 2.466\,l$$

The critical case is shown here.

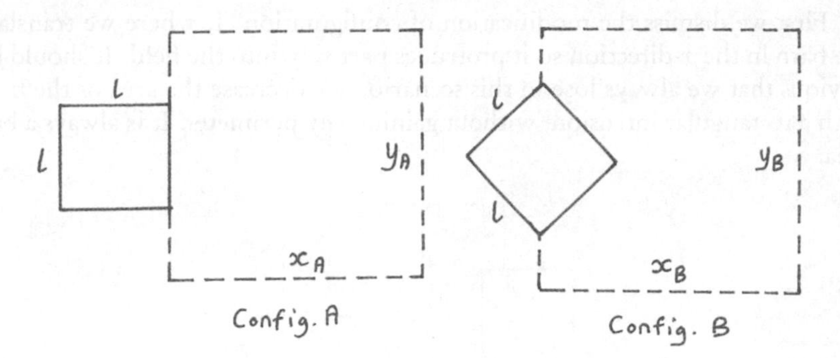

So it seems that both Farmer Haystack and Farmer Pitchfork are correct. The solution depends on the ratio of the length of the fence to the length of the side of the barn. We may or may not have noticed that these solutions are valid for fence lengths in the restricted range $3\sqrt{2} < L/l < \infty$. Let's consider this point a little further in the discussion.

Further discussion

We have defined the best solution for long fences. We said that for $L > L_c$ (where $L_c = 7l/\left[2\left(\sqrt{2}-1\right)\right]$), configuration B was optimum. We said that for fences shorter than this configuration, A was optimum, but we have not looked rigorously at this region. What happens if the fence becomes very short, for example? We now look at shorter fences in more detail.

For both configurations A and B we found that the optimum area was for a *square field*, which requires the following minimum lengths for configurations A and B.

$$L_{A,\,\text{min}} = 3l \text{ and } L_{B,\,\text{min}} = 3\sqrt{2}l$$

Below these lengths it is impossible to have a square field. It is apparent that both $L_{A,\,\text{min}}$ and $L_{B,\,\text{min}}$ are well below the critical length. That is, $L_{A,\,\text{min}} < L_c$ and $L_{B,\,\text{min}} < L_c$. Noting that $L_{A,\,\text{min}} < L_{B,\,\text{min}}$, we see that in the range $L_c > L > L_{B,\,\text{min}}$ (where both solutions apply), configuration A is best. Which solution is best in the range $L_{B,\,\text{min}} > L > L_{A,\,\text{min}}$, however? We consider this now.

First we note that in the range $L_{B,\,\text{min}} > L > L_{A,\,\text{min}}$, a square field in configuration B which uses all of the longest diagonal of the barn is not possible (the fence is too short). It *would* be possible, however, to keep the alignment of the barn the same (at 45°, that is) but use a section of the barn smaller than the longest diagonal. We'll analyse configurations of this sort now. What follows is not an entirely rigorous approach, but it goes some way to testing other possible configurations.[25] There are further configurations that we do not discuss.

[25] We restrict our approach to square field solutions, for example, and an additional proof is required to show that these are optimum.

First we dismiss the modification of configuration A, where we translate the barn in the x-direction so it protrudes part-way into the field. It should be obvious that we always lose in this scenario. We decrease the area of the field with a rectangular intrusion, without gaining any perimeter. It is always a bad idea.

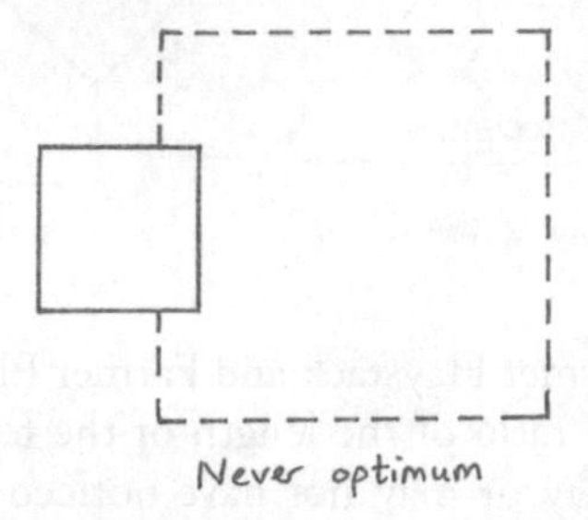

What if we take configuration B and translate it in the negative x-direction? Is it possible that some optimum exists for a triangular notch out of the field that extends the perimeter by a length X which is smaller than the length of the longest diagonal ($\sqrt{2}l$)? We now examine this in two parts: first the case $l < X < \sqrt{2}l$ and $X < L/3$, and second the case $X \leq l$ and $X < L/3$.

- CONFIGURATION C: DIAGONAL NOTCH OUT OF FIELD EXTENDING PERIMETER BY $l < X < \sqrt{2}l$ AND $X < L/3$.

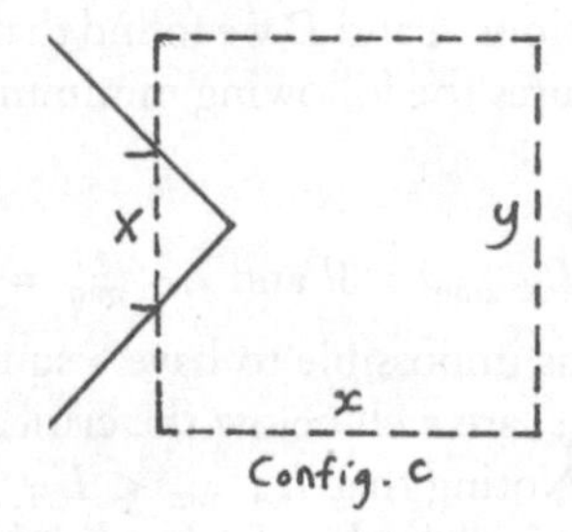

One way to determine whether a configuration C could be better than a configuration B ($X = \sqrt{2}l$) is to find the value of X that maximises the area of a square field for a given L. The area is given by

$$A_C = \frac{L^2}{16}\left[1 + 2\frac{X}{L} - 3\left(\frac{X}{L}\right)^2\right]$$

Making the substitution $X' = X/L$, differentiating with respect to X', and setting $dA_C/dX' = 0$, we have $X' = 1/3$. That is, $X = L/3$. We have shown, for a square field at least, that the optimum size of intrusion is equal to the length of a full side of the square field. The form of $A_C = f(X/L)$ is *quadratic*, so we should know that there exists only a single maximum, which is for $X = L/3$. For a barn of side l the maximum exists for $X = \sqrt{2}l = L/3$, or $L = 3\sqrt{2}l$. As we move away from this maximum the area of the field gets smaller. But we have determined that $A_A > A_B$ for $L < L_c$. It is apparent (although it may take some thought to appreciate it) that $A_C < A_B$ *always*, because an intrusion of length $l < X < \sqrt{2}l$ is always further from the maximum area than an intrusion of length $X = \sqrt{2}l$.

- CONFIGURATION D: DIAGONAL NOTCH OUT OF FIELD EXTENDING PERIMETER BY $X \leq l$ AND $X < L/3$.

  For this case it is obvious by inspection that we can achieve the same increase in perimeter, but with no loss of area, by using configuration A. So configuration D can never be better than configuration A.

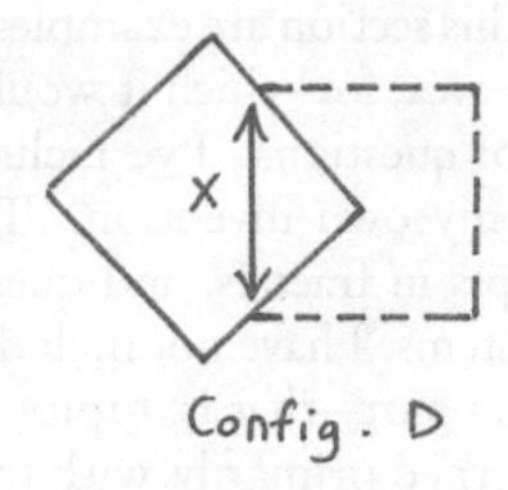

It seems that for $L_{B,\,\min} > L > L_{A,\,\min}$, that is, $3\sqrt{2}l > L > 3l$, configuration A is optimal. We also state, with reference to the previous question, but without further proof, that for $2l < L < 3l$, rectangular fields with aspect ratios $1/2 < x/y < 1$ are optimal. For these situations, one of the long sides of the field, y, is provided by the wall of the barn. For $L \leq 2l$, fields with aspect ratios $x/y = 1/2$ and fences built alongside the barn (which is used for one of the long sides, y) are optimal.

Chapter 2

Mathematics

Most of the questions in this section are examples of what we might call *applied* mathematics. It is a huge area, for which it would be impossible to give a representative distribution of questions. I've included a few well-known, easier questions, and a few of my own invention. Topics covered include simple logic, probability, concepts in fractals, and questions which ask us to set up equations for physical systems. I have not included many problems that mathematicians would regard as *pure*—that is, topics which are are intrinsically abstract. Here we are concerned primarily with the application of mathematics to real-world problems.

2.1 Human calculator ★

There are only a few examples of people with such exceptional skills in arithmetic (addition, subtraction, multiplication and division) that we can legitimately call them 'human calculators'. It is thought that the earliest academic account of what we now refer to as *savantism* was by Benjamin Rush in 1789.[1] He described an African slave called Thomas Fuller, who had developed an ability to do lightning calculations in his head. The subject of savantism has now been studied extensively. A relatively recent article by the Berkeley educational psychologist Arthur Jensen[2] starts with the words:

> It seems hard to believe, but the following is reported in the Guinness Book of Records (1982), which has a reputation for the authenticity of its claims: 'Mrs. Shakuntala Devi of India demonstrated the multiplication of two 13-digit numbers of 7,686,369,774,870 × 2,465,099,745,779 picked at ran-

[1] Rush, B., 1789, "Account of a wonderful talent for arithmetical calculation in an African slave, living in Virginia," American Museum, 1789, pp. 62–63.

[2] Jensen, A.R., 1990, "Speed of information processing in a calculating prodigy," Intelligence, 14, pp. 259–274.

> dom by the Computer Department of Imperial College, London on 18 June 1980, in 28 s. Her correct answer was 18,947,668,177,995,426,462,773,730.'

The literature contains many examples of savantist feats,[3] by individuals most commonly with a prodigious talent for music, drawing, calendar calculating, arithmetic and precise distance measurement. Most savants have some form of autism, in many cases to a severely disabling extent, but in rarer cases high-functioning autism which allows them to lead quite conventional lives. A recent memoir[4] of the young English arithmetic savant Daniel Tammet gives a fascinating account of how he perceives calculation. He describes a process of calculation that is incredibly visual, rather than mechanical. The individual numbers, which he says all have distinct shapes and colours in his mind, merge together to form a new number which is the result of the calculation. The separate tendency to associate numbers and words with colours is described as *synesthesia*, making the experience of the process he describes almost unique and extremely vivid.

The feats of these human calculators attract derision from some quarters of the scientific community. It is true that calendar calculating, or the ability to quickly give the day of the week corresponding to a particular random date many years in the past or future, is made possible by simple algorithms which can be committed to memory. One of the first of these was *Zeller's congruence*, an algorithm devised by Christian Zeller in 1883.[5] Five years later, in March 1887, the rather more famous Lewis Carroll[6] wrote a letter to Nature entitled *To find the day of the week for any given date.*[7] The letter opened with these words:

> Having hit upon the following method of mentally computing the day of the week for any given date, I send it you in the hope that it may interest some of your readers. I am not a rapid computer myself, and as I find my average time for doing any such question is about 20 seconds, I have little doubt that a rapid computer would not need 15.

---

[3]Treffert, D.A., 2009, "The savant syndrome: an extraordinary condition. A synopsis: past, present, future," Philosophical Transactions of the Royal Society, B—Biological Sciences, 364 (1522), pp. 1351–1357.

[4]Tammet, D., 2007, "Born on a blue day: the gift of an extraordinary mind," Hodder Paperbacks.

[5]Zeller, C., 1883, "Problema duplex calendarii fundamentale," Bulletin de la Société Mathématique de France, 11, pp. 59–61.

[6]Lewis Carroll's real name was Charles Lutwidge Dodgson (1832–1898). He is most famous for writing the children's book Alice's Adventures in Wonderland, but he performed academic work in mathematics and logic, fields which informed even his non-academic writing. He also wrote some fantastic poems, including The Hunting of the Snark and Jabberwocky.

[7]Carroll, L., 1887, "To find the day of the week for any given date," Nature (Letter), March 31, 1887, p. 517.

This seems to have been the inspiration for what is now probably the best known method for date calculation, the Doomsday algorithm.[8] Developed by John Conway[9] in 1973, it enables the user, with practice, to calculate the day of the week for any date in the past or future within about two seconds! So this particular savantist feat is demystified. Or perhaps not. Research performed at the Department of Neurosciences at the University of California[10] suggests that the explanation may not be quite so simple:

> Despite several investigations of calendar savants, the methods used by savants to determine dates remain poorly understood. Several algorithms exist for determining the day of the week for any given date. However, the idea that these are widely used has been largely dismissed because the impaired arithmetic ability of the typical calendar savant makes it unlikely that the task could be approached in this way. Furthermore, many savants are able to answer "reverse" calendar questions (e.g., "What is the date of the second Monday of March 1983?"), which would require either a substantial modification to these algorithms or a slower method based on generating possible answers and then working toward the correct date. A more likely alternative is that the savant uses either memory alone or a combination of memory and simple calculation to take advantage of regularities in the structure of the calendar.

This would be consistent with the other abilities of many arithmetic savants, which often include extraordinary powers of retention. On 14 March 2004, for example, Daniel Tammet repeated the number π to 22,514 digits in just over five hours without making a single mistake.

Most normal people need to develop strategies for doing even modest arithmetical tasks in their heads. This question is designed to encourage students to think about possible ways of making a computational task manageable for someone with average ability.

DEVISE A SIMPLE method to multiply 993 and 1,007 in your head without resorting to pen and paper.

Answer

Over dinner once I was told what I believe is a true story about the principal of a Cambridge college taking out a calculator to multiply a number by 100.

---

[8]Conway, J.H., 1973, "Tomorrow is the day after doomsday," Eureka, Volume 36, pp. 28–31.

[9]John Conway (1937–) is an influential British mathematician who works on finite groups, number theory, game theory, knot theory and coding theory. In popular mathematics Conway is well-known for inventing the Game of Life, a *cellular automaton* on a two-dimensional grid, in which complex patterns can emerge (and in some cases self-replicate) from very simple underlying rules.

[10]Kennedy, D.P., Squire, L.R., 2007, "An analysis of calendar performance in two autistic calendar savants," Learning Memory, 14, pp. 533–538.

In a rare moment of lucidity I quipped, "was it a difficult number that was being multiplied?" Only the scientists got the joke.

To multiply even relatively small numbers in my head I need to resort to strategies to break the problem down. The purpose here is to see if a student can do the same. One approach is to see that 993 and 1,007 are both related to the number 1,000. Specifically,

$$993 = 1,000 - 7 = a - b$$

$$1,007 = 1,000 + 7 = a + b$$

A *method* that could be used is to note—without paper, of course—that

$$993 \times 1,007 = (a - b)(a + b) = a^2 - b^2$$

This leads to something that most of us can handle fairly easily in our heads.

$$a^2 - b^2 = 1 \times 10^6 - 49 = 999,951$$

As a final footnote, I showed this question to a Japanese friend who I studied physics with. I expected him to say it was easy, but he thought it was *laughably* easy. He then went on to explain that, in Japan, young children still learn to do arithmetic on an abacus,[11] and quickly learn to visualise the process of multiplying modestly large numbers in their heads. Apparently it is not at all unusual for Japanese children to be able to multiply three- to five-digit numbers in their heads in a matter of seconds. Crumbs!

2.2 Professor Fuddlethumbs' reports ★

PROFESSOR FUDDLETHUMBS HAS forgotten the names of his students, and unfortunately all six of them are boys. At the end of term meeting he has to read out the reports from their tutor and is in the sticky position of having to do this at random. Calculate the chance that exactly five of the students receive the correct report.

Answer

I am not a great fan of trick questions, but I suppose a purist would argue that there is no trick here. In fact the question should be totally straightforward for someone who is good at questions involving *chance*. It is impossible for *exactly* five students to receive the correct report, because if five students receive the correct report the sixth must also receive the correct report. The chance is zero.

[11] The abacus, or counting frame, is a system of beads that slide on a wire frame (or in troughs in a board), and which is used to perform arithmetical calculations. It has been in use since about 2700 BC.

We might equally have been asked to calculate the chance of *at least* five students receiving the correct report. In this case we would have computed the number of outcomes that satisfy the condition (there is only one) and the total number of possible outcomes (the number of ways of arranging the reports), and taken the ratio.

The answer to that question would have been

$$P = \frac{1}{6!} = \frac{1}{720}$$

Professor Fuddlethumbs would have had to have been very lucky indeed.

2.3 More of Professor Fuddlethumbs' reports ★

PROFESSOR FUDDLETHUMBS HAS forgotten the names of his students again. Thank heavens that this year four of them are girls and three are boys. From the names on the seven reports he can at least work out whether they belong to girls or boys. Unfortunately they are highly contrasting reports, either excellent or poor. In fact, they simply read either "excellent work this term" or "poor work this term". There are two excellent reports for the boys, and two excellent reports for the girls. Calculate the chance of all the students getting an appropriate report.

Answer

This question can be broken down into relatively simple parts. Because we can sort the reports into those for girls and those for boys, the chance of the girls all getting an appropriate report, $P(A)$, and the chance of the boys all getting an appropriate report, $P(B)$, are *independent* of each other. The chance of all the girls and all the boys getting appropriate reports is therefore

$$P(A \cap B) = P(A)\,P(B)$$

Let us calculate $P(A)$ and $P(B)$ individually.

Consider the number of *unique ways* the four reports for girls might be arranged. The number of unique ways is equal to the total number of ways all four reports might be arranged (4!) divided by the number of indistinguishable ways the two good reports can be arranged (2!), divided by the number of indistinguishable ways the two bad reports can be arranged (2!). The number of unique arrangements is

$$\frac{4!}{2!2!} = 6$$

The probability of all the girls getting appropriate reports is the reciprocal of the number of unique ways the reports can be arranged:

$$P(A) = \frac{1}{6}$$

The possibilites are 1100, 1010, 1001, 0110, 0101 and 0011, where we use 1 to signify an excellent report, and 0 to signify a poor report.

Let us now consider the three boys. There are 3! ways of arranging the reports. The are 2! indistinguishable ways of arranging the good reports and 1! indistinguishable ways of arranging the bad reports. The number of unique arrangements is

$$\frac{3!}{2!1!} = 3$$

The probability of all the boys getting appropriate reports is the reciprocal of the number of unique ways the reports can be arranged:

$$P(B) = \frac{1}{3}$$

The probability of all the girls *and* all the boys getting appropriate reports is therefore

$$P(A \cap B) = P(A)\,P(B) = \frac{1}{6} \times \frac{1}{3} = \frac{1}{18}$$

Professor Fuddlethumbs still needs to be very lucky to get them all right, but the situation is not quite as bleak as before.

2.4 Ant on a cube I ★★

AN ANT STARTS on one vertex of a cube and walks randomly along the edges of the cube, from vertex to vertex. Whenever the ant encounters a vertex it decides at random which edge to walk along, including returning along the edge it just traversed. Calculate the probability that the ant will end up at the vertex diagonally opposite the starting vertex after exactly seven moves, having also visited all vertices.

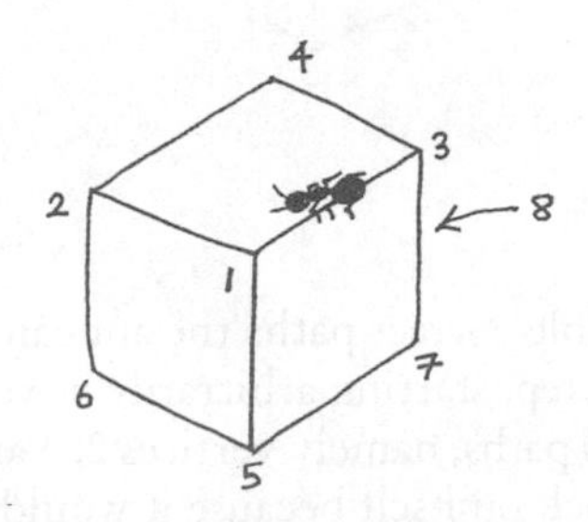

Answer

Consider the possible correct paths the ant can take. Starting arbitrarily at vertex 1 and ending at vertex 8, the paths must involve no doubling back and must take in all vertices en route to vertex 8. We will call these paths *good paths*. For the first step, starting at vertex 1, the ant has three possible options for good paths, namely vertices 2, 3 and 5. From each of these vertices there are two second steps, which are enumerated in the table. From each of these paths there is only one possible route that takes in all the vertices on the way to vertex 8 with no doubling back. The total number of good paths is 6.

$$1\left\{\begin{array}{l} 2\left\{\begin{array}{l}437568\\657348\end{array}\right.\\ 3\left\{\begin{array}{l}426578\\756248\end{array}\right.\\ 5\left\{\begin{array}{l}624378\\734268\end{array}\right.\end{array}\right.$$

If the ant makes seven steps, the total number of possible paths is $3^7 = 2,187$. The probability of reaching vertex 8 in seven steps that take in all the vertices is

$$\frac{6}{2187} = \frac{2}{729}$$

2.5 Ant on a cube II ★★★

AN ANT STARTS on one vertex of a cube and walks randomly along the edges of the cube, from vertex to vertex. Whenever the ant encounters a vertex it decides at random which edge to walk along, including returning along the edge it just traversed. Calculate the probability that the ant will have visited each vertex once after seven moves.

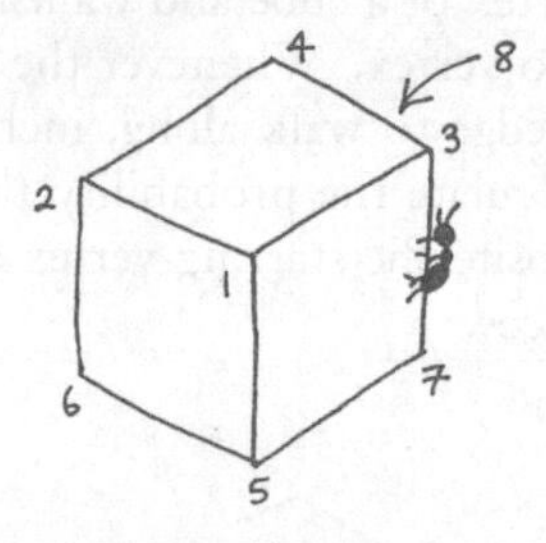

Answer

Let us consider the possible correct paths the ant can take, which we will call *good paths*. For the first step, starting arbitrarily at vertex 1, the ant has three possible options for good paths, namely vertices 2, 3 and 5. At the second step, the ant cannot double-back on itself because it would not have enough moves available to then complete the route to all vertices in the required seven steps. The ant therefore has only two good paths open to it. These are enumerated in the tree of good paths shown below. At the third step the ant again has two good paths open to it. Each of these paths leads to one of two fourth steps that have either a single solution, or two solutions. So there are 18 good paths in total.

$$1\left\{\begin{array}{l} 2\left\{\begin{array}{l} 4\left\{\begin{array}{l} 37\left\{\begin{array}{l}568\\ 865\end{array}\right.\\ 86573\end{array}\right.\\ 6\left\{\begin{array}{l} 57\left\{\begin{array}{l}843\\ 348\end{array}\right.\\ 84375\end{array}\right.\end{array}\right.\\ 3\left\{\begin{array}{l} 4\left\{\begin{array}{l} 26\left\{\begin{array}{l}578\\ 875\end{array}\right.\\ 87562\end{array}\right.\\ 7\left\{\begin{array}{l} 56\left\{\begin{array}{l}248\\ 842\end{array}\right.\\ 84265\end{array}\right.\end{array}\right.\\ 5\left\{\begin{array}{l} 6\left\{\begin{array}{l} 24\left\{\begin{array}{l}378\\ 873\end{array}\right.\\ 87342\end{array}\right.\\ 7\left\{\begin{array}{l} 34\left\{\begin{array}{l}268\\ 862\end{array}\right.\\ 86243\end{array}\right.\end{array}\right.\end{array}\right.$$

If the ant makes seven steps, the total number of possible paths is $3^7 = 2,187$. The number of good paths is 18. The probability of visiting all eight vertices in seven steps when taking a random walk is

$$P = \frac{18}{2187} = \frac{2}{243}$$

2.6 Ant on a cube III ★★★★

This is a difficult question. I recommend that you try both of the easier variants before attempting this one.

An ant starts on one vertex of a cube and walks randomly along the edges of the cube, from vertex to vertex. Whenever the ant encounters a vertex it decides at random which edge to walk along, including returning along the edge it just traversed. Calculate the probability that it will end up at the vertex diagonally opposite the starting vertex after exactly seven moves.

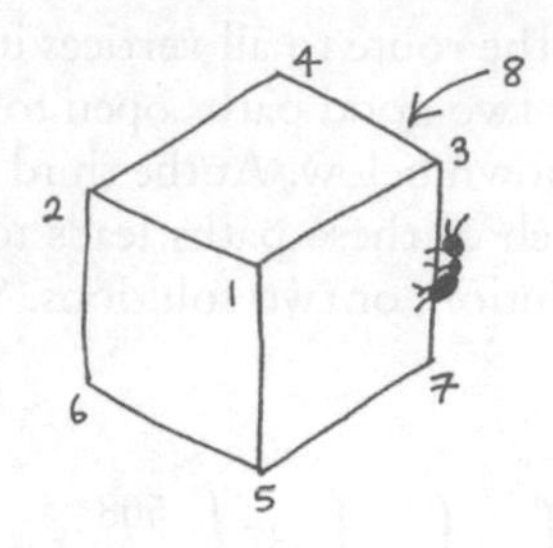

Answer

Place the cube in an x, y, z co-ordinate system so that the starting point is at the origin $(0, 0, 0)$ and the goal is at $(1, 1, 1)$. We need to calculate the probability of changing the state from $(0, 0, 0)$ to $(1, 1, 1)$ in exactly seven steps. At each step, the ant can change the state of one of the co-ordinates (either the x, y or z co-ordinate). We now represent in tabular form the number of times the ant has changed the state of each of the x, y and z co-ordinates, n_x, n_y and n_z respectively, and the resulting final state. We first represent the number of times as either *even* or *odd*, because this uniquely determines the final state.

n_x	n_y	n_z	Result
odd	odd	odd	Takes $(0, 0, 0)$ to $(1, 1, 1)$
odd	even	even	This (and permutations) cannot take $(0, 0, 0)$ to $(1, 1, 1)$
even	odd	odd	Sum of steps even, so not possible in exactly seven steps
even	even	even	Sum of steps even, so not possible in exactly seven steps

Consider the number of ways that 7 can be divided by odd numbers. Only the following ways are possible

$$7 = 3 + 3 + 1 \text{ and its three permutations}$$
$$7 = 5 + 1 + 1 \text{ and its three permutations}$$

Consider the number of *unique ways* the ant can perform seven steps with the restriction that there are indistinguishable groups of 3, 3 and 1. It is the total number of ways, 7!, divided by the product of the numbers of ways each indistinguishable group can be arranged. We get 7!/(3!3!1!). Taking each way of dividing 7 into odd numbers, and each permutation of that division, the total number of ways of getting odd, odd, odd from seven steps is therefore given by

$$\frac{7!}{3!3!1!} \times 3 + \frac{7!}{5!1!1!} \times 3 = 546$$

At each vertex the ant can take three possible routes, so the total possible number of paths it can take in seven steps is 3^7. The probability of getting odd, odd, odd in each of the x, y and z directions is given by

$$\frac{546}{3^7} = \frac{182}{729} \approx 0.2497$$

Because odd, odd, odd *always* leads to changing the state from $(0, 0, 0)$ to $(1, 1, 1)$, and because no other odd/even combination leads to $(1, 1, 1)$, the probability of the ant arriving at $(1, 1, 1)$ in a random walk of seven steps is $182/729$. It is interesting to note that this is very close, but not equal, to $1/4$.[12]

2.7 A falling raindrop ★★

Clouds are formed by the cooling of air which is saturated with water vapour. Because the *saturated vapour pressure* (a measure of the amount of water vapour air can contain) increases strongly with temperature, as saturated air cools water vapour condenses out of the air to form very small water droplets. These droplets can range in size from about 1 to 100 μm, are small enough that they

---

[12]With an odd number of steps, the ant can only change state from $(0, 0, 0)$ to one of $(1, 0, 0)$, $(0, 1, 0)$, $(0, 0, 1)$ or $(1, 1, 1)$. That is, there are four possible outcomes. The chance of achieving each outcome depends less on the starting position as we increase the number of steps. As $n \to \infty$ (for odd n), the probability of the ant finishing at $(1, 1, 1)$ approaches $1/4$. We demonstrate this by considering the cases of five steps and nine steps. We first note the magnitude of the remainder for the case of seven random steps. We have $(1/4) - (182/729) = 1/2916$.

- Case of Five Steps. The number of ways of dividing by odd numbers is

$$5 = 3 + 1 + 1 \text{ and its three permutations}$$

The total number of ways of getting odd, odd, odd from five steps is therefore

$$\frac{5!}{3!1!1!} \times 3 = 60$$

The total possible number of paths the ant can take in five steps is $3^5 = 243$. The probability of the ant arriving at $(1, 1, 1)$ in a random walk of five steps is therefore $60/243$. The remainder is $(1/4) - (60/243) = 1/324$. As expected, this is larger than the remainder for seven steps.

- Case of Nine Steps. The number of ways of dividing by odd numbers is

$$9 = 3 + 3 + 3 \text{ (this is the only permutation)}$$
$$9 = 5 + 3 + 1 \text{ and its six permutations}$$
$$9 = 7 + 1 + 1 \text{ and its three permutations}$$

The total number of ways of getting odd, odd, odd from nine steps is therefore

$$\frac{9!}{3!3!3!} \times 1 + \frac{9!}{5!3!1!} \times 6 + \frac{9!}{7!1!1!} \times 3 = 4,920$$

The total possible number of paths the ant can take in nine steps is $3^9 = 19,683$. The probability of the ant arriving at $(1, 1, 1)$ in a random walk of nine steps is therefore $4920/19683$. The remainder is $(1/4) - (4920/19683) = 1/26244$. As expected, this is smaller than the remainder for seven steps.

rarely collide, and are light enough that they remain aloft on air currents for a very long period. Raindrops form by the gradual coalescence of small droplets, until the larger droplet that has been formed can overcome air currents and fall through the remainder of the cloud, coalescing with smaller droplets as it goes.

A SPHERICAL RAINDROP falls through a cloud with a uniform density of small condensed droplets. After falling 1 km, the radius of the raindrop is 5 mm. What volume fraction of the cloud is made up of condensed water droplets? We can assume that the raindrop starts with negligible size, and accumulates any small condensed droplets that it passes through.

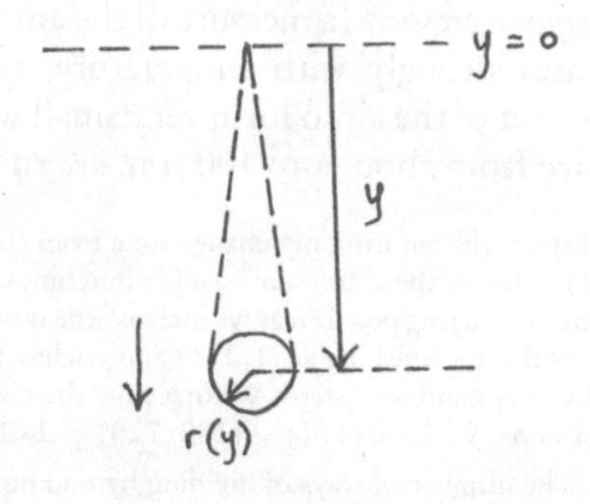

Answer

Let the raindrop have radius r after falling a distance y through the cloud. Letting the swept volume of the cloud be V_c, the rate (with respect to y) at which the raindrop sweeps out volume, dV_c/dy, is equal to the cross-sectional area of the raindrop, πr^2. We write

$$dV_c/dy = \pi r^2$$

Letting the raindrop have volume V_d, the rate (with respect to y) at which the raindrop changes volume, dV_d/dy, is equal to the rate at which the raindrop sweeps out volume multiplied by the fraction of the swept volume that is condensed water droplets. We define this *volume fraction* as

$$\lambda = \frac{\text{Vol}_{\text{water}}}{\text{Vol}_{\text{air+water}}}$$

We write

$$\frac{dV_d}{dy} = \frac{dV_c}{dy}\lambda = \pi r^2 \lambda$$

The volume of a spherical raindrop is given by $V_d = (4/3)\,\pi r^3$. Combining this with the relationship, and using the chain rule, we have

$$\frac{dV_d}{dy} = \frac{d\left((4/3)\,\pi r^3\right)}{dy} = 4r^2\pi\frac{dr}{dy} = \pi r^2\lambda \quad \text{or} \quad \frac{dr}{dy} = \frac{\lambda}{4}$$

Taking the raindrop to be negligibly small at $y = 0$, we integrate y between 0 and Y, and r between 0 and R

$$\int_0^R dr = \frac{\lambda}{4}\int_0^Y dy$$

to obtain

$$R = \frac{\lambda}{4}Y$$

We see that R is proportional to Y. The relationship between the increase in radius and the distance the raindrop falls is a linear one.

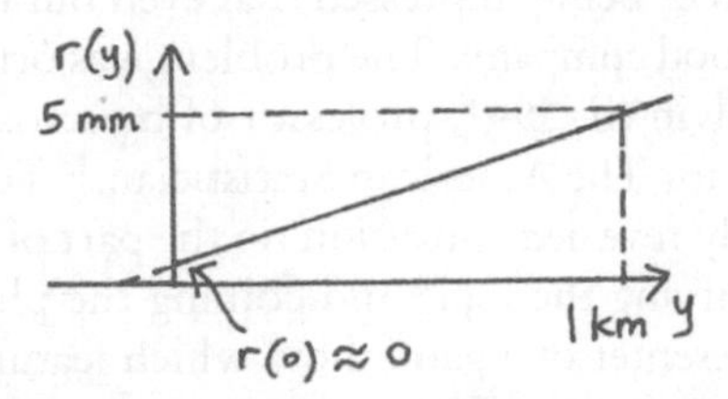

This should be no surprise. The rate at which volume accumulates is proportional to cross-sectional area, which is itself proportional to r^2, and a small change in volume can be expressed by $\delta V = (4\pi/3)\left[(r+\delta r)^3 - r^3\right] \approx 4\pi r^2\delta r$, which has the same form with r.

If after falling 1 km ($Y = 1{,}000$ m) the raindrop has radius $R = 5$ mm, we obtain

$$\lambda = \frac{4R}{Y} = 2 \times 10^{-5}$$

The volume fraction of the cloud that is condensed water is $\lambda = 2 \times 10^{-5}$.

At approximately 22 °C the *specific volume* of saturated water vapour is $v_g = 50\ \text{m}^3\ \text{kg}^{-1}$. This gives a density of $\rho_g = 1/v_g = 0.020\ \text{kg m}^{-3}$. If water vapour of this density were to condense out to form water droplets, the volume the condensed water would occupy is $2 \times 10^{-5}\ \text{m}^3$ (taking the density of condensed water to be approximately $1{,}000\ \text{kg m}^{-3}$). This corresponds to a volume fraction, $\lambda = 2 \times 10^{-5}$. It seems we are dealing with a tropical rainstorm.

2.8 The Three Door Problem ★★

I was asked this question when I came for a physics interview at Oxford. I still remember it vividly, but not with much pleasure. I did abysmally. It is a long time ago now, but if I remember correctly I not only got the answer wrong, I also did not immediately understand the explanation when it was offered. I nodded in agreement and said something vague about it being interesting, hoping desperately that it wouldn't be followed up. It wasn't. But at the same time, I doubt I got away with it. A young and slightly over-enthusiastic graduate student had asked the question–fortunately I never saw that person again. On the train on the way home it suddenly *clicked*, and I felt very stupid indeed. The thing was, I had never seen a probability paradox before, not even a very simple one, which this undoubtedly was. It was useful to learn the lesson from personal experience that very simple ideas can be quite confusing when you meet them for the first time. I try and bear this in mind when teaching. Some ideas just take *time* to settle in.

I took the problem back to my school and shared it with our A-level maths class. Not only could nobody answer it, but very few people believed the explanation. I remember being impressed that even our teacher was confused! It seems we were in good company. The problem was originally posed in 1975 by Professor Steve Selvin (b. 1941, professor of biostatistics at UC Berkeley) in an article he wrote for The American Statistician.[13] Following criticisms of his logic, which merely revealed confusion on the part of the readers, he wrote a further letter[14] clarifying the logic and coining the phrase the Monty Hall problem, after the presenter of a gameshow which featured a similar scenario to the one posed in the question. The problem was popularised by Marilyn vos Savant,[15] who ran the problem in her Parade magazine column in September 1990. She met with angry and vocal criticism from many readers. In the days that followed, over 10,000 letters arrived at the Parade offices! They flooded in from people who maintained not only that her answer was incorrect, but that she was encouraging mathematical illiteracy. Among her detractors was a statistician from the National Institute of Health, and professors from at least seven US universities, including a Nobel Prize winner. The controversy continues today. There have been countless newspaper and magazine articles dealing with the problem and its aftermath, and at least two books are devoted solely to the history of the problem.[16] If there is one probability paradox you should be familiar with it is certainly this one.

---

[13]Selvin, S., 1975, "A problem in probability (letter to the editor)," The American Statistician 29 (1): 67.

[14]Selvin, S., 1975, "On the Monty Hall problem (letter to the editor)," The American Statistician 29 (3): 134.

[15]Marilyn vos Savant (b. 1946) is an American columnist famous for the Ask Marilyn column in Parade magazine, which often features mathematical puzzles. From 1986 to 1989 Savant was listed in the Guinness Book of World Records in the category Highest IQ.

[16]Rosenhouse, J., 2009 "The Monty Hall problem: The remarkable story of math's most contentious brain teaser," Oxford University Press, ISBN–10: 0195367898.

The curious thing from the perspective of someone who *is* now familiar with the problem is that it is quite hard to imagine not being able to see the solution. You can form your own opinion on this once you have solved it.

YOU ARE INVITED to play a gameshow. There are three closed doors. Behind them, in an order known to the gameshow host, but unknown to you, are two goats and a bag of gold. You are asked to pick a door. At the end of the game you get to keep whatever is behind your chosen door. But the game isn't over yet. Once you have picked a door the host announces that he is going to open one of the other two doors and reveal what is behind it. He opens a door and reveals a goat. Before the game ends and you open a door, you are then asked if you want to swap to the other unopened door, or stick with the door you originally picked. What should you do and why?

Answer

The correct answer is that you double your chance of winning by switching. If you stick, your chance of winning is $1/3$, and if you switch your chance of winning is $2/3$.

Let us deal first with the situation if you stick. When you first pick a door, you have no information about which door the bag of gold is behind. It is obvious that your chance of picking the correct door at random is $1/3$. If you stick with this door, no matter what new information the gameshow host gives you, your probability of winning is unchanged.

Now let's look at the situation if you switch. By opening a door to reveal a goat, the gameshow host has injected new information into the problem. We now know that the bag of gold is behind either the door you originally picked—call the probability of this $P(A)$—or the remaining unopened door—call the probability of this $P(B)$. If the gold is behind one or the other we have $P(A) + P(B) = 1$. We know, however, that $P(A) = 1/3$. This gives $P(B) = 2/3$. The chance of winning if you switch is $2/3$. If you switch you double your chances of winning.

If this still doesn't *feel* right, consider the following example. We go on a trip to the city of London together. I take something from you—say, your watch—and head off to hide it behind one of the doors in the city. Let's say there are a million doors. Hours later I come back. I could have gone a very long way or I could have been waiting in the next street. You don't even know which suburb I visited to hide your watch, let alone which part of that suburb, or which street, or which house. I then invite you to head off into London to pick a door. You come back with the address of the door you picked. I then write down one other address, so you have two addresses in London, your one and my one. I tell you that your watch is behind the door at one of those two addresses. I think it is fairly obvious which address you should go to first.

If you found that challenging, you were not alone. The cognitive psychologist Massimo Piattelli-Palmarini wrote of the Monty Hall problem "... no other statistical puzzle comes so close to fooling all the people all the time...even Nobel physicists systematically give the wrong answer, and they insist on it, and they are ready to berate in print those who propose the right answer."

2.9 Dr Bletchley's PIN ★★★

DR BLETCHLEY HAS forgotten the four-digit PIN[17] for his bank card. The digits can take the values 0, 1...9. All he can remember is that there was at least one zero in the PIN. What is the maximum number of PINs he will need to try?

Answer

The easy way of solving this problem is to compute the number of PINs that do not include a zero, and subtract that from the total number of PINs. If zero is not permitted there are 9 options $(1, 2...9)$ for each of the four numbers. The number of PINs that *do not* include a zero is $9 \times 9 \times 9 \times 9 = 9^4$. By the same logic, the total number of PINs is 10^4. The maximum number of PINs Dr Bletchley will need to try is

$$N = 10^4 - 9^4 = 3,439$$

[17] Personal Identification Number, also referred to redundantly as a PIN number, or a pincode.

This is certainly the preferred method for problems of this type, and illustrates the power of considering the inverse problem when tackling combinatorics puzzles.

We can solve the problem by another, slightly less direct route, in which we consider the possibility that the known number (0) appears as each of the first, second, third or fourth digits. We consider the problem in four steps.

- STEP 1. Consider first the number of PINs that take the form

  where $\star$ is a number between 0 and 9. The number of PINs with this form is $1 \times 10 \times 10 \times 10 = 1,000$.

- STEP 2. Consider PINs that take the form

$$\star' \quad 0 \quad \star \quad \star$$

  where $\star$ is a number between 0 and 9, and $\star'$ is a number between 1 and 9. We distinguish between $\star'$ and $\star$ because in step 1 we already counted every possible PIN with 0 in first place, and do not wish to re-count them. The additional number of PINs step 2 generates is $9 \times 1 \times 10 \times 10 = 900$.

- STEP 3. Consider PINs that take the form

$$\star' \quad \star' \quad 0 \quad \star$$

  We have already considered every PIN that starts with either 0 or 0 0, so the number of unique PINs generated in step 3 is given by $9 \times 9 \times 1 \times 10 = 810$.

- STEP 4. Consider PINs that take the form

$$\star' \quad \star' \quad \star' \quad 0$$

  We have already considered every PIN that starts with either 0, 0 0 or 0 0 0, so the number of unique PINs generated in step 4 is given by $9 \times 9 \times 9 \times 1 = 9^3 = 729$.

The total number of PINs is given by the sum of the unique PINs generated in steps 1 to 4. That is

$$1,000 + 900 + 810 + 729 = 3,439$$

We see that the answer is the same as that acheived using the simpler method.

2.10 Mr Smith's coins ★★★

A friend was asked this question as part of a university physics entrance test in the mid-1980s. It is a well-known puzzle which has appeared in a number of forms over the years. The question normally comes in three parts, with a third, harder variant which introduces concepts in information and distribution selection. The easier parts of the problem are as follows.

ANSWER THE FOLLOWING questions.

1. Mr Smith tossed a coin twice. The second toss yielded a head. What is the probability that both tosses yielded heads?

2. Mr Smith tossed a coin twice. At least one toss yielded a head. What is the probability that both tosses yielded heads?

Answer

This is a well-known paradox in probability theory, referred to as the Two Child Problem, or the Boy or Girl paradox. The first popular example of it is credited to Martin Gardner, who published it in his column in Scientific American in 1959.[18] The reason it is a well-known paradox is that instinct leads us to the wrong answer. We assume, incorrectly, that the probability of two heads is the same in both cases. Let us consider the cases in turn.

- CASE 1: Mr Smith tossed a coin twice. The second toss yielded a head. What is the probability that both tosses yielded heads?

  Consider first all possible combinations. There are four in total which could in principle occur with equal likelihood.

[18] Martin Gardner (1914–2010) is without doubt the biggest name in *recreational mathematics*: mathematics for fun, and fun alone. He has done more to popularise mathematics than perhaps anyone else, having written over 100 books, many of them on mathematical topics. He is probably most famous for his Mathematical Games column in Scientific American, which he wrote from 1956 to 1981. If you have any interest in recreational mathematics his books are unsurpassed.

$$\begin{matrix} H & T \\ H & H \\ T & T \\ T & H \end{matrix}$$

Based on the information provided, the only possible options for Mr Smith's two tosses are $H\,H$ and $T\,H$. They occur with equal likelihood. The probability that both tosses yielded heads is $P_1 = 1/2$.

- CASE 2: Mr Smith tossed a coin twice. At least one toss yielded a head. What is the probability that both tosses yielded heads?

  Based on the information provided, the possible options are $H\,T$, $H\,H$ and $T\,H$. They occur with equal likelihood. The probability that both tosses yielded heads is $P_2 = 1/3$.

We see that $P_1 \neq P_2$. The apparent paradox is explained by the fact that the information given in the questions is quite different. In the first case we are given information about specifically one event. In the second case we are given information that relates to either or both events.

2.11 The three envelope problem ★

I GIVE YOU an envelope. You open it and find it contains a certain amount of money. I put half the amount of money into a second envelope, and twice the amount into a third envelope. The envelopes are indistinguishable once I have shuffled them. To maximise your *expected return*,[19] should you stick with the envelope you have, or swap?

Answer

You should swap. Let us say that your envelope contains X. You know that the second and third envelopes therefore contain $X/2$ and $2X$. Consider the two scenarios.

[19]The *expected return* is the average value of a random variable when a process is repeated an infinite number of times. It is a measure of the centre of the distribution of the variable. The expected return $E[R]$ is given by $E[R] = \sum_{i=1}^{n} R_i P_i$, where R_i is the return given outcome i (of a possible set of n outcomes), and P_i is the probability of outcome i.

- SCENARIO 1: YOU STICK. The only possible outcome is a return $R_1 = X$, with probability $P_1 = 1$. Your expected return is $E[R] = \sum_{i=1}^{n} R_i P_i = X.1 = X$.

- SCENARIO 2: YOU SWAP. There are two possible outcomes. A return $R_1 = X/2$, with probability $P_1 = 1/2$, and a return $R_2 = 2X$, with probability $P_2 = 1/2$. Your expected return is $E'[R] = \sum_{i=1}^{n} R_i P_i = (X/2) . (1/2) + 2X\,(1/2) = (5/4)\,X$.

We see that $E'[R] > E[R]$, so our expected return is greater if we swap. It is perhaps a little counter-intuitive, but not nearly so counter-intuitive as the *two envelope paradox*, which we now discuss.

A note on the *two envelope paradox*

There is another, much harder, problem known as the two envelope paradox. A question is first phrased as follows.

You have two indistinguishable envelopes, and you know that one contains twice as much money as the other. You pick an envelope and open it. Once you have done so, I ask if you would like to stick with the envelope you have, or swap to the other envelope. To maximise your expected return, what should you do?

Common sense says that swapping shouldn't make any difference. The *paradox* is illustrated by the difficulty we have in finding the flaw in an apparently logical argument in favour of always swapping the envelope to maximise our expected return. The argument is as follows.

1. The amount of money in your envelope is X.

2. There is a fifty-fifty probability that X is the smaller amount, and a fifty-fifty probability that X is the larger amount.

3. Therefore the second envelope contains either $2X$ or $X/2$, each with a probability of $1/2$.

4. The expected return on switching to the second envelope is therefore $(1/2)\,(2X) + (1/2)\,(X/2) = (5/4)\,X$.

5. This is greater than the current return, X, so you will gain by swapping.

The problem is attributed to the Belgian mathematician Maurice Kraitchik (1882–1957), and it first appeared as The Paradox of the Neckties in his 1930 La Mathématiques des Jeux ou Recréations Mathématiques. It also appears in his 1953 Mathematical Recreations.[20] Kraitchik's version, as we might expect from a Belgian, is rather more elegant: "B and S each claim to possess the

[20]Kraitchik, M., 2006, "Mathematical recreations," Dover Recreational Math, Dover Publications Inc., 2nd edition, ISBN-10: 0486453588 ISBN-13: 978-0486453583.

better necktie..." Martin Gardner's American re-phrasing has two rich men comparing the contents of their wallets![21]

Every year there is at least one academic paper devoted to the two envelope paradox. So while the problem is not suited to a full discussion here, it is interesting to discuss briefly because it *looks* so elementary. It appears that no simple complete solution to this problem exists, but it is insightful to give an *incomplete explanation* of the flaw in logic.

> The incomplete explanation of the flaw in logic is that we have used X to represent two different things, both the smaller amount and the larger amount (see steps 2 and 3). We then perform an expected value calculation in which X represents two contradictory things. In the first term X represents the smaller amount, and in the second term the larger amount. The expectation value (step 4) is therefore meaningless, and this leads to an apparent paradox.
>
> A proposed resolution is to use X to represent a single quantity. *Before* any envelope is opened we can state that the envelopes contain X and $2X$. That is, we *define* X to represent only the *smaller* amount. If we pick an envelope at random, we have a fifty-fifty chance of picking X, and a fifty-fifty chance of picking $2X$. The expected value calculation is
>
> $$\frac{1}{2}2X + \frac{1}{2}X = \frac{3}{2}X$$
>
> According to this logic, the average expected value (if we perform the selection a large number of times) is $(3/2)\,X$, which is midway between the values in the two envelopes X and $2X$. By swapping we stand to gain as much as we stand to lose. There is no advantage in swapping.

The problem with this incomplete explanation is that it does not deal satisfactorily with the *distribution* of possible values X. Is X picked from an infinite set of possible values? Is X infinitely divisible? And so forth. These questions, and many more, are raised in papers dealing with this topic.

2.12 A card game ★★

Professor Plemis has invited Professor E. Lucid to dinner. He has been struggling with a problem of *chance*, and intends to ask Professor Lucid to enlighten him. As they attempt to find meat on a miniature fowl, Plemis sets the scene.

"My dear Lucid," he begins, "I really am very confused. Allow me to illustrate with these..." Plemis draws a pack of cards from his pocket and lays them on the table. Both men look at the cards. "Fifty-two cards, Lucid. If I

[21]Gardner, M., 1982, "Aha! Gotcha: paradoxes to puzzle and delight," W.H. Freeman & Co Ltd., ISBN-10: 0716714140 ISBN-13: 978-0716714149.

take a shuffled pack and deal a hand of one card, there are fifty-two possible combinations. If I deal a hand of two cards there are almost twice that number of combinations. The number of combinations goes up as I deal a hand with more and more cards."

"Not quite," says Lucid, "but do continue."

"Well that is exactly my problem," says Plemis. "When I deal a hand of fifty-two cards, there is only one possible combination. What I want to know is the number of cards which has the maximum number of combinations. I really am quite confused."

"My dear Plemis," says Lucid, "I am glad you are professor of Egyptology and not mathematics. This really is utterly elementary. Allow me to explain..."

Lucid then outlines a proof for the number of cards that gives the maximum number of combinations.

FOR A PACK of n unlike cards, what number of cards k maximises the possible number of combinations that can be dealt?

Answer

For a pack of n unlike cards, if we deal a hand of k cards there are ${}^nC_k$ combinations where

$${}^nC_k = \frac{n!}{k!\,(n-k)!}$$

We seek to maximise this number. We note that the numerator $n!$ is the same for all k for a particular pack of cards. The problem reduces to minimising the denominator, $k!\,(n-k)!$. We write this out explicitly.

$$k!\,(n-k)! = [1 \times 2 \times \ldots \times (k-1) \times k]\,[1 \times 2 \times \ldots \times (n-k-1) \times (n-k)]$$

If we deal one more card, we must consider the following product of terms

$$(k+1)!\,(n-(k+1))! = (k+1)!\,(n-k-1)!$$

or

$$(k+1)!\,(n-k-1)! = [1 \times 2 \times \ldots \times k \times (k+1)]\,[1 \times 2 \times \ldots \times (n-k-1)]$$

The ratio of these two products of terms is given by

$$\frac{k!\,(n-k)!}{(k+1)!\,(n-k-1)!} = \frac{n-k}{k+1}$$

We see that the ratio ${}^nC_{k+1}/{}^nC_k$ is given by

$$\frac{{}^nC_{k+1}}{{}^nC_k} = \frac{(k)!\,(n-k)!}{(k+1)!\,(n-k-1)!} = \frac{n-k}{k+1}$$

We now examine this ratio, noting that if ${}^nC_{k+1} > {}^nC_k$, *increasing* k would increase the number of combinations, and if ${}^nC_{k+1} < {}^nC_k$, *decreasing* k would increase the number of combinations.

We see that if $n-k = k+1$, or equivalently $k = (n-1)/2$, then ${}^nC_{k+1} = {}^nC_k$. For $n \in \mathbb{Z}^+$ and $k \in \mathbb{Z}^+$, this can only occur for odd n. Thus, for odd n, the solution is $k = (n-1)/2$. For $n - k > k + 1$, or equivalently $k < (n-1)/2$, we have ${}^nC_{k+1} > {}^nC_k$. Thus, for even n the solution is $k = n/2$.

As an example, consider the case of 52 cards. Here we have $n = 52$. Using the solution for even n we have $k = n/2 = 26$. The number of combinations of 26 cards we can deal is ${}^{52}C_{26} = 52!/(26! \times 26!) = 495,918,532,948,104$, or approximately 5×10^{14} combinations.

After explaining all this to Professor Plemis, Lucid sits back in his chair and goes quiet for a moment. "That number, Plemis," he said, finally, "is roughly equal to the number of grains of rice that would take up the same volume as the Great Pyramid of Giza.[22]

[22] Taking a grain of rice to be approximately 8 mm^3.

Chapter 3

Statics

In this section we look at problems in *statics*. By definition, *statics* is the study of systems that are stationary, and non-accelerating. In these systems the forces acting on individual masses must sum to zero, and the *torques* (or *moments*) acting on a body must sum to zero. These are requirements for zero acceleration (both linear and rotational). We generally look at the systems at a particular instant, and *resolve*[1] the forces so that the conditions for zero acceleration are satisfied. We should state a few definitions more formally.

- Torque or Moment. The magnitude of the torque (or moment) about a particular point (or pivot) due to a force is the product of the force and the perpendicular distance between the line of action of the force and the point of interest. We refer to this distance as the *lever arm* of the force. Torque has units of N m.

- Equilibrium Requirement for Zero Linear Acceleration. For an object of finite mass, for the linear acceleration in a particular direction to be zero, the forces in that direction must sum to zero. For the simplest two-dimensional Cartesian system, with axes x and y, we write $\sum F_x = \sum F_y = 0$. This condition must hold in every direction.

- Equilibrium Requirement for Zero Angular Acceleration. Taking an object of finite moment of inertia[2] about a particular pivot, A, for the angular acceleration to be zero, the sum of torques, T_{Ai}, about that pivot must be equal to zero. We write $\sum_i T_{Ai} = 0$. By the same logic, given that every real body has a finite moment of inertia, the sum of torques about *any point* in a system must be equal to zero, whether or not that point lies on the body.

[1]To resolve, take the sum of vector components in a particular direction.

[2]In the same way as mass is a measure of the resistance of a body to linear acceleration, the moment of inertia is a measure of the resistance of a body to angular acceleration. The moment of inertia I of a body is defined about a particular point. For a body composed of mass elements m_i at distances r_i from a point of interest A, the moment of inertia is defined by $I_A = \sum_i m_i r_i^2$.

- Centre of Mass (Gravity). The centre of mass (or gravity, for a uniform gravitational field) of a rigid body is the point at which an applied force would cause linear acceleration but zero angular acceleration (no rotation). For a body with uniform density, the centre of mass is at the centroid of the body. It does not need to lie on the body (consider a doughnut shape, for example).

Now we are ready to try a few problems in statics. Most of these questions are my own invention, but a couple are modifications of well-known problems.

3.1 Sewage worker's conundrum ★

A sewage worker is inside a large underground aqueduct (diameter $\gg$ a man's height) of circular cross section, and so smooth that friction is negligibly small. He has a ladder which is the same length as the diameter of the aqueduct. He wishes to inspect something in the roof. He mounts the ladder and continues to climb until he reaches the other end. What happens?

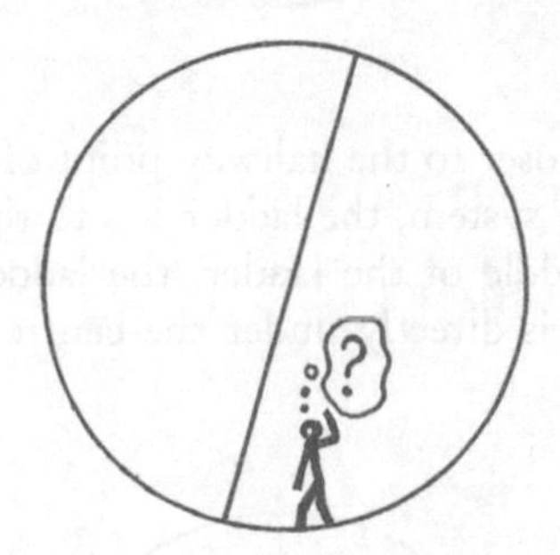

Answer

It hardly matters as far as solving the problem is concerned, but we might start with the realization that the sewage worker must be at the lowest point of the aqueduct. Were he anywhere else he would slide to the bottom because the aqueduct is perfectly smooth. The ladder, interestingly, is equally happy in any rotational position despite there being no friction. We could argue the case by drawing forces on the ladder. Or we could simply note that it has the same potential energy in all rotational positions. This must be so because the centre of mass of the ladder is always at the centre of the tunnel: the length of the ladder is equal to the diameter of the tunnel. Energy arguments can be powerful, and we will use them in this question.

But what *does* happen next?

Well, as the man takes hold of the ladder and mounts it, the very first thing that happens is that the ladder tips back slightly. Because the *system* is free to move, it moves to the position with the lowest potential energy. Here we are using the *principle of minimum total potential energy*.[3] When we talk about the system, we mean the ladder and the man. The potential energy of the ladder is the same in every rotational position. The man's potential energy varies with the angle he makes with the vertical, and is minimised when his centre of mass is below the centre of the ladder, which can be thought of as a virtual pivot.[4] The man's centre of mass hangs slightly away from the plane of the ladder, and his centre of mass must be under the centre of rotation of the system for it to be in equilibrium.

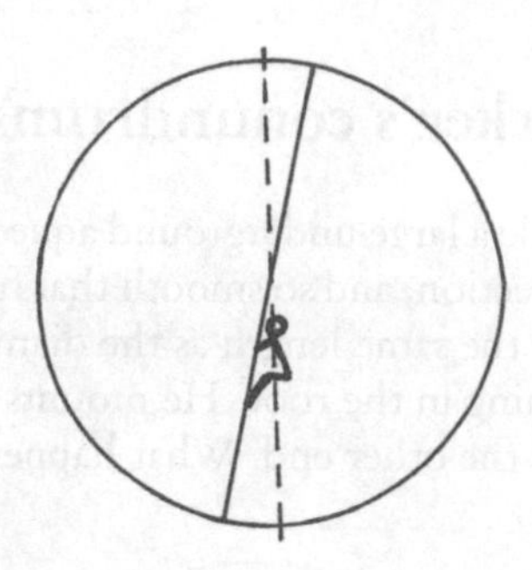

As he climbs and gets closer to the halfway point of the ladder, which is the centre of rotation of the system, the ladder has to tip back more. When the man approaches the middle of the ladder, the ladder must be almost horizontal, until his weight is directly under the centre of the cylinder and the ladder *is* horizontal.

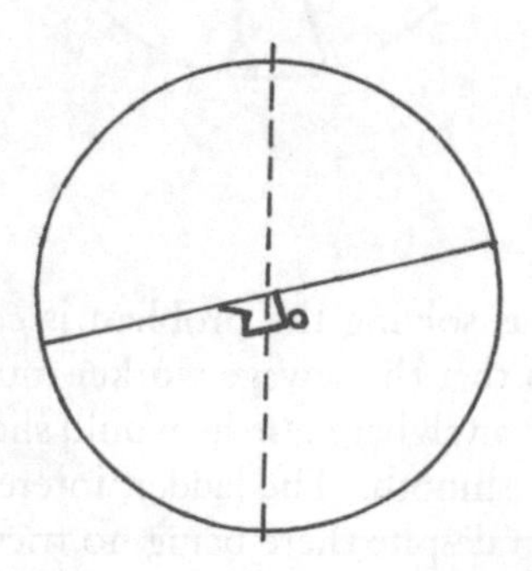

[3]This rather impressive sounding principle simply tells us that systems that are free to move do so until they are at the point of minimum potential energy. The *lost* energy is dissipated, as heat, for example. A ball placed on the top of a hill will roll to the bottom of the hill. Water will move until the free surface is perpendicular to the direction of gravity. The most stable configuration is the one with minimum potential energy. This is the *equilibrium position* for the system.

[4]If this is not perfectly obvious, consider that when the man is a given distance, say r, from the midpoint of the ladder, his possible positions (for every angle θ his centre of mass makes with respect to the vertical) are defined by a circle of radius r centred on the midpoint of the ladder. The lowest point on that circle, or the point with the minimum potential energy, is below the midpoint of the ladder.

As the man continues to climb along the ladder from the midpoint, the ladder rotates past the horizontal position. The man climbs down to the bottom of the cylinder, this time upside-down.

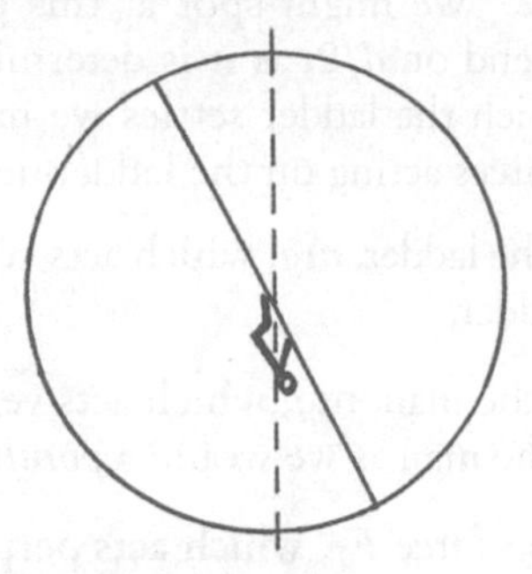

Eventually he is back to where he started but the wrong way up. It seems his inspection of the roof will have to wait.

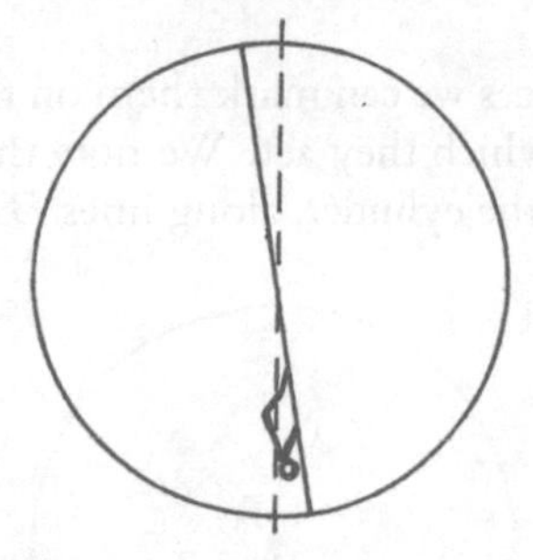

3.2 Sewage worker's escape ★★

A SEWAGE WORKER is inside a large and smooth underground aqueduct of circular cross section and diameter d. To escape a tide of effluent, he balances on the end of a ladder of length $d/2$ and mass m. If the sewage worker also has mass m, at what angle does the ladder settle? Assume that the length of the ladder is very much greater than the height of the man, so he acts as a *point mass* on the end of the ladder.

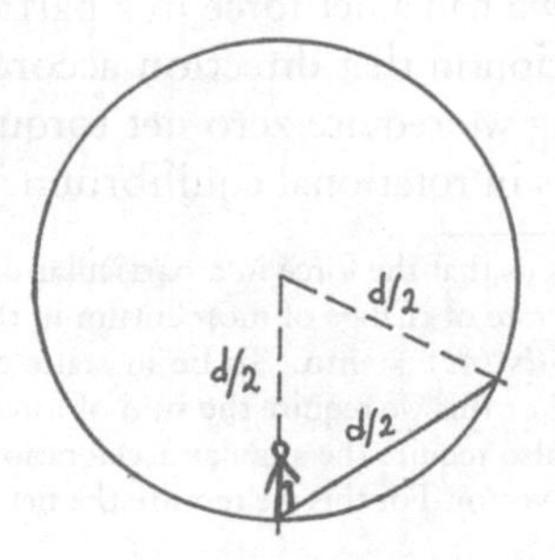

Answer

We first need to consider the geometry of the problem. Triangle ABC is equilateral, with side $d/2$. We might spot at this point, however, that the answer is unlikely to depend on $d/2$, as it is determined by the angles of the system. The angle at which the ladder settles we mark as θ in the diagram. There are four external forces acting on the ladder-man system.

- The weight due to the ladder, mg, which acts vertically downwards from the centre of the ladder.
- The weight due to the man, mg, which acts vertically downwards from point B (we treat the man as we would a *point mass* at B).
- The normal reaction force F_1, which acts perpendicular to the smooth wall at point B.
- The normal reaction force F_2, which acts perpendicular to the smooth wall at point C.

Now we have the four forces we can mark them on the diagram, paying attention to the directions in which they act. We note that forces F_1 and F_2 both act through the centre of the cylinder, along lines BA and CA respectively.

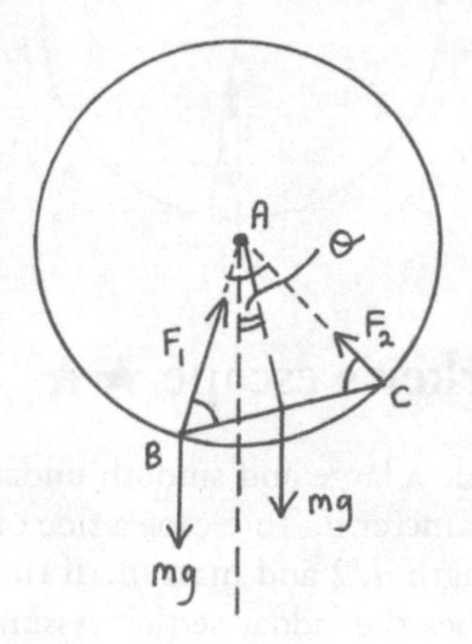

For the man-ladder system to be in static equilibrium, the forces in all directions have to balance. If we had a net force in a particular direction we would have a non-zero acceleration in that direction according to Newton's Second Law. By similar reasoning we require zero net torque (moment) on the man-ladder system, so that it is in rotational equilibrium.[5] We are interested in the

[5]Netwon's Second Law tells us that the force in a particular direction $\boldsymbol{F}$ (where we note this is a vector) is equal to the time rate of change of momentum in that direction, $\boldsymbol{F} = d\boldsymbol{p}/dt$. For constant m we have $\boldsymbol{F} = m\,(d\mathbf{v}/dt) = m\mathbf{a}$. To be in static equilibrium we must have zero acceleration, $d\mathbf{v}/dt = \mathbf{a} = \mathbf{0}$. For this we require the sum of forces in every direction to be zero. To be in static equilibrium we also require the angular acceleration to be zero: $\boldsymbol{\alpha} = d\boldsymbol{\omega}/dt = \mathbf{0}$, where $\boldsymbol{\omega}$ is the angular velocity vector. For this we require the net torque on the object to be zero, giving $\boldsymbol{\tau} = I\boldsymbol{\alpha} = 0$.

vector forces, and the points of action of the forces, on the man-ladder system. As long as we preserve the relative magnitudes, angles, and distances we can consider the diagram in any rotated position. A slightly simpler representation of the man-ladder system is shown below. Note that the weight vectors make an angle θ to the ladder, as they did in the original representation.

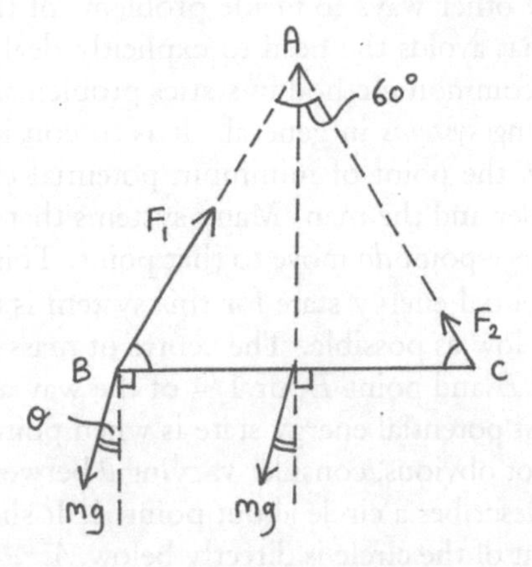

The standard procedure from here, which should be the mantra of any student of statics, is to *resolve, resolve, take moments*. Let us do just that.

- Resolve in the horizontal direction: $2mg \sin\theta = F_1 \cos 60° - F_2 \cos 60°$
- Resolve in the vertical direction: $F_1 \sin 60° + F_2 \sin 60° = 2mg \cos\theta$
- Take moments about point B: $2F_2 \sin 60° = mg \cos\theta$

Now we just need to solve these equations for θ, the angle of the ladder to the horizontal. We should check that we have enough equations: we need as many independent equations as we have unknowns. The unknowns are F_1, F_2 and θ—only three. We have three independent equations, so a solution should be possible. I've left the algebra as an exercise, as it should be fairly straightforward from here. If you do it correctly you should get

$$\theta \approx 16.10°$$

The answer feels about right: the ladder is at a shallow angle to the horizontal. Of course, if we wish, we can then also solve explicitly for F_1 and F_2, giving

$$F_1 = \sqrt{3}mg \cos\theta \approx 1.66\,(mg)$$
$$F_2 = \frac{\sqrt{3}}{3} mg \cos\theta \approx 0.555\,(mg)$$

Again, the answers feel about right. F_1 is approximately three times larger than F_2, giving greater reaction force at the end where the man is balancing, as we'd expect.

Further discussion

We did well to solve the last problem. We took a completely new situation, reduced it to the underlying equations, then solved the equations quite simply to get an exact result. The *resolve, resolve, take moments* method worked well. But there are many other ways to tackle problems of this sort, and here we demonstrate one that avoids the need to explicitly deal with forces and moments. It is a very common method in statics problems, and a very common technique in studying *systems* in general. It is to consider the *energy* of the system—specifically, the point of minimum potential energy for the system comprising the ladder and the man. Many systems that are free to move to a lower potential energy point *do* move to that point. This is one such system.

The lowest potential energy state for this system is the one in which the centre of mass is as low as possible. The centre of mass of the system is midway between point B and point E, or $1/4$ of the way along the ladder from point B. The lowest potential energy state is when point D is directly below point A. If this is not obvious, consider varying θ between $0°$ and $360°$ in the diagram. Point D describes a circle about point A. It should now be obvious that the lowest point of the circle is directly below A.

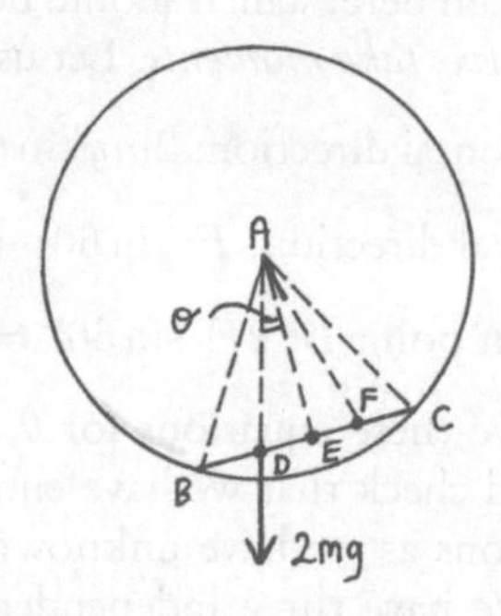

It is now a simple exercise in geometry to define the angle θ that line AE makes to the vertical. This is the same angle that BC makes to the horizontal. Consider the geometry of the situation.

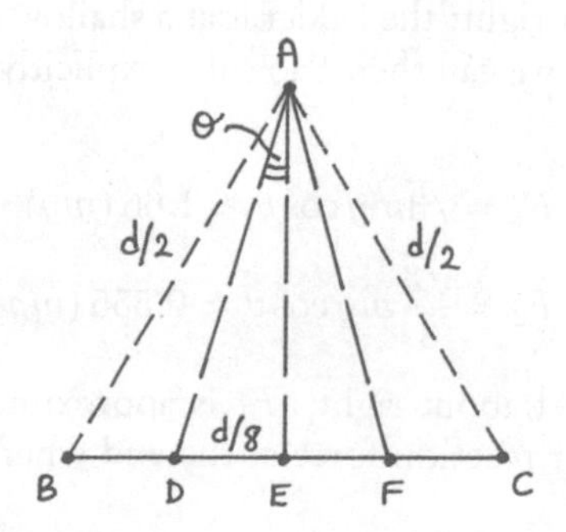

We have

$$\theta = \arctan \frac{(d/8)}{|AE|} = \arctan \frac{1}{2\sqrt{3}} \approx 16.10^\circ$$

This is the same answer as before. By using the principle of minimization of potential energy we have considerably simplified the problem.

Assumptions

We started by saying that the aqueduct was smooth, and we implicitly suggested that the man could get onto the ladder in a position that was not the point at which the ladder ultimately settles. If we wanted to be pedantic, we could argue that if there was no friction *at all* the ladder would simply oscillate indefinitely, and never settle. Likewise, getting onto the ladder could be problematic. If the expected answer to "at what angle does the ladder settle?" was either of these I think we would feel we had been somewhat misled. We therefore assume that the walls are *smooth enough* that the reaction forces are always normal to the wall, but not quite so smooth that the ladder cannot eventually settle.

An important skill in addressing physics problems is identifying the fundamental subject of interest, the nub of the question, and filtering out everything that is unimportant. In that latter category are an almost infinite number of things which we should know we need to discount: the material of the ladder—we know it is smooth; air resistance—we are considering an equilibrium/stationary solution; the name of the man; the day of the week; etc... If we feel the need to introduce these latter things, we are probably more interested in riddles than physics problems.

3.3 Sewage worker's resolution ★★★

A SEWAGE WORKER needs to use a ladder inside a large and perfectly smooth underground aqueduct of circular cross section. The ladder is of the same length as the diameter of the aqueduct. Draw the forces on the ladder when the ladder is placed: i) vertically; ii) horizontally. Discuss the diagrams.

Answer

We consider the answer in two sections. First the vertical ladder, then the horizontal ladder.

i) VERTICAL LADDER

This should seem, and is, very simple. There are only two external forces acting on the ladder:

- The weight of the ladder, mg, which acts vertically downwards from the centre of mass of the ladder.
- The reaction force, F, which acts normal to the smooth wall of the cylinder at the point of contact.

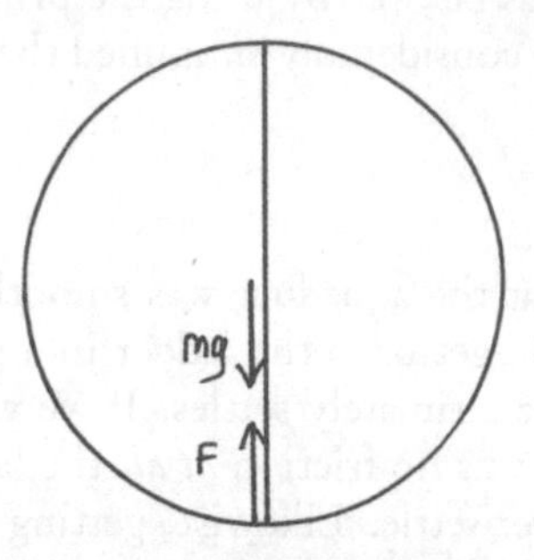

If the ladder is in static equilibrium the two forces must be in balance. That is, $F = mg$.

ii) Horizontal Ladder

Our first reaction might be that this is equally simple. We can have three external forces acting on the ladder, one due to the weight of the ladder, and one at each end of the ladder. The ladder is symmetrically disposed about the y-axis, so we expect the two forces at the ends of the ladder to be equal in magnitude and opposite in sign. We know that the wall of the cylinder is smooth, so the forces at the ends must be normal to the wall of the cylinder. The arrangement of forces is shown in the diagram below.

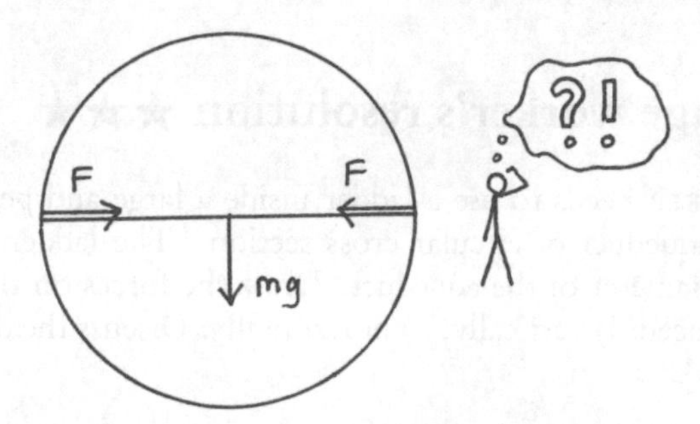

A problem is immediately apparent. We can balance forces in the horizontal direction, but what about the vertical direction? We seem to have a net downward force due to the weight, mg. The ladder cannot be in static equilibrium with this arrangement of forces. We require upward forces at the walls to balance the weight of the ladder. As the boundary conditions are stated, and, making assumptions about the ladder having negligible width in the vertical direction, it is impossible to achieve a balance of forces.

There are a number of ways we can modify the boundary conditions, or relax the assumptions, that would allow us to have a more realistic arrangement of forces.

- We could relax the smooth wall condition, and have a coefficient of friction slightly greater than zero.
- We could allow the ladder to have finite width in the vertical direction, so that the points of contact are just above and just below the centre of the cylinder. This would allow an upward component of reaction force at the lower points of contact.
- We could relax the condition that the ladder is exactly the same length as the diameter of the cylinder, and allow it to be infinitesimally shorter. This would allow it to drop a small distance below the centre of the cylinder, allowing an upward component of force at the points of contact. In practice this would happen due to both bending and compression.

Let us consider the option of the ladder being infinitesimally shorter. This could be on account of a tiny amount of compression, or due to very slight bending, or both. Whatever the mechanism, the points of contact with the cylinder move just below the line through the centre of the cylinder. If we still make the assumption that the cylinder is perfectly smooth, the symmetrical reaction forces, F, are still normal to the wall, and are therefore directed through the centre of the cylinder. This arrangement of forces allows us to have zero net horizontal and vertical force on the ladder. The conundrum is resolved. It is interesting to note that for a small amount of compression we require $F \gg mg$, so the mechanism for providing measurable compression of even a fairly rigid ladder is also apparent.

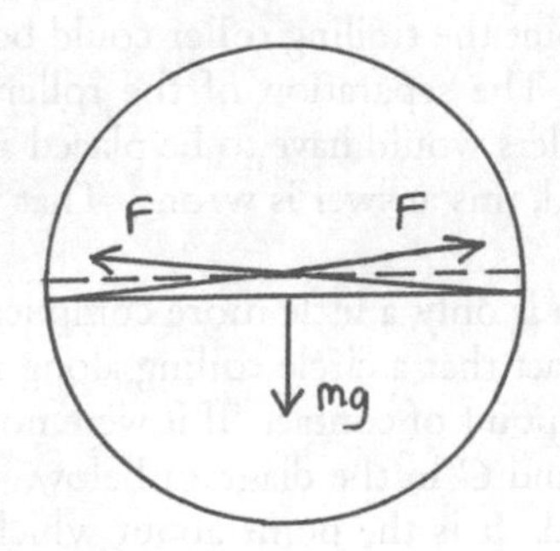

3.4 Aztec stone movers ★★

There is some uncertainty about how the Aztecs and Egyptians moved huge stones great distances to their temples and pyramids, even if we assume that a practically unlimited supply of slaves must have been available. There is even more uncertainty about how our Neolithic ancestors behind such structures as Stonehenge could have accomplished what even today would have been quite a feat of engineering. The sarsen stones that form the outer circle of the monument weigh 50 tonnes each. To lift such a stone would take the strength of about 1,000 men—unfortunately only about 20 men fit round the perimeter of each stone. The nearest known source of sarsen stone is 25 miles from the monument. And the stones were transported in about 2000 BC.

We have almost no direct evidence of how such a feat was accomplished, but most theorists have it that the stones were moved using sleds and ropes. The sleds were made of tree trunks, and the ropes of leather. The sleds were loaded onto rollers, also made from tree trunks. It has been estimated that using this relatively sophisticated system would have required 500 men to pull the load and a further 100 men to place the rollers.

IF A BLOCK of stone 2 m long is to be pulled 1 km, how many times do rollers have to be placed in the path of the stone?

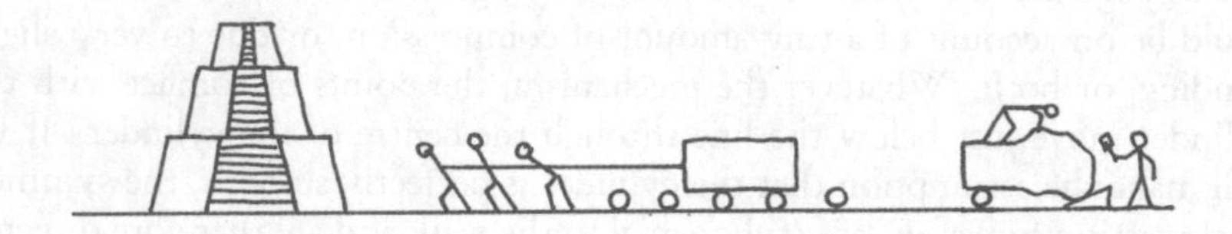

Answer

The *obvious* answer to this goes as follows. The block of stone is 2 m long and can only balance on a single roller when that roller is exactly in the middle of the block. At this point the trailing roller could be removed and a leading roller could be added. The separation of the rollers is therefore 1 m, and approximately 1,000 rollers would have to be placed along the 1 km path. As you have already guessed, this answer is wrong. That would be too easy to be interesting.

The correct solution is only a little more complicated, however. The key observation lies in the fact that a circle rolling along the ground is instantaneously stationary at the point of contact. If it were not, slipping would occur. Consider points A, B and C in the diagram below. Point A is the point of contact with the ground. It is the point about which—instantaneously—the body appears to be rotating. We conclude that $v_A = 0$. Neither is the ground moving, so $v_E = 0$. Points B (the centre) and C (the top) lie vertically above point A. Thus, B and C can have *only horizontal components of velocity* if we consider that the body is rotating instantaneously about A. This is an important point, which is worth thinking about. If the wheel is in solid-body rotation about point A, the velocity of C must be twice the velocity at B, so $v_C = 2v_B$. This is easy to see because B and C are, respectively, one radius and two radii from A. The velocity distribution for points on a line between A and C is linear; that is, a triangle of velocities is formed.[6]

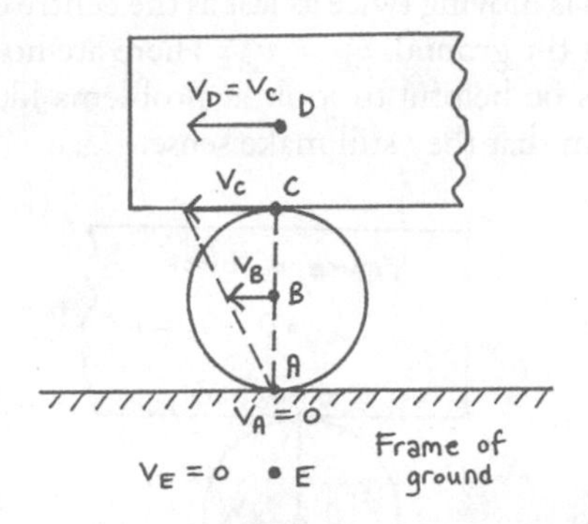

We can apply a similar argument to the block of stone. If the roller *rolls* against the block of stone, rather than slipping, the roller and the stone must have the same velocity at the point of contact, so $v_D = v_C$. The important result is that the speed of advance of the stone is twice the speed of advance of the roller. We will return to this in a moment.

For interest's sake, let's consider the problem in the frame of reference of the block, that is, the frame of reference in which the block appears to be stationary, $v'_D = 0$. This is the frame we would be observing from if we were sitting on top of the block as it was being slid along by the workers. In our *block frame* we will mark all velocities v', to distinguish them from velocities measured in the *ground frame*, v. In our block frame it looks like the ground is being slid along on rollers underneath the block. The ground has to go in the opposite direction to the block when observed from this frame. Because there is no slipping at the point of contact with the block, $v'_C = v'_D = 0$.

[6] A popular, related puzzle is how to derive the equation of a *cycloid*, or the path traced by a point on the rim of a wheel as it rolls without slipping along a plane. For a wheel of radius r, which has rolled about its centre through an angle θ, it is possible to show using relatively simple geometry that the *parametric equations* (that is, the equations for x and y as a function of θ) are $x = r(\theta - \sin\theta)$ and $y = r(1 - \cos\theta)$. By eliminating θ it is possible to show that the Cartesian equation ($x = f(y)$) is given by

$$x = r\cos\left(1 - \frac{y}{r}\right) - \left(2ry - y^2\right)^{1/2}$$

This is valid for $y = 0$ to $y = 2r$, and gives the first half of the first hump (from the origin to P in the diagram).

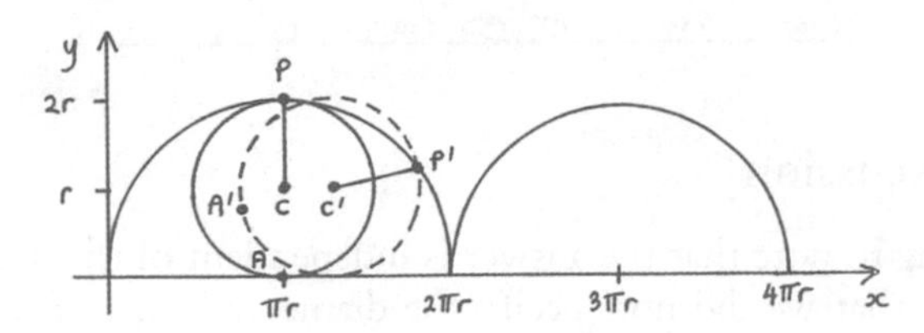

Once again, the velocity distribution along a vertical line of points through the centre of a roller is linear in form, giving $v'_A = 2v'_B$. In this frame of reference, the ground is moving twice as fast as the centre of the roller. Because there is no slipping at the ground, $v'_E = v'_A$. There are no great surprises here, but it can sometimes be helpful to look at problems like this from another perspective to confirm that they still make sense.

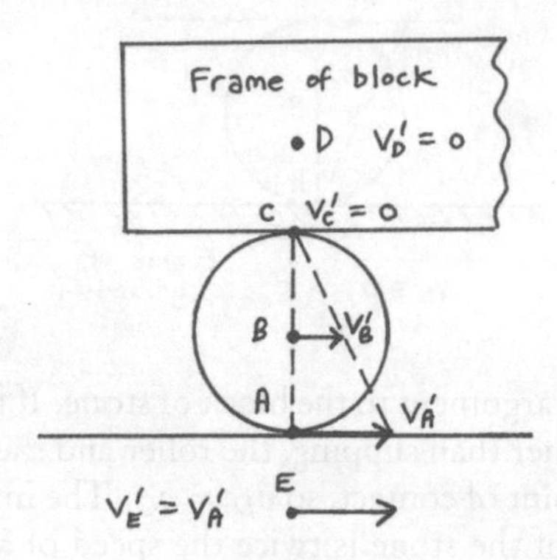

The key result of this analysis is that for every 1 m the block is pulled forward, the roller advances 1/2 m. We decided at the outset that we needed the rollers to be spaced exactly 1 m apart under the block. But this would be achieved by having them spaced every 2 m apart on the ground, because as the roller is pulled forward 2 m, the rollers already underneath the block go back only 1 m relative to the block. The process is illustrated below.

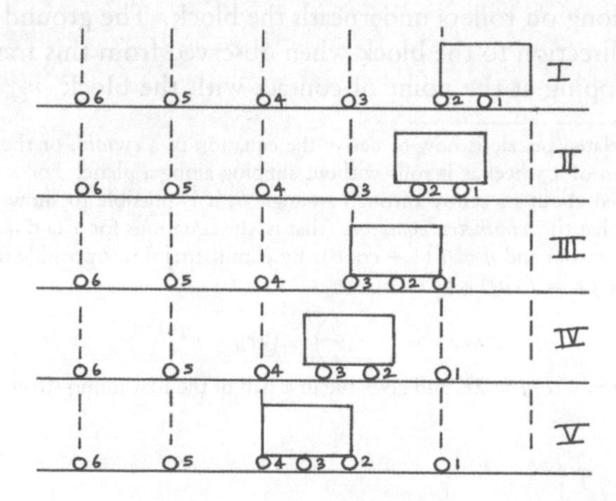

Further discussion

It is interesting to note that the answer is independent of the diameter of the rollers. Note that we did not specify the diameter in the analysis. In fact, as we will see in questions relating to imperfect wheels, the force required to rotate an imperfect wheel on a smooth surface also depends only on *shape*,[7] and does not depend on diameter. When rolling on a rough surface, however, the diameter of the wheel is important, and on rough surfaces larger wheels, or larger rollers, are clearly better.

There is a convincing case that the development of wheels was inspired by observations about the relative speed of rollers and load. In this version of history, men started with the observation that pushing or pulling great loads was made easier by using rollers made from tree trunks.

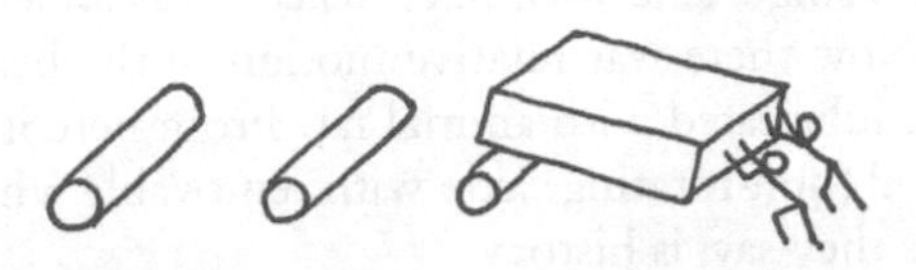

Around the same period in man's technological development, people started to use primitive sleds to haul loads, because the smooth runners had lower friction on most surfaces than the uneven loads did. One day, a particularly clever person realized that combining the sled and the roller was more efficient than using either in isolation.

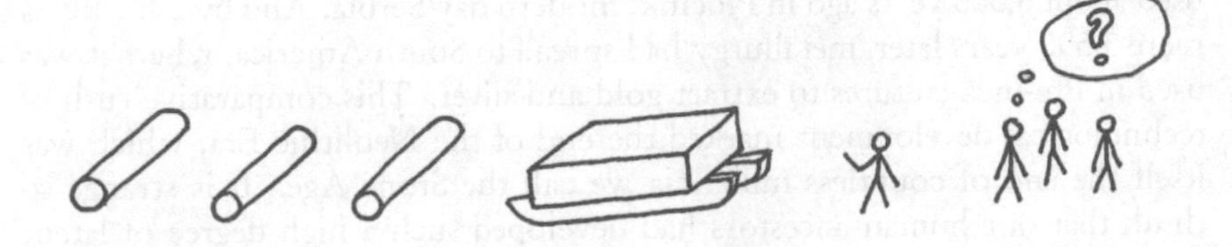

Many many years passed in which the roller-sled system was adopted more routinely for heavy loads. When objects were hauled very long distances the sleds cut into the rollers, creating deep grooves in them. This served to guide the sled, which was helpful, but there was an even more important effect. As the grooves got deeper, the rollers needed to be placed less frequently. With very deep grooves you might need to place the rollers, say, ten times less frequently than with un-grooved rollers.[8] This made moving loads over long distances much more efficient.

[7] By *shape* we mean how circular the wheel is. In a question that follows, we model imperfections in the wheel, or facets, by representing the wheel as an n-sided polygon.

[8] We require a diagram to explain this interesting effect fully, but I've left that as an exercise for the reader. Consider, however, that when the grooves get very deep, the radius r of the remaining axle of wood (at the very bottom of the groove, that is) is very much less than the radius of the roller, R. We write $r \ll R$. We should be able to show that the forward progression of the load in the lab frame is proportional to $R + r$, and the backward progression of the roller with respect to the load (in the load frame) is proportional to r. The distance we can travel before having to replace a log is therefore proportional to $(R + r)/r$, and goes up quickly as r becomes smaller. Prove it for ★★★!

Neolithic man then started pre-grooving the rollers, creating what we would now recognize as axles with fixed wheels, which were placed periodically under the load. Later a single axle with fixed wheels was attached through fixed holes in the sled. Now there was relative motion of the bearing and axle, so the holes had to be lubricated with animal fat. From here it was only a small step to design a fixed (non-rotating) axle with removable wheels.

And the rest, as they say, is history.

3.5 The Wheel Wars I ★★★

About five thousand years ago, tens of thousands of years of somewhat uncertain prehistory were ending rather abruptly in almost all corners of the world, with the almost simultaneous beginnings of Copper Ages, Bronze Ages and Iron Ages. The historical evidence suggests that smelting techniques were first used about 5,500 years ago in Pločnik, modern day Serbia. And by 2000 BC, a mere 3,500 years later, metallurgy had spread to South America, where it was used in Pre-Inca cultures to extract gold and silver. This comparative rush of technological development marked the end of the Neolithic Era, which was itself the end of countless millennia we call the Stone Age. It is strange to think that our human ancestors had developed such a high degree of latent sophistication, which would go on to suddenly manifest itself in all the great inventions during the few thousand years between then and now.

One of the most important inventions in human history, along with the tools to make fire (fire itself had been around for a bit longer), and much later the printing press and penicillin, was the wheel.

The first wheels seem to have been invented at similar times in several places: Mesopotamia (a region between the Tigris and Euphrates rivers which includes a large part of modern-day Iraq), the Indus Valley (modern day Afghanistan and Pakistan), the North Caucasus (home to the Maykop culture) and Central Europe. The very oldest intact wheel is thought to be the Ljubljana Marshes Wheel, which is believed to be 5,150 years old. Some people believe that the oldest known representation of a wheeled vehicle is on the Bronocice pot, dug up in the 1970s in Poland, and which dates back to 3500 BC.

As with all interesting topics in history, there are naysayers, those who believe the repeated motif is not a cart at all. You can decide for yourself. From around 3500 BC onwards, however, there is increasing evidence that people were using the wheel for pottery-making, transportation, and many other applications besides. The fanciful cartoon images of cavemen with stone wheels are just that. Strangely, however, huge stone wheels were, and to some extent still are, used on the island of Yap (in Micronesia) as a form of currency. The wheel truly has wide application.

The wheel's ability to roll is a strong function of how *smooth* it is—how far it deviates from a perfect circular form. Of course, it also depends on how flat the road is—a problem in ancient times.

During the Wheel Wars, a very brief period of Paleolithic history between 1,000,000 years ago and 1,000,001 years ago—the exact dates are disputed by some scientists—there was a sudden surge in wheel research, and wheel makers were selling wheels with three, four, and five sides, or even special made-to-order n-sided wheels. Wheel sellers claimed that it was worth spending more caveman currency (probably mammoth tusk or some such) on wheels with more sides, and it certainly seemed to be the case that large-n wheels rolled very smoothly indeed. But it would be another million years until mathematics was advanced enough to prove or disprove these marketing claims.

HOW MUCH HORIZONTAL force, F, applied at the frictionless axle of a regular n-sided polygon of mass m, is required to cause it to roll without slipping?[9] If we wish, we can take r to be the distance from the axle to the vertices of the polygon, defined so that in the limit $n \to \infty$, r defines the radius of a circular wheel.

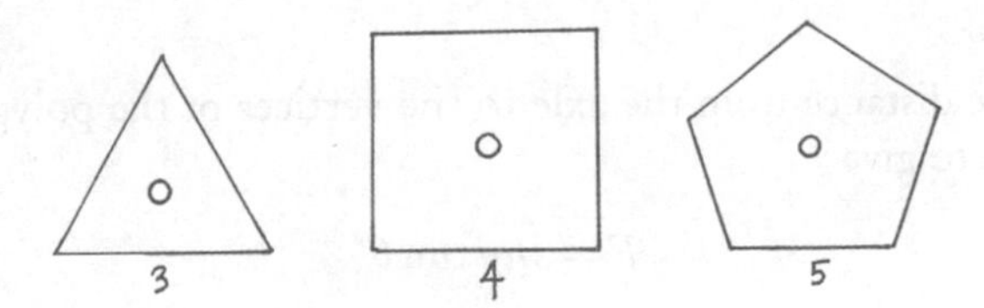

Answer

It may have been beyond the average Neolithic man, but this is actually relatively easy to answer. Let's start with a polygon with an arbitrary number of sides: $n = 4$, say (a square). In a moment we will see that it is a trivial extension to generalize the answer.

First we consider the point about which the square rotates. If we apply a force towards the left, and if the square rolls without slipping, the point about which the square rotates is the bottom left corner. When the object is on the point of overbalancing—that is, when it is about to rotate—all the force from

[9]This implies that we must assume that the coefficient of friction is high enough that this motion can occur.

the ground is transmitted through the point of rotation. If this is not obvious, consider the object a moment later.

If we apply a horizontal force F at the centre of the square (the axle), it must be balanced by a force equal in magnitude and opposite in sign which acts at the point about which the square rotates. Similarily, the downward force due to weight, mg, must be balanced by an upward force, mg, acting at the point of rotation. These two conditions are required for the object to be in static equilibrium–in other words, not accelerating in either the horizontal or vertical direction. We can now take moments about the point of rotation. When the applied force, F, is just large enough to cause rotation to occur, the moments about the pivot due to this force, and the weight mg, must be equal and opposite.

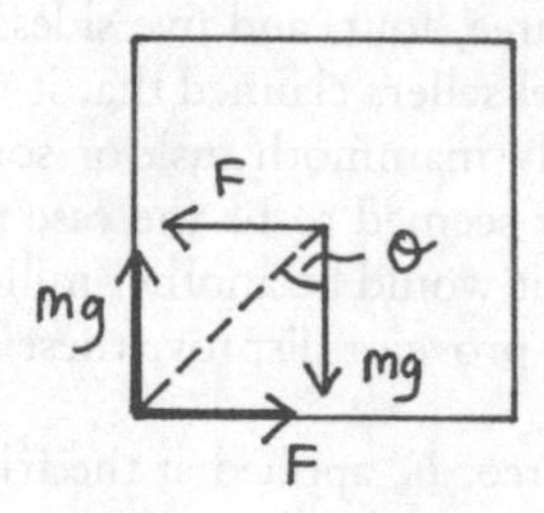

We have

$$Fr\cos\theta = mgr\sin\theta$$

where r is the distance from the axle to the vertices of the polygon. We can rearrange this to give

$$F = mg\tan\theta$$

where, for a square, $\theta = 2\pi/8$. In general for an n-sided polygon, we have $\theta = 2\pi/2n = \pi/n$. So the required force becomes

$$F = mg\tan\left(\frac{\pi}{n}\right)$$

Because the side of a polygon subtends a smaller angle at the axle as n increases, the required force reduces rapidly as n increases. As $n \to \infty$, $\theta \to 0$, so the required force also approaches zero. A perfectly circular wheel (with no friction at the axle) requires no force to rotate on flat ground. A million years later we have proved the cavemen right–it was worth spending a bit more mammoth tusk on a wheel with lots of sides.

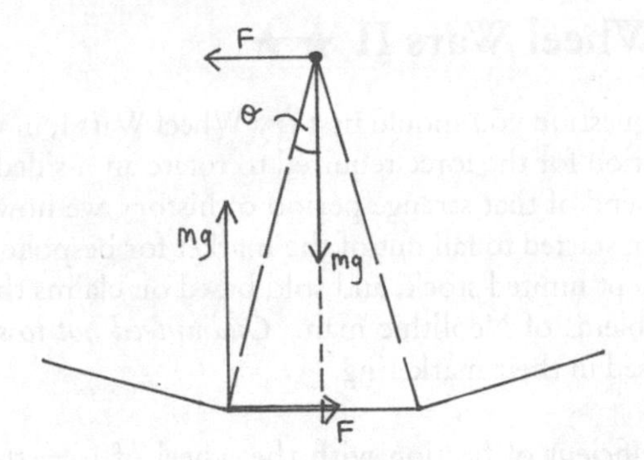

If we plot the solution we see that for $n = 4$ the force required to rotate the object is equal to mg. For a twenty-sided polygon $F = 0.16mg$. The solution *asymptotes* to zero for $n \to \infty$, which represents a perfect circle.

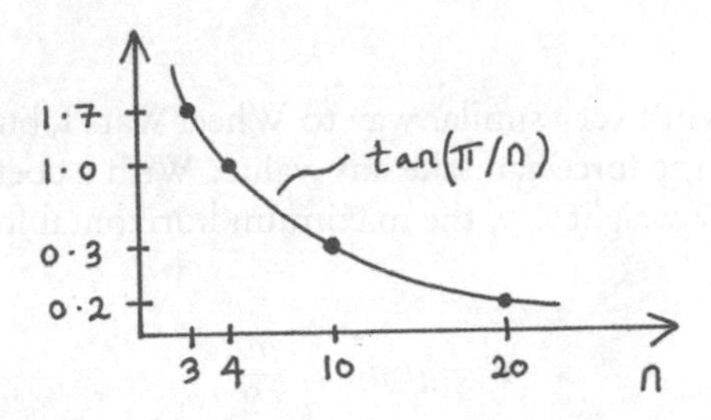

Further discussion

You will probably have noticed that in the analysis we have just performed the solution for the force required to rotate a wheel does not depend on diameter. Yet we know from practical experience that larger wheels give a smoother ride than smaller wheels. Bikes are less shaky than push scooters. There are a number of things at play here.

- Imperfections in the flatness of the ground act in the same way as facets on the wheel, and lead to large starting forces, and, by the same argument, a bumpy ride.
- Manufacturing tolerances become proportionally more important as wheels become smaller.
- Larger wheels often have additional mechanisms to reduce the peak force. These could include pneumatic tyres, or some form of suspension, or just the *compliance*[10] of the overall structure of the wheel.

[10] *Compliance* is the inverse of stiffness, or the deformation of a structure under a given load. The compliance of a structure is often responsible for the redistribution of stress, so that peak stress within the structure is reduced.

3.6 The Wheel Wars II ★★

To attempt this question you should first try Wheel Wars I, in which we solved the general equation for the force required to rotate an n-sided polygon.

Towards the end of that strange period of history we now call the Wheel Wars, the bottom started to fall out of the market for bespoke n-sided wheels. Manufacturers kept limited stock, and sold based on claims that appealed directly to the problems of Neolithic man. *Guaranteed not to slip on ice* was a popular claim used in their marketing.

IF ICE HAS a coefficient of friction with the wheel of $\mu = 0.5$, calculate the minimum number of sides that a regular n-sided polygon must have to *turn before slipping* when we apply a slowly increasing horizontal force F at the axle.

Answer

We can answer this in a very similar way to Wheel Wars I, but we need to relax the condition that the force can take any value. With a coefficient of friction $\mu = 0.5$, and with a weight mg, the maximum horizontal force at the ground is limited to

$$F = \mu mg = \frac{mg}{2}$$

To prevent slipping we are therefore limited to the range $0 \leq F \leq mg/2$.

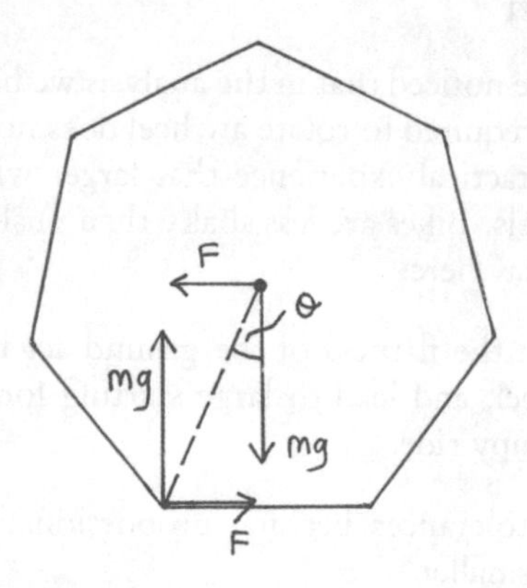

Our previous result

$$F = mg \tan\left(\frac{\pi}{n}\right)$$

becomes

$$\frac{mg}{2} = mg \tan\left(\frac{\pi}{n}\right)$$

giving

$$n = \frac{\pi}{\tan^{-1}(1/2)} \approx 6.78$$

n must be an integer, giving $n \geq 7$. So our wheels must have at least seven sides to prevent them slipping on ice.

3.7 Obelisk raiser ★★

The Egyptians, Assyrians and Romans were all great fans of the obelisk. In ancient Egypt, under the instruction of the Pharaohs, many tens of these monolithic structures were hewn from the bedrock and transported over land and water to adorn the entrances to temples. They appear to have been associated with the cult of sun worship—many scholars believe that they represent petrified rays of sunlight, in tribute to the deity Ra. This god was so powerful that he called into existence not only all forms of life, but also numerous other gods. Ra was truly one to be venerated. The centre of sun worship was the Heliopolis, where the Temple of Re-Atum, another god identified with Ra, was built by Senusret I. Senusret I was the second pharaoh of the Twelfth Dynasty of Egypt: Egyptian history really is quite complicated. At the Re-Atum temple, most of which now sits about 15 m below the busy streets of Cairo, Senusret I erected two 21 m-high granite obelisks, each weighing 120 tonnes. He did this in about 1950 BC. Quite remarkably, almost four thousand years later, one of these obelisks still stands on the original spot.

The largest standing ancient obelisk is the Lateran Obelisk at the Archbasilica of St John Lateran in Rome. It is 37 m high, and is estimated to weigh about 450 tonnes. The Roman emperors were keen on acquiring these ancient monuments, and using specially designed obelisk ships transported many of them up the Nile, and across a huge tract of the Mediterranean sea to Rome. The engineering feat involved is remarkable.

It seems that those obelisks plundered from Egypt and re-erected by Romans were put up with vast teams of men and horses, with the aid of sophisticated wooden scaffolds.

There are several engravings of the comparatively recent operation to raise the obelisk in St Peter's Square in 1586. The enterprise involved nine hundred men and over one hundred horses. This team was probably very small by Roman standards, and even smaller by ancient Egyptian standards, as the effort of the men was multiplied by great steel levers arranged around capstans. There is no evidence that such sophisticated technology was used by the Pharaohs.

As with most things Egyptian and many things ancient, there is much speculation about how the Egyptians managed to raise such vast monoliths. Sadly the Egyptians were better at documenting administrative trivia than they were their engineering methods, which seem for the most part to have vanished into the desert sand. A few clues have survived, however. The base stone, or pedestal, of every obelisk raised in ancient Egypt had a radiused groove at least ten centimeters deep carved into its top. This ran the full width of the stone, very close to one edge. There seems little doubt that this served as a retaining groove to locate the heel of an obelisk while it was raised to the vertical position. The heel would somehow—and we will deal with this *somehow* in a moment—be located in the groove, before the obelisk was raised by an army of men pulling ropes attached with a frame to the apex of the monolith.

A fascinating article entitled "An ancient Egyptian mechanical problem. Papyrus Anastasi I. About 1300 BC" appears in the December 1912 edition of The Open Court, a monthly magazine "devoted to the science of religion". The author of the article, F. M. Barber, begins as follows.

> So far as I am aware this is the only ancient Egyptian Papyrus that has ever been found which makes even a remote reference to the apparatus used or methods employed in the installation of their gigantic monuments, and even here the account is so fragmentary as to seem at first sight merely to excite curiosity rather than to offer a satisfactory solution.

Given the importance of the obelisks to ancient Egyptians, and the vast effort that must have been devoted to their construction, by whatever means, the scarcity of documentary evidence of the Egyptians' methods is more than a minor mystery. Over the course of 11 pages Barber examines interpretations of the Anastasi I papyrus, which to this day seems to be the most technologically explicit. The purpose of the document seems to be the subject of some debate, with many believing it to be a work of literature rather than an administrative record. What it describes in some detail, however, is the cutting and transportation of a huge obelisk.

> Let there be made a new obelisk, sculptured in the name of the Lord Royal, of 110 cubits in height...The periphery of its foot will be 7 cubits on each side...Thou hast placed me as chief of those who haul it...

The measurements correspond to a stone which is 59 m high and 3.7 m on a side. Barber goes on to estimate its weight at about 1,400 tonnes. The Unfinished Obelisk, so called because it is still bonded to the bedrock at Aswan, would have weighed 1,200 tonnes when finished, making it the largest (unfinished) obelisk of ancient times, considerably larger than the 37 m high, 450 tonne Lateran Obelisk. The papyrus then describes the construction of an inclined roadway or ramp, retained within 120 vast caissons. It suggests that the obelisk would have been dragged up this ramp until it rested high above the base stone, which was buried 20–30 m below under a great mass of sand. The biggest clue of all in the Anastasi I papyrus translates as follows.

> Thou sayest I need the great box which is filled with sand with the colossus of the Lord Royal thy master which was brought from the red mountain...Empty the box which is loaded with sand under the monument of your master...

The colossus, or obelisk to venerate the Lord Royal, has been brought from Aswan (the red mountain), dragged up the inclined road (built no doubt at unfathomable cost) to rest at the top of a sand box. The sand box is then emptied, lowering the obelisk that thousands of workers have toiled to bring into position. And it is this single clue that seems to unravel the enigma of obelisk construction. Barber explains with a diagram.

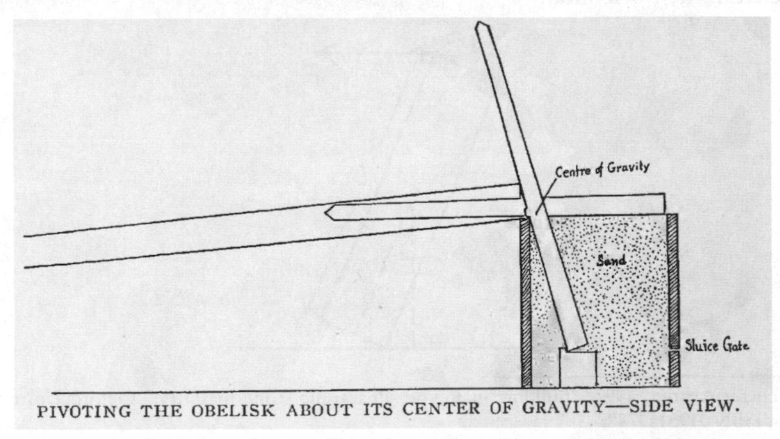

PIVOTING THE OBELISK ABOUT ITS CENTER OF GRAVITY—SIDE VIEW.

The obelisk is pivoted about its centre of mass by slowly emptying sand from the caisson, thereby lowering the heel towards the groove in the base stone. There is only one chance to get this operation right. After years of work, the obelisk is successfully raised or broken in the process. Barber hypothesizes that the obelisk could be guided using this method to 15 degrees from the vertical. The last step in raising the obelisk would be done by teams of men pulling on ropes. Of all the surviving hieroglyphics from ancient Egypt, very few represent engineering methods in any detail. There is a single relief on a temple decorated by Ptolemy XII,[11] however, that symbolically illustrates a pharaoh raising an obelisk from an off-vertical position. This suggests that the final part of the operation at least was performed as Barber maintains.

Let us now turn our attention to calculating the number of men required, according to Barber's method, for the final pull into the vertical position.

FOR A REGULAR obelisk (of uniform cross section) of mass m, height l and base width d, calculate the minimum horizontal force, F, required to raise it incrementally when it makes an angle θ to the vertical.

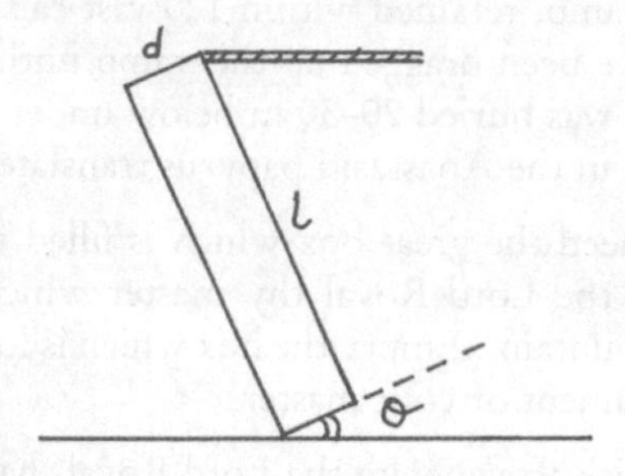

Answer

We should start with a diagram of the obelisk with the given dimensions l, d and θ marked. If θ is the angle from the vertical then it is also the angle of the base from the horizontal.

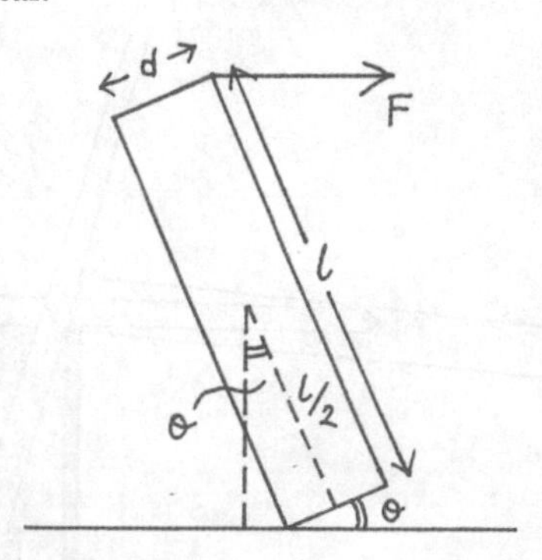

[11]Dieter Arnold, 1997, "Building in Egypt: Pharaonic stone masonry," Oxford University Press, ISBN 0195113748.

The minimum horizontal force, F, required to raise the obelisk will clearly be achieved when the force is applied at the very top of the obelisk. To apply the force purely horizontally we would need either a very long rope or a frame to direct the force at the appropriate angle. We assume that this is possible. In total, then, there are four[12] external forces acting on the obelisk.

- The weight, mg, which acts vertically downward from the centre of mass, point C.
- The force due to tension in the rope, F, which acts horizontally from the top of the obelisk, point D.
- The reaction force from the ground, R, which acts vertically upwards at the point of contact with the ground, A. This has to be equal to mg to balance the forces in the vertical direction.
- The frictional force at the ground, point A, which, when the obelisk is in equilibrium, is equal in magnitude and opposite in sign to the force due to tension in the rope, F. If we assume that the heel of the obelisk is located in a turning groove, we do not need to worry about the value of the coefficient of friction.

We now need to consider the geometry of the problem. We intend to take moments about A to find the force F which allows the obelisk to be rotationally stable at a particular angle θ. With reference to the diagram, the rope tension force, F, acts at a distance $y_1 + y_2$ above the pivot, A, where from simple geometry $y_1 = l\cos\theta$ and $y_2 = d\sin\theta$.

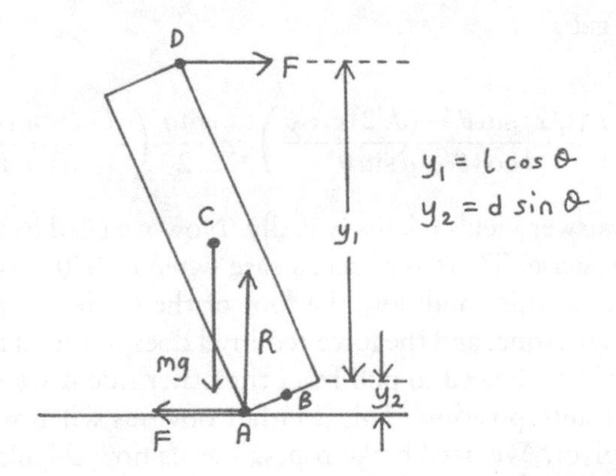

With reference to the diagram, the weight force, mg, acts with an equivalent moment arm $x_1 = x_3 - x_2$ about the pivot A, where, from the geometry of the diagram, $x_3 = (l/2)\sin\theta$ and $x_2 = (d/2)\cos\theta$.

[12]We regard the upward reaction force and the friction force as separate, to simplify the analysis. Clearly we could represent them as a single force vector if we wanted to, because they act at the same point.

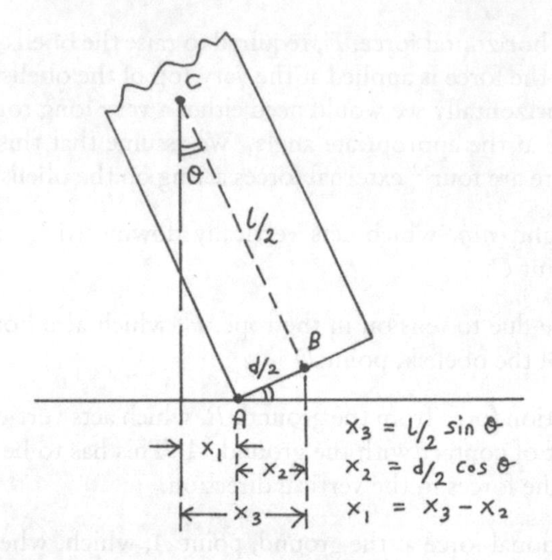

Taking moments about A we have

$$mg(x_3 - x_2) = F(y_1 + y_2)$$

where

$$\begin{aligned} x_3 - x_2 &= \frac{l}{2}\sin\theta - \frac{d}{2}\cos\theta \\ y_1 + y_2 &= l\cos\theta + d\sin\theta \end{aligned}$$

Combining, we get

$$F = mg\left(\frac{(l/2)\sin\theta - (d/2)\cos\theta}{l\cos\theta + d\sin\theta}\right) = \frac{mg}{2}\left(\frac{(l/d)\tan\theta - 1}{(l/d) + \tan\theta}\right)$$

We see that the answer yields relatively easily. Now we need to test the answer to see if it makes sense. There is a special case when $\theta \to 0$. You may already have realized that at this condition, the foot of the obelisk is about to come flat against the base stone, and the force required does not tend to zero. In fact it becomes negative: we need to pull from the other side if we want to gently lower the obelisk into position. This is rather obvious when we consider the location of the pivot. We need brake ropes. Let us now calculate the tension required in those brake ropes when $\theta \to 0$.

Our equation becomes

$$F_{\theta=0} = -\frac{mgd}{2l}$$

Take the Lateran Obelisk, which has $m = 450 \times 10^3$ m, $l = 37$ m and $d = 2.5$ m. We can take $g \approx 10$ ms^{-2}. This gives $F_{\theta=0} = 1.52 \times 10^5$ N. If each worker pulled with a force equivalent to 20 kg or 200 N, it would take 760 of them to hold the brake ropes. To raise the obelisk from $\theta = 15°$ would take 2,234 workers, a major feat of both engineering and logistics. There is a further special case at $\theta = \tan^{-1}(d/l) \approx 3.865°$, where the force required to hold the obelisk in position is zero. At this point, the obelisk's centre of mass is over the pivot.

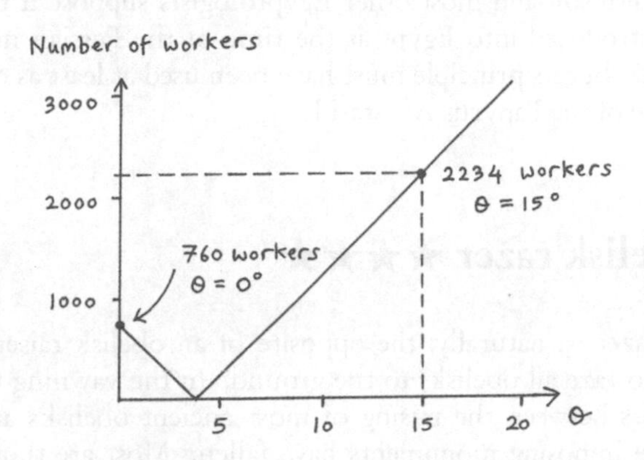

The ancient Egyptians were sound accountants and experts at logistics, as many of their legacies will testify. It is interesting to think that despite their relative sophistication at engineering and accounting, however, their mathematics was not nearly as advanced as even that of today's average high-school student. Three thousand three hundred years ago, the scribe of the Anastasi I papyrus is being asked, almost in the form of a riddle, to calculate how many men would be required to perform the task of transporting and raising the obelisk. Something that would have relied on considerable practical experience to execute. Today, with relatively basic concepts in statics and algebra we can circumvent the requirement for years of practical understanding, and perform accurate calculations for any situation of this type. The Papyrus Anastasi I scribe would probably have been glad of a modem high-school student when performing his calculations.

Barber estimated that 21,600 men would have been required to drag the obelisk mentioned in the papyrus into position at the top of the sand box. He speculates that the practical logistics of this would have been almost insurmountable, and suggests that the ancient Egyptians relied on technology like the capstan. This idea is attractive, but if we accept the hypothesis we have to contend with the problem that no evidence of any kind has ever been found in support of the argument. Either way we are left with something of a mystery. I'll leave the last word to Barber, and leave you to ponder which of two slightly inadequate versions of obelisk history is the more compelling.

> I once made a calculation to ascertain how many men would be required to drag the Karnak obelisk which weighs 374 tons. It proved to be 5,585 men harnessed in double rank to four drag

ropes and covering a space of 1,400 feet. The obelisk of the papyrus [Anastasi I] would therefore require 21,600 men, and they would cover a space on the road for over a mile. Nobody could drill such a body of men to pull together. Capstans must therefore have been employed. The sakiya or geared wheel and water buckets worked by cattle embodies the principle of the capstan, and Wilkinson and most other Egyptologists suppose it to have been introduced into Egypt at the time of the Persian invasion B.C. 527; but its principle must have been used at least as early as the time of the Papyrus Anastasi I.

3.8 Obelisk razer ★★★★

An obelisk razer is, naturally, the opposite of an obelisk raiser. He is one who wishes to raze all obelisks to the ground. In the yawning three to four thousand years between the raising of most ancient obelisks and today, almost all these imposing monuments have fallen. Most are thought to have succumbed to earthquakes, or been deliberately destroyed during conflicts, or—both in Roman times and, more recently, colonial times—plundered by invading powers. A few are thought to have been pulled over by local inhabitants. The prolific British Egyptologist Sir Ernest Alfred Wallis Budge speculated in the 1920s that the fallen Heliopolis obelisk (the other of the pair is still standing) was deliberately pulled down by people who were searching for treasure hidden at the top.[13] In this version of history there was no treasure, of course, as the remaining obelisk survives to testify.

AN OBELISK RAZER wishes to pull down an obelisk. The obelisk has a height L and a base width very much smaller than L. He has a strong, light, inextensible rope also of length L with an additional lasso on the end, which we do not count in L. The man can position the lasso at any height h from the base of the obelisk. We can assume that the lasso does not slip once positioned. Both the obelisk and the man are free standing on flat ground, and have the same coefficient of friction with the ground. We can assume that the obelisk is very much taller than the man, and has very much greater mass. Estimate the optimum value of h to maximise the chance of sucessfully razing the obelisk.[14]

[13] Wallis Budge, E.A., 1926, "Cleopatra's needles and other Egyptian obelisks; a series of descriptions of all the important inscribed obelisks, with hieroglyphic texts, translations, etc.," Religious Tract Society, (OCoLC) 614632306.

[14] It is worth noting that this question is somewhat more open-ended than most of the questions in this book, and it may be more suited to a discussion around the solution method and assumptions than to blind attempts. To estimate the optimum value of h, we need to make a number of approximations and simplifications. This is not, therefore, a clean textbook-type question.

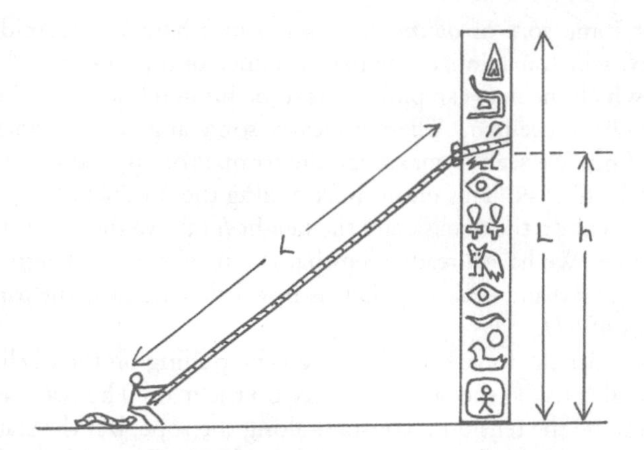

Answer

If we have tried the question on obelisk raising we will already know that we need an awful lot of men to pull an obelisk from a slightly off-vertical position to the true vertical. This is something of a reverse problem, but with a few complications. One of the Heliopolis obelisks is, in fact, still standing on its original spot after four thousand years. This provides a clue as to the sort of force required to topple such a structure.

The first thing we need to note is that the base width of the structure is not given in the question, nor is the mass of the man or the obelisk. In fact this makes complete sense, because this is an *optimisation* problem. We know this because we are being asked to find the *optimum* value of h. Of course, h scales with L, which we can regard as the *length scale* for the problem. Let us take a closer look at *why* such an optimum height for toppling might exist.

What would happen if the man attached the lasso to the very base of the obelisk, $h/L = 0$? He would pull with an essentially horizontal force (notwithstanding the slight inclination of the rope). The maximum force he could exert would be equal to the limiting friction force with the ground. His reaction force is mg, so at most he can exert a horizontal force equal to μmg, where μ is the coefficient of friction with the ground. It happens that we do not need to know the value of any of these new variables to solve the problem. We will discuss this more in a moment. The obelisk also stands on flat ground. It is very much heavier than the man, so—recalling that the coefficients of friction with the ground for the man and the obelisk are the same—it will not slide under the action of such a horizontal force applied at the base.

Let us now consider what would happen if the man attached the lasso to the very top of the obelisk at $h/L = 1$. Because the rope and obelisk are both of length L the rope would hang almost vertically downwards and the man could at most exert his own weight, mg, almost exactly downwards on the obelisk. This is likely to simply plant it more firmly onto the ground. The torque, or moment, the man could exert would be very small indeed.

Perhaps some sort of *optimum* exists somewhere in the middle–some value of h/L which maximizes the man's chance of toppling the obelisk. The force with which the man can pull seems to be limited both by sliding friction and the weight of the man. There is clearly some angular dependence which we need to find. We seek to maximize the torque (or applied moment) about the base of the obelisk. This means maximizing the product of the horizontal force component on the obelisk and the height h (above the base) at which we exert the force. We have already seen that the torque goes to zero at the two extremes $h/L = 0$ and $h/L = 1$. Let us now calculate how the torque varies for all values of h/L.

First consider the force caused by the rope pulling on the obelisk. In due course we will consider the opposing force on the man. The two are equal and opposite because the tension is constant along the rope, but the static equilibrium of the man limits the magnitude of the force. To enable us to calculate the torque on the obelisk we will introduce a new variable θ, which is the angle the rope makes with the ground. We define θ by $\sin\theta = h/L$. We are assuming that the height of the man is negligible in comparison to the length of the rope, so that the *end effect* is neglected. These kinds of approximation, though rare in textbook problems, are very common in real-world calculations. It is important to feel comfortable making them, but this only comes with experience. We treat the obelisk as very thin compared to its height, so we are not going to worry about the detail of how the lasso interacts with the column. We could, of course, deal with both more formally by setting up the problem more rigorously, or simply by adding the complexity into the analysis that follows. If we did that we would find that it has only a small impact on the answer for slender obelisks.

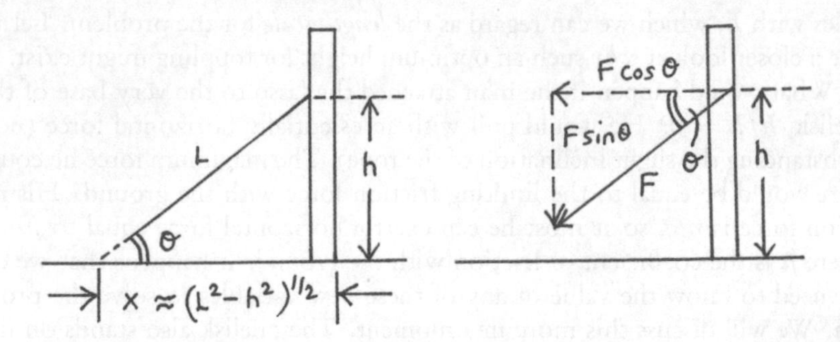

Now consider the torque on the obelisk, T. As we said earlier, this is simply the horizontal force component multiplied by the height h. We write $T = F\cos\theta h$. We recall that we are seeking to maximise T.

We now consider the forces on the man. They are:

- His weight, mg, which acts downwards from the centre of mass.
- The force due to tension in the rope, F, which acts at an angle θ above the horizontal.

- The normal reaction force from the ground, which acts vertically upwards at the point of contact. This needs to be equal to $(mg - F \sin\theta)$ to balance forces in the vertical direction.

- The frictional force with the ground, which acts horizontally at the point of contact with the ground. To balance forces in the horizontal direction, this needs to be equal in magnitude but opposite in direction to the horizontal component of force due to tension in the rope, which has magnitude $F\cos\theta$. The frictional force is limited to being less than or equal to the coefficient of friction multiplied by the reaction force, otherwise the man will slip forwards. We can write this as the inequality $F\cos\theta \le \mu\,(mg - F\sin\theta)$. When we have *limiting friction*—that is, when the man is on the point of slipping and static friction has reached a maximum—we can write the equality $F\cos\theta = \mu\,(mg - F\sin\theta)$. This gives us an equation for the maximum value that F can take:

$$F = \frac{\mu m g}{\mu \sin\theta + \cos\theta}$$

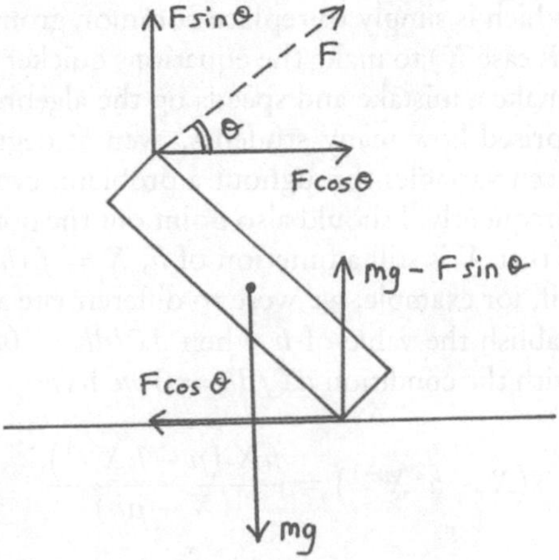

We have discussed all the forces, so all that remains is to solve our equations to find the maximum torque the man can exert on the obelisk. First we substitute for F in the equation for T, above, giving

$$T = F\cos\theta h = \frac{\mu m g h}{\mu \tan\theta + 1}$$

We seek the maximum T for any h, so we put the expression above in a form we can differentiate to find this maximum. If we are observant we should notice that θ and h are interrelated. Specifically,

$$\tan\theta \approx \frac{h}{\left(L^2 - h^2\right)^{1/2}}$$

This is an approximation because there is a small end effect due to the finite height of the man, which we have ignored in the analysis. Substituting for $\tan\theta$ gives an expression for T as a function of h, $T = f(h)$. The expression is

$$T = f(h) \approx \frac{\mu mgh \left(L^2 - h^2\right)^{1/2}}{\mu h + \left(L^2 - h^2\right)^{1/2}}$$

We note that T is now purely a function of h. The other terms are constants (μ, m, g and L). So we can differentiate T with respect to h to find the maximum. The maximum is found by setting $dT/dh = 0$. We can perform the differentiation fairly easily using the chain rule, though it requires many lines of algebra. We get

$$\frac{dT}{dh} \approx \frac{\mu mg \left(X - h^2 X^{-1}\right)}{X + \mu h} - \frac{\mu mghX \left(\mu - hX^{-1}\right)}{(X + \mu h)^2}$$

where $X = \left(L^2 - h^2\right)^{1/2}$. Here I have used a little trick I like to use in problems of this type, which is simply to replace common groups of terms with a new variable (in this case X) to make the equations quicker to write. It means I am less likely to make a mistake and speeds up the algebraic manipulations. You would be surprised how many students, even at degree-level study, insist on using the given variables throughout a problem, even when groups of terms appear very frequently. I should also point out the potential peril of this simplification. It is that X is still a function of h, $X = f(h)$, a fact we would need to remember if, for example, we were to differentiate a second time.

We wish to establish the value of h when $dT/dh = 0$. Rearranging the expression above with the condition $dT/dh = 0$ we have

$$\left(X - h^2 X^{-1}\right) = \frac{hX \left(\mu - hX^{-1}\right)}{(X + \mu h)}$$

This leads to

$$X^3 = h^3 \mu$$

Recalling that $X = \left(L^2 - h^2\right)^{1/2}$, the condition for the maximum torque reduces to

$$h = \frac{L}{\left(\mu^{2/3} + 1\right)^{1/2}}$$

This is quite an interesting dependence on μ. If we consider the case $\mu = 1$, we get

$$h = \frac{L}{\sqrt{2}}$$

Considering the geometry of the problem, this gives us $\theta = 45°$, or $\pi/4$ rad. This is satisfyingly simple to remember in the heat of obelisk toppling, but only applies when the rope has the same length as the obelisk's height.

It might also be interesting to work out the force the man can exert at this maximum torque condition for the case $\mu = 1$. Recalling the equation for T and substituting $\mu = 1$ and $h = L/\sqrt{2}$, we get

$$T = \frac{mgh}{2}$$

The torque was given by $T = F \cos\theta h$, so

$$F = \frac{T}{\cos\theta h} = \frac{mg}{\sqrt{2}}$$

So at the maximum torque condition, the man can exert $1/\sqrt{2}$ times his own weight before he starts to slide.

Further discussion

There is one little subtlety that we neatly glossed over when we drew the free-body diagram of the man. If you noticed it you have a mind for even the most tiny detail. If you did not, you can definitely be forgiven, and this further discussion may encourage you to check this aspect of problems in future. The original question (what height h is best for obelisk toppling?) was posed in quite an open-sounding way. We needed to work out that our aim was to maximize the torque on the obelisk. We also needed to realize that for a given coefficient of friction there was a maximum horizontal force the man could exert on the rope without slipping, and that this force depended on the angle of the rope. The problem is that as θ increases the reaction force between the man and the ground decreases for the same force on the rope. What we failed to check was whether the man was *rotationally stable* for the solution $h = L/\sqrt{2}$, or $\theta = 45°$. Let's consider that now.

In the solution above we balanced the forces on the man in the horizontal and vertical directions, but we did not consider the *net torque* on the man. That is what we will do here. We represent his feet as A, his centre of mass as C, and the top of his body as B. By considering the mass distribution of the average person, it is possible to argue that the man's centre of mass is about halfway between A and B, so we let $AC = CB = l/2$. Here we are only concerned with rotational stability, so we consider only the force F due to tension in the rope, and the force mg due to gravity. We ignore forces passing through the point of contact P, which is the pivot for the system. We now *assume* that the man will be rotationally stable at an angle β to the horizontal, and test this assumption by attempting to solve for β. The forces are shown below.

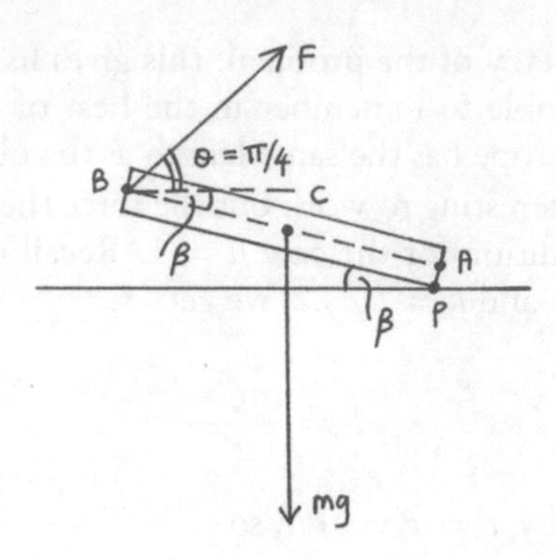

The man will be rotationally stable (that is, have an angular acceleration of zero) when the net torque about P is zero. Taking the distance AP to be very much smaller than AC (that is $AP \ll AC$) and balancing moments about P, we have

$$mg\frac{l}{2}\cos\beta \approx Fl\cos\left(\frac{\pi}{4}-\beta\right)$$

The force the man can exert at the maximum torque condition is $F = mg/\sqrt{2}$. We have

$$\cos\beta \approx \sqrt{2}\cos\left(\frac{\pi}{4}-\beta\right)$$

This condition is satisfied for $\beta = 0$, which is the limit of the man lying parallel to the ground. We can represent forces in the limit as $\beta \to 0$. From the diagram, it should be obvious that for $\theta = \pi/4$ and $F = mg/\sqrt{2}$ we have zero net torque about A. If it isn't obvious from the diagram we should repeat our analysis to check.

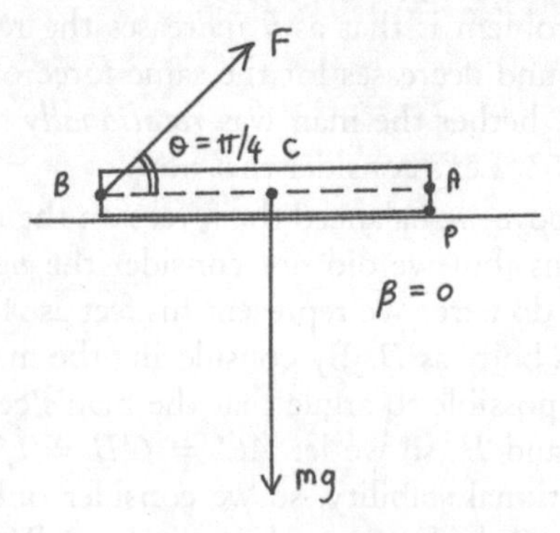

We conclude that the man can exert the force required to achieve the maximum torque on the obelisk without toppling over himself. However, the system is limited both by the toppling force on the man, and by friction at his point of contact with the ground.

3.9 The Ravine of (Not Quite) Certain Death ★★★

In a famous scene from Monty Python and the Holy Grail, five knights, led by King Arthur, are working their way along a precipitous cliff, chainmail and swords clanking through the mist. In the distance there is a dilapidated rope bridge over a gorge, guarded by an old man, the keeper of the bridge. They have arrived at The Bridge of Death. "Oh, great," says Sir Robin.

The fact that Sir Robin is a little put out by the whole bridge thing is explained by the fact that they are about to cross The Gorge of Eternal Peril, "which no man has ever crossed". So the odds of survival are looking fairly bleak from the outset. King Arthur explains that, to be allowed across, the aspirant bridge-crosser must answer three questions posed by the bridge-keeper. Sir Robin is still a little ill at ease, and asks what happens if they get an answer wrong. King Arthur—very much a straight talker—answers, "then you are cast into The Gorge of Eternal Peril". Sir Robin looks a little pale. And with good reason, because he is one of two knights who are eventually cast into the Gorge. Then the bridge-keeper asks King Arthur his question: "what is the air-speed velocity of an unladen swallow?"

At this point King Arthur is a little confused, no doubt trying to recall the exact velocities for different species. Eventually he replies, "what do you mean? An African or European swallow?" The keeper is cast into the Gorge. One of the two remaining knights asks King Arthur how he knows so much about swallows. "Well, you have to know these things when you're a king, you know," King Arthur replies.

King Arthur was rather ahead of his time. It was not until 2001 that this great mystery of science eventually started to unravel. First to shed light on the problem was Dr Park at the Department of Animal Ecology at Lund University in Sweden, who carried out wind tunnel experiments on barn swallows.[15]

> Two barn swallows (Hirundo rustica) flying in the Lund wind tunnel were filmed using synchronised high-speed cameras to obtain posterior, ventral and lateral views of the birds in horizontal flapping flight.

Later, Dr Taylor led theoretical developments at the University of Oxford.[16] I have had lunch with Dr Taylor on a few occasions, so know a little about his work. I recently learnt that he has performed theoretical estimates of air-speed for both African and European swallows, based on his work on optimum

[15]Park, K.J., Rosén, M., Hedenström, A., 2001, "Flight kinematics of the barn swallow (Hirundo rustica) over a wide range of speeds in a wind tunnel," The Journal of Experimental Biology, 204, pp. 2741–2750.

[16]Taylor, G.K., Nudds, R.L., Thomas, A.L.R., 2003, "Flying and swimming animals cruise at a Strouhal number tuned for high power efficiency," Nature 425, pp. 707–711.

Strouhal number[17] flight characteristics for flying animals. The air-speed of an unladen swallow, it turns out, is about 8.9 m s^{-1} for both species.

Although I am fairly sure I was not inspired by Python at the time, it is now difficult to see how the following question could not have been directly based on that brilliant scene. All the students I have tried this question on seemed to enjoy it, and it gives good scope for discussion with relatively simple mathematics. When asking the question, I made it clear that it was a straightforward problem in statics (no tricks, that is), not a river-crossing puzzle of the wolf, goat and cabbage variety.[18] The question is as follows.

TWO MEN, EACH of mass m, are on opposite sides of a gorge of width w. Each man has a tree trunk of length l and mass $5m$. What is the widest ravine that can be crossed by one of them (so that they are both on the same side)?

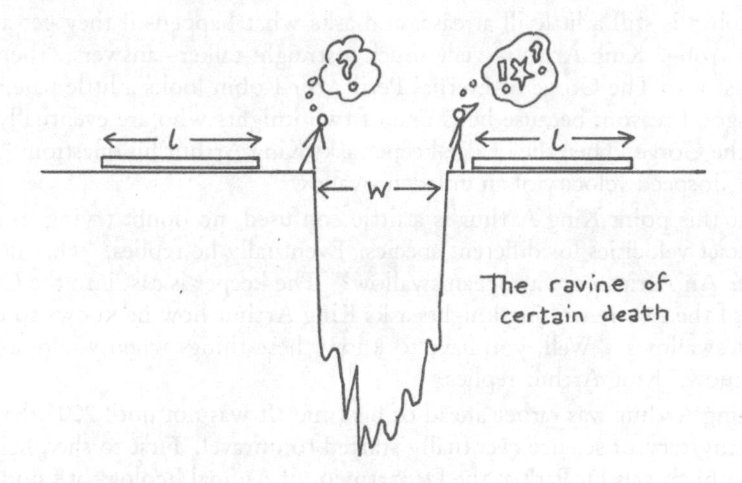

Answer

There are in principle a number of possible ways of interpreting the implicit *rules* of the question. The simplest answer, and therefore the preferred one, is now given. We will invoke some possible complications in the discussion section.

Let's consider the problem in two parts. We will assume that the man on the right (R) will join the man on the left (L).

[17] The Strouhal number is a dimensionless quantity (group of variables) that describes the vortex-shedding frequency of objects in fluid flow.

[18] Transport puzzles, or so-called river-crossing puzzles, are among the oldest forms of popular logic puzzle, and date back to the ninth century or earlier. One of the most well-known examples is the wolf, goat and cabbage problem. You have probably heard it already, but it goes like this. A farmer needs to cross a river with a wolf, a goat and a cabbage. His boat will only hold him plus one of the wolf, goat or cabbage. If the goat and the cabbage are left together, however, the goat will eat the cabbage. If the wolf and the goat are left together, the wolf will eat the goat. How can the farmer safely transport the wolf, goat and cabbage across the river?

Consider first the man on the R. He will need to slide his tree trunk out to the maximum distance that will support his weight when he is standing on its end. We define this distance as x, and then calculate x by taking moments around the edge of the ravine. There are only two forces, the force mg due to the weight of the man, and a force $5mg$ due to the weight of the tree trunk. The clockwise moment is equal to the anticlockwise moment when the tree trunk is in balance. The moment equation is as follows.

$$mgx = \left(\frac{l}{2} - x\right) 5mg$$

That is

$$x = \frac{5l}{12}$$

We now consider the L side of the problem. When the man on the R steps on to the tree trunk on the L, he will be counterbalanced by the man on the L, so, by symmetry, the tree trunk can overhang the gorge by $l/2$.

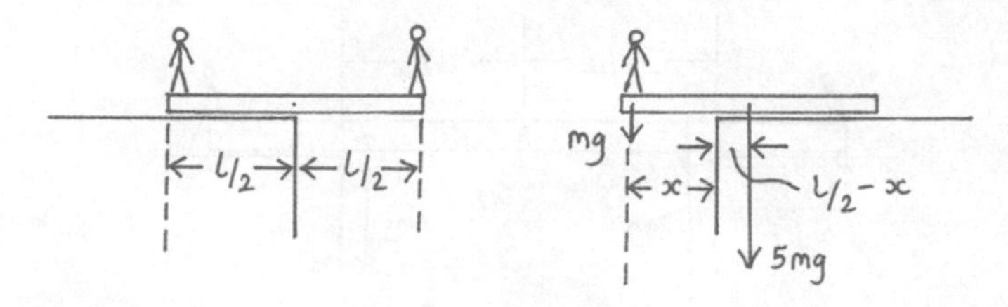

Thus, the widest ravine the men can cross is $w = (l/2) + (5l/12) = (11/12)\,l$.

Further discussion

When I devised this puzzle, I worried a little that some students might come up with more esoteric approaches, so I worked out a number of alternative solutions to include in the discussion. In fact, I probably had nothing to fear, Having tried the problem on quite a number of students, only one proposed a solution different from that above. I took this to be a good sign. In general, problems should be solved in the simplest possible way, without invoking the sort of extra complexity I present now. These two alternative solutions may amuse, however, and are interesting examples that give slightly different twists to the problem, although not within the spirit of the original wording.

- Tree trunks one on top of the other. If we allow the tree trunks to be placed one on top of the other (which would require some sort of overlapping joint, when the problem of *assembly* is considered), it should be apparent that it is now possible to cross a ravine of width $w = l$. We note that the man on the L provides a counterbalancing moment for the man on the R, which increases linearly from 0 to $mgl/2$ as the man on the R moves from $x = 0$ to $x = l/2$ from the R edge of the ravine.

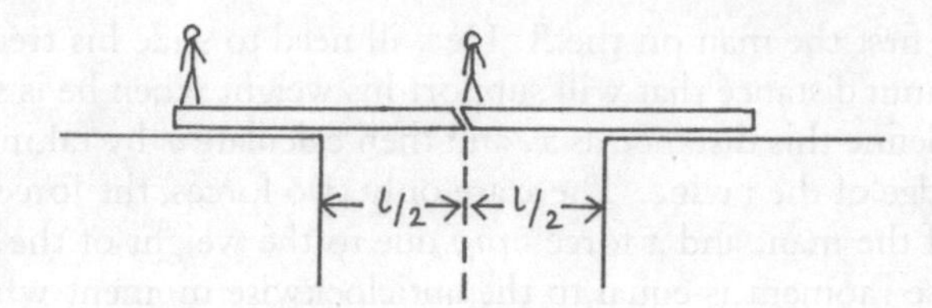

- Tree trunks rigidly jointed. Here we assume that we can assemble the tree trunks in such a way that they form a rigid joint that is stiff and able to take a load. This places a new upper bound on the possible width of the ravine, $w = (14/12)\,l$. To check this answer you would need to take moments about the edge of the gorge on each side, and consider the maximum distance the tree trunks can be extended over the precipice (in static equilibrium) without toppling. When the counterbalancing force due to the weight of the man is taken into account, the maximum distance the end of each tree trunk can extend is $(7/12)\,l$.

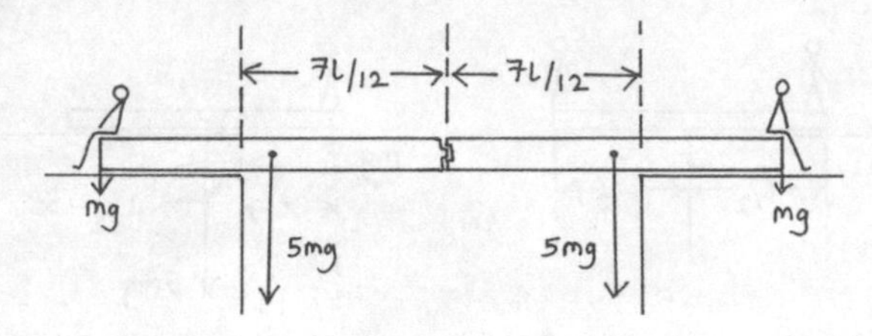

Chapter 4

Dynamics and collisions

In this chapter we look at dynamics and collisions. We consider collisions in terms of the velocities of the interacting bodies, the forces on the bodies during a collision, and the energy of the bodies before and after a collision. In the simplest case, we deal with head-on collisions—that is, one-dimensional collisions. The mathematics for two- and three-dimensional collisions is not much more complicated, but we need to consider the vector components (in the x, y and z directions, or in polar co-ordinates) when solving the equations. The basic principles remain the same.

To solve problems in dynamics and collisions we need a number of concepts to hand. We won't elaborate on all the equations here, because you should be able to *derive* them as you solve these problems. If we have the basic principles correct, however, the rest should follow naturally. Let us consider some of the concepts that come up in dynamics problems, and the notation that is used. If any of these are unfamiliar to you, it may be worth consulting a textbook before trying the problems.

- VELOCITY. When we consider interacting bodies, we are generally interested in their *initial* and *final* velocities—that is, the velocities before and after the collision. We need a notation to distinguish these velocites. For example, we can use $\boldsymbol{v}_1$ and $\boldsymbol{v}_2$ for the initial velocities, and $\boldsymbol{v}'_1$ and $\boldsymbol{v}'_2$ for the final velocities. The corresponding scalar *speeds* are v_1, v_2, v'_1 and v'_2. In general we will use the prime symbol ($'$) to indicate the state following a collision.

- ENERGY. The kinetic energy of a body is given by KE $= \frac{1}{2}mv^2$ (see proof[1]).

[1]This proof of the formula for kinetic energy is a popular problem that encourages students to think more deeply about an equation we normally take for granted. The incremental work dW performed on a particle is defined as the force F acting on the particle multiplied by the displacement in the direction of the force ds. In vector notation we write $W = \int_{s1}^{s2} \boldsymbol{F}.d\boldsymbol{s}$, where we take the dot product of vectors $\boldsymbol{F}$ and $\boldsymbol{s}$. If we consider motion in a single direction—the

- LINEAR MOMENTUM. The momentum of a body is given by $\boldsymbol{p} = m\boldsymbol{v}$. Because velocity is a vector quantity, momentum must also be a vector quantity. According to the law of conservation of momentum, in all collisions, whether elastic or inelastic, the linear momentum of the system is conserved. For two bodies we write

$$\begin{aligned} m_1\boldsymbol{v}_1 + m_2\boldsymbol{v}_2 &= m_1'\boldsymbol{v}_1' + m_2'\boldsymbol{v}_2' \\ \boldsymbol{p}_1 + \boldsymbol{p}_2 &= \boldsymbol{p}_1' + \boldsymbol{p}_2' \end{aligned}$$

- ELASTIC AND INELASTIC COLLISIONS. In an *elastic* collision the total energy of the bodies is *conserved*: the sum is the same before and after a collision. In an *inelastic* collision energy can be lost. By *lost* we mean that it is converted from kinetic energy, which is the useful form that interests us, to less useful forms of energy like heat.[2] For two bodies we write

$$m_1v_1^2 + m_2v_2^2 \geq m_1\left(v_1'\right)^2 + m_2\left(v_2'\right)^2$$

where the equality applies for elastic collisions.

- COEFFICIENT OF RESTITUTION. The coefficient of restitution e is defined as the ratio of relative velocities after and before a collision. We write

$$e = \frac{\text{speed of separation}}{\text{speed of approach}} = -\frac{|\boldsymbol{v}_1' - \boldsymbol{v}_2'|}{|\boldsymbol{v}_1 - \boldsymbol{v}_2|}$$

where, in general,[3] $0 \leq e \leq 1$. For an *elastic* collision $e = 1$, for an *inelastic* collision $e < 1$, and for a *perfectly inelastic* collision $e = 0$. Later in this chapter we prove that, for elastic collisions, saying there is zero kinetic energy loss is the same as saying that $e = 1$.

- IMPULSE AND FORCE. Newton's Second Law tells us that force on a body is equal to the rate of change of momentum of the body. We write $\boldsymbol{F} = d\boldsymbol{p}/dt$, where the momentum $\boldsymbol{p} = m\boldsymbol{v}$. This reduces to

---

x-direction, say—this reduces to $W = \int_{x1}^{x2} F_x dx$. The acceleration of a particle of mass m is related to the force F_x by Newton's Second Law, $F_x = ma_x$. We write $W = \int_{x1}^{x2} ma_x dx$. The acceleration a_x is defined in terms of the velocity v_x as $a_x = dv_x/dt$. We write

$$W = \int_{x1}^{x2} m\frac{dv_x}{dt}dx = \int_{x1}^{x2} m\frac{dv_x}{dx}\frac{dx}{dt}dx = \int_{x1}^{x2} m\frac{dv_x}{dx}v_x dx$$

Simplifying, we write $W = \int_{v1}^{v2} mv_x dv_x$, where we change the limits to be in terms of velocity, the variable in which we are integrating. Integrating we get

$$W = \frac{1}{2}mv_2^2 - \frac{1}{2}mv_1^2$$

which we define to be the kinetic energies of the initial and final states.

[2]One measure of the usefulness of a particular form of energy is the ease with which it can be converted to another form. In this context kinetic energy can readily be converted to heat, but heat cannot as readily be converted to kinetic energy.

[3]For *superelastic* collisions, or collisions in which stored energy is released, we can have $e > 1$.

$$F = m\frac{dv}{dt} + \frac{dm}{dt}v = ma \quad \text{for} \quad \frac{dm}{dt} = 0$$

We can integrate this relationship to get the impulse on a body due to a force, which is simply defined as the integral of the force over a period of time. So, impulse I is defined by

$$I = \int_{t}^{t'} F dt = \int_{p}^{p'} dp = p' - p$$

If the force is constant during the period of time over which we are integrating, we get

$$I = F\Delta t = \Delta p = p' - p$$

So to achieve a particular change of momentum of a body—to bring it to rest, for example—we require a fixed impulse. The magnitude of the force required to achieve that impulse depends on the period of time over which the force acts. If we fall over on a concrete floor, in the resulting collision we are brought to rest very quickly indeed and the force is high. It hurts. If we fall over and hit a springy surface we are brought to rest rather slower and the peak force is lower.[4] In each case the total impulse is the same.

- REFERENCE FRAME. According to the principle of *Galilean invariance*, Newton's laws apply in all inertial frames of reference. An inertial frame of reference is one for which the acceleration is zero. This means that the laws of conservation of momentum and energy (for elastic collisions) apply equally in any non-accelerating frame. In 1632 Galileo used the example of an observer in the belly of a ship sailing over a smooth sea. According to the invariance principle, when performing experiments the observer would be unable to detect whether the ship was in constant motion or stationary. For all practical purposes this statement is true, but we should note that the Earth's surface is not quite an inertial frame. This is because it rotates about its own axis, and about the Sun, which itself rotates around the centre of the galaxy, and so on. We will ignore the secondary non-inertial effects, which are small.

  In addition to the laboratory frame of reference, which we take as being the frame in which the observer is stationary, there are two special frames of reference we might choose for convenience during our calculations. Whichever frame we choose we should get the same result. The frames are:

[4]In collisions between springy/deformable objects, the peak pressure is also lower, because the force tends to be distributed over a larger surface area.

- A frame in which one body is stationary, allowing us to simplify the equations.
- The *centre-of-momentum frame*, in which momentum is zero both before and after the collision (by the law of conservation of momentum). This frame has the special property that in an elastic collision the bodies rebound with the same speeds as before the collision (but with the signs of the velocities reversed).

4.1 Pulleys ★

Many students find this straightforward question (in the sense that there are no tricks) harder than they expect.

A LIGHT, INEXTENSIBLE rope connecting 5 kg and 10 kg masses is suspended over a light frictionless pulley. The system is stationary, and has just been released from rest at the instant shown. Answer the following questions:

1. Are the tensions T_1 and T_2 the same?
2. What is the magnitude and direction of the force on the pulley?

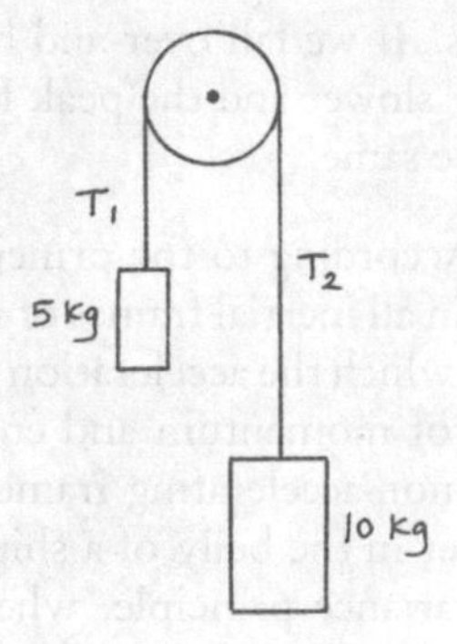

Answer

Let us consider the two parts of the problem in turn.

1) ARE THE TENSIONS T_1 AND T_2 THE SAME?

The simple answer is yes, they are the same. They have to be the same because both the rope and the pulley are *light*, so have no inertia.[5] Likewise, the rope is *inextensible* so has no ability to store energy. If we pull on one end, the force must be transmitted to the other end.

[5]Here we can think of *mass* as being the *translational inertia*. The *rotational inertia* of the pulley is defined by $I = \sum mr^2$, the sum of the mass elements multiplied by their distance from the centre of rotation.

1) WHAT IS THE MAGNITUDE AND DIRECTION OF THE FORCE ON THE PULLEY?

For this part we need to solve the forces on all the objects. To do this we need to draw free-body diagrams of all three components. To simplify the diagram we let the smaller weight have mass m, and the larger weight have mass M. We have established that the tensions in the ropes are equal. We represent the tensions by T.

Let's deal first with the pulley. It should be immediately apparent that to achieve zero acceleration of the pulley, we require an upward force equal to $2T$ acting from the axle of the pulley. We now need to establish the value of T.

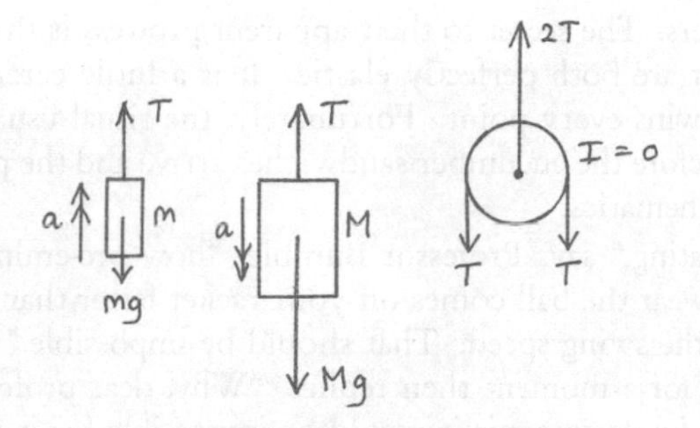

Consider the smaller mass m. Resolving forces *upwards positive*, and allowing the mass to have a *positive upward* acceleration a, we have

$$T - mg = ma$$

Consider the larger mass M. Resolving forces *downwards positive*, and allowing the mass to have a *positive downward* acceleration a, we have

$$Mg - T = Ma$$

Adding these equations we have

$$(M - m)\,g = (M + m)\,a$$

Rearranging for a we have

$$a = \frac{(M - m)}{(M + m)}g$$

Putting in values ($m = 5$ kg, $M = 10$ kg) we have $a = (1/3)\,g$. We can now solve either of the equations for the masses to find T. Doing this gives $T = (20/3)\,g$. The upward force on the pulley is equal to $2T$, so is therefore $(40/3)\,g$. Taking $g \approx 10$ m s^{-2} gives a force of approximately 133 N.

We can compare this value to the upward force on the pulley that would be required if the system were prevented from rotating (and the rope were prevented from slipping over the pulley). In this case we would need to support the entire weight of the masses, and the upward force would be $15g$, or approximately 150 N. By allowing the centre of mass of the system (which comprises both masses and the pulley[6]) to accelerate downwards, we reduce the force requirement on the pulley axle.

4.2 Dr Lightspeed's elastotennis match ★★

Dr Lightspeed and Professor Bumble are lining up for another elastotennis match. Viewed from a distance it looks an awful lot like ordinary tennis with very good players. The secret to their apparent prowess is that the elastoball and elastoracket are both perfectly elastic. It is a futile ceremony in which Dr Lightspeed wins every point. Fortunately, the ritual usually only lasts a couple of sets before the cucumber sandwiches arrive and the pair tackle some problem in mathematics.

"It is fascinating," says Professor Bumble, "how pre-eminently fast your returns are. I swear the ball comes off your racket faster than the sum of the ball speed and the swing speed. That should be impossible." Dr Lightspeed looks confused for a moment then replies, "Why, dear professor, surely the reverse is true. In elastotennis it would be impossible for it *not* to come off my racket faster than the sum of those speeds." They return to the match, in which Dr Lightspeed wins every point.

Later, over more sandwiches, the professor is rather subdued. They sit nibbling in silence. Eventually the professor speaks. "I still don't understand, Dr Lightspeed, how you can return so fast. You really must explain."

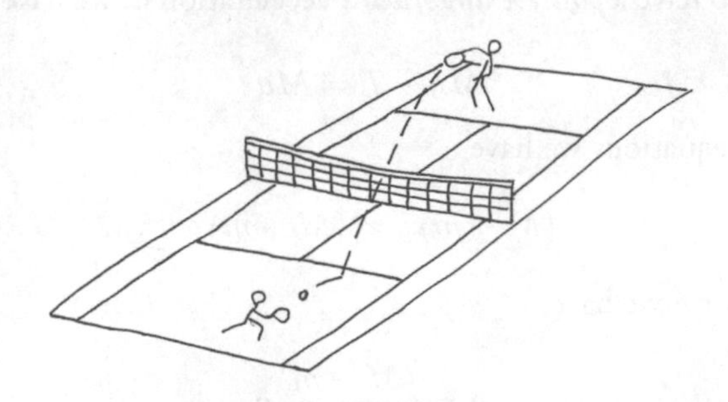

IN ELASTOTENNIS (TENNIS in which both the ball and the racket are entirely elastic), what is the maximum return speed of the ball in terms of the original ball speed and swing (racket) speed?

[6]Though note that, in this case, the pulley has no moment of inertia.

Answer

First we will take a brute-force approach to the problem, the standard method. We'll solve the momentum and energy equations in the global frame of reference (the frame of reference of the tennis court). I've also given an alternative answer, which exploits the centre-of-momentum frame, and which is more elegant.

In the global frame of reference we define the velocities of the elastoball and the elastoracket as v_1 and v_2 respectively.

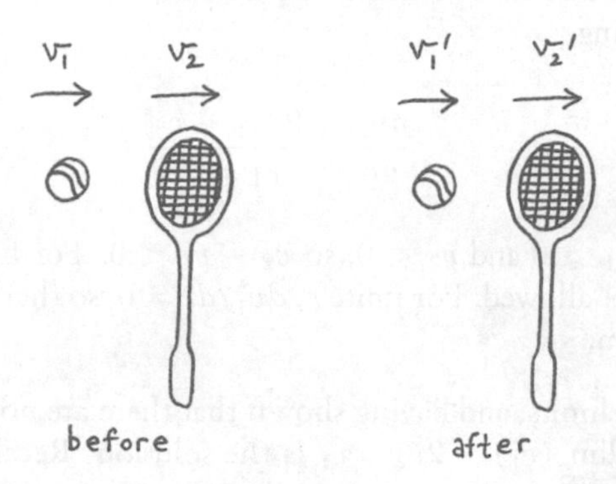

We arbitrarily define a positive value of velocity as corresponding to motion to the right, and a negative value of velocity as motion to the left. Letting the masses of the ball and racket be m_1 and m_2, we can write the momentum equation

$$m_1 v_1 + m_2 v_2 = m_1 v_1' + m_2 v_2'$$

or

$$v_1 + r v_2 = v_1' + r v_2'$$

where $r = m_2/m_1$. For a completely elastic collision we have

$$e = -\left(\frac{|v_1' - v_2'|}{|v_1 - v_2|}\right) = 1 \quad \text{or} \quad v_1 - v_2 = v_2' - v_1'$$

where v_1' and v_2' are the velocities after the collision. Eliminating v_2' we have

$$v_1' = \frac{v_1(1-r) + 2rv_2}{(1+r)}$$

Taking v_1 to be positive (the ball moving to the right) and v_2 as negative (the racket moving to the left), we seek to make v_1' as negative as possible. We therefore look at both limits, $\lim_{r\to 0}(v_1')$ and $\lim_{r\to\infty}(v_1')$, and investigate all possible minima. Let's consider each of the three in turn.

- CONSIDER $\lim_{r\to 0}(v_1')$. We see that $\lim_{r\to 0}(v_1') = v_1$. This is because m_1 is so heavy (relative to m_2) that it continues unimpeded after the collision.

- CONSIDER $\lim_{r\to\infty}(v_1')$. We see that $\lim_{r\to\infty}(v_1') = 2v_2 - v_1 < 0$. In this case m_2 is so heavy (relative to m_1) that it continues unimpeded after colliding.[7] The smaller mass m_1 (the ball) bounces off the unaffected larger mass m_2 (the racket).

- CONSIDER ALL POSSIBLE MINIMA. To do this, we first differentiate to get dv_1'/dr, giving

$$\frac{dv_1'}{dr} = \frac{2\,(v_2 - v_1)}{(1+r)^2}$$

  However, $v_1 > 0$ and $v_2 < 0$, so $v_2 - v_1 < 0$. For finite m_1 and m_2, $r = \infty$ is not allowed. For finite r, $dv_1'/dr \neq 0$, so there is no minimum or maximum.

Considering both limits, and having shown that there are no minima or maxima, we see that $\lim_{r\to\infty}(v_1') = 2v_2 - v_1$ is the solution. Recalling that $v_1 > 0$ and $v_2 < 0$ we see that the maximum speed of return $|v_1'|$ is equal to the ball speed plus twice the racket speed. We write

$$|v_1'| = 2\,|v_2| + |v_1|$$

Dr Lightspeed is correct. In elastotennis it is impossible for the elastoball *not* to come off the racket faster than the sum of the ball speed and the racket speed.[8] No wonder Professor Bumble is struggling to return his shots.

An elegant approach

We now consider an approach that invokes the centre-of-momentum frame of reference.

We are interested in determining the *maximum* return speed. This will occur when the change in velocity of the ball is greatest, no matter what inertial frame of reference we consider the problem in. Consider the frame of reference in which the elastoracket is stationary. It *should* be obvious that in an elastic collision, the biggest change in velocity of the ball occurs when the change in velocity of the racket during the collision approaches zero. This condition occurs when the racket has a much greater mass than the ball ($m_2 \gg m_1$). If this doesn't feel right, even in the frame of reference of the

[7] To show this, we consider the symmetry of the problem (that is, what would happen if we interchanged 1 for 2 throughout the analysis). Alternatively we can solve directly for v_2'.

[8] For an individual shot this is true even for realistic ball and racket weights (see further discussion).

racket, try and invent some examples to convince yourself.[9] In the discussion we solve the more general problem (m_1 and m_2 having the same order of magnitude[10]).

Because we can perform our calculations in any inertial frame of reference, and because for $m_2 \gg m_1$ the racket will experience no velocity change during the collision, we consider the problem in the frame of reference of the racket. This frame is travelling at v_2. The velocities v_1 and v_2 in the global frame of reference become $v_1 - v_2$ and 0, respectively, in the racket frame. We represent the velocities in both frames below, and show the velocities of the frames with respect to the global frame.

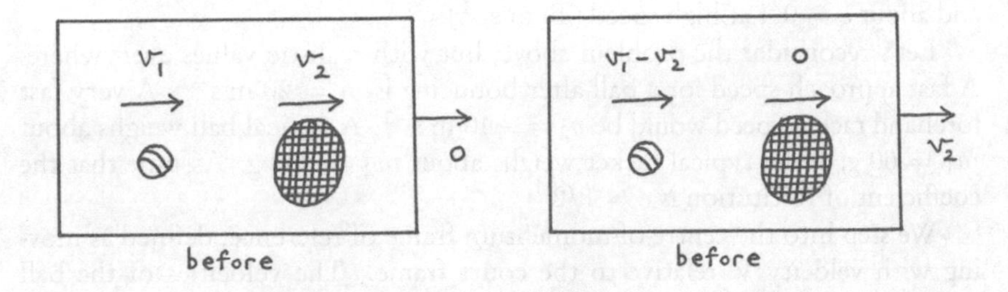

Considering the collision in the racket frame, for $e = 1$, and for $m_2 \gg m_1$, the velocity of the ball is reversed. A ball velocity before collision, $v_1 - v_2$, becomes a velocity after collision, $v_2 - v_1$. Because $m_2 \gg m_1$, the velocity of the racket after the collision is still zero. We now step back into the global frame by adding the racket frame velocity, v_2, to the velocities of the ball and racket within the racket frame. For $m_2 \gg m_1$, the velocities of the ball and racket after collision are $v_1' = 2v_2 - v_1$ and $v_2' = v_2$.

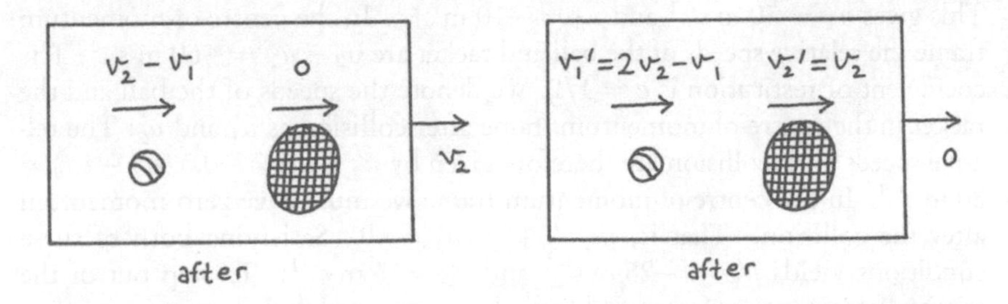

[9]Imagine floating in a space station with a superball (a highly elastic toy ball) and a number of hard steel plates of various thicknesses. We throw the superball at a particular floating stationary plate. If the plate is very heavy, the ball rebounds with approximately the same speed as it had when it arrived. This is like bouncing the ball off a hard floor. As we make the plate lighter, the ball rebounds more slowly. When the plate is very much lighter than the ball, it can no longer impede the ball's progress, and the ball does not return.

[10]The *order of magnitude* of a number is its size to the nearest power of 10, and is used to make approximate comparisons of the sizes of two numbers. The order of maguitude of a number a is $\log_{10}(a)$ rounded to the nearest integer. If two numbers a and b have the same order of maguitude we write $a \sim b$.

We see that by using this approach we achieve the same result with considerably less complexity. The centre-of-momentum frame can be a powerful tool.

A realistic case ★★★

Believe it or not there are quite a number of academic papers on the physics of tennis. Specifically ball-racket interactions, and ball-surface interactions.[11] In both cases the coefficients of restitution are reasonably similar, and both are functions of the impact speed. The latter defines the deformation of the tennis ball, which is the main source of energy loss during the collision. The coefficients of restitution vary between about $e = 0.7$ at low speed (10 m s^{-1}), and about $e = 0.4$ at high speed (45 m s^{-1}).

Let's reconsider the problem above, but with realistic values everywhere. A fast approach speed for a ball after bouncing is $v_1 = 20 \text{ m s}^{-1}$. A very fast forehand racket speed would be $v_2 = -40 \text{ m s}^{-1}$. A typical ball weighs about $m_1 = 60$ g, and a typical racket weighs about $m_2 = 300$ g. Assume that the coefficient of restitution is $e = 1/2$.

We step into the centre-of-momentum frame of reference, defined as moving with velocity V relative to the court frame. The velocities of the ball and racket relative to the centre-of-momentum frame are $u_1 = v_1 - V$ and $u_2 = v_2 - V$. In this frame, by definition, the momentum is zero. That is, $m_1 u_1 + m_2 u_2 = 0$. Expanding this we find that

$$m_1 (v_1 - V) + m_2 (v_2 - V) = 0$$

or

$$V = \frac{m_1 v_1 + m_2 v_2}{m_1 + m_2} = -30 \text{ m s}^{-1}$$

This gives $u_1 = 50 \text{ m s}^{-1}$ and $u_2 = -10 \text{ m s}^{-1}$. In the centre-of-momentum frame the relative speeds of the ball and racket are $u_2 - u_1 = -60 \text{ m s}^{-1}$. The coefficient of restitution is $e = 1/2$. We denote the speeds of the ball and the racket in the centre-of-momentum frame after collision as u_1' and u_2'. The relative speeds after collision are therefore given by $u_2' - u_1' = -0.5\,(u_2 - u_1) = 30 \text{ m s}^{-1}$. In the centre-of-momentum frame we must have zero momentum after the collision. That is, $m_2 u_2' + m_1 u_1' = 0$. Satisfying both of these conditions yields $u_1' = -25 \text{ m s}^{-1}$ and $u_2' = 5 \text{ m s}^{-1}$. To step out of the centre-of-momentum frame and into the court or global frame, we need to add the frame velocity. The velocities of the ball and racket in the court frame are therefore $v_1' = u_1' + V = -55 \text{ m s}^{-1}$ and $v_2' = u_2' + V = -25 \text{ m s}^{-1}$.

In this more realistic case we achieve a ball velocity of 55 m s^{-1} (123 mph), for racket and ball speeds of 40 m s^{-1} and 20 m s^{-1} respectively. For a per-

[11] For further discussion see: Miller, S., 2006, "Modern tennis rackets, balls, and surfaces," British Journal of Sports Medicine, May 2006, 40(5), pp. 401–405, doi: 10.1136/bjsm.2005.023283; and Brody, H., 1997, "The physics of tennis—III—The ball-racket interaction," American Journal of Physics, 65, pp. 981–987.

fectly elastic collision the ball speed after collision would have been 80 m s^{-1}, and the racket speed 20 m s^{-1}.

In the last few years recorded tennis serves have been getting steadily faster. In 2010, the record stood at about 140 mph. By 2012 it was 163.4 mph, an enormously powerful shot by the Australian player Samuel Groth at a match in South Korea. But even this is not nearly as fast as the sport of Basque pelota, which is the fastest ball sport in the world. Competitors hurl small goatskin balls at a wall (rather like the game of squash) using curved wicker baskets. Ball speeds of 188 mph have been recorded. Catch that!

4.3 Accelerating matchbox ★★★

A MATCHBOX OF height h and uniform density stands on its end on a table with coefficient of friction μ. We wish to move it with initial horizontal acceleration a. What value of y should we choose to keep the matchbox upright?

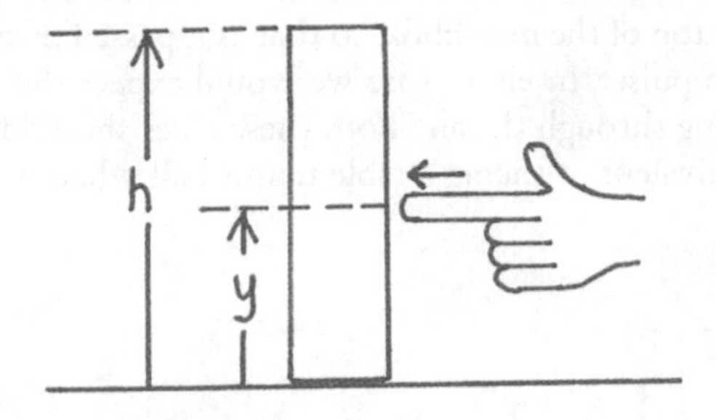

Answer

This is a perfectly innocuous-looking question which I am almost certain most high-school students would struggle with if they came to it cold. In many problems at school in which we have to resolve forces, we may be asked to find, say, the force required to hold an object in static equilibrium. Often there is a single solution for a given geometry. In this example, the value of y required for the matchbox to slide but not topple depends on the value of a we choose. To help illustrate this, let's consider some limiting cases before solving the problem. If our intuition is right we can then use the limiting cases to cross-check the solution we derive.

- CASE 1: ACCELERATION SLIGHTLY GREATER THAN ZERO, $a = 0 + \delta a$. To achieve this we need to just overcome the friction force that acts horizontally at the point of contact with the table. It feels appropriate to push horizontally and very gently as close as we can to the bottom of the matchbox until it's just starting to slide. If this does not feel right, I hope we can agree that it definitely feels *wrong* to push gently at either the centre or the top of the matchbox. I think it is quite obvious that both of those actions would simply topple the matchbox.

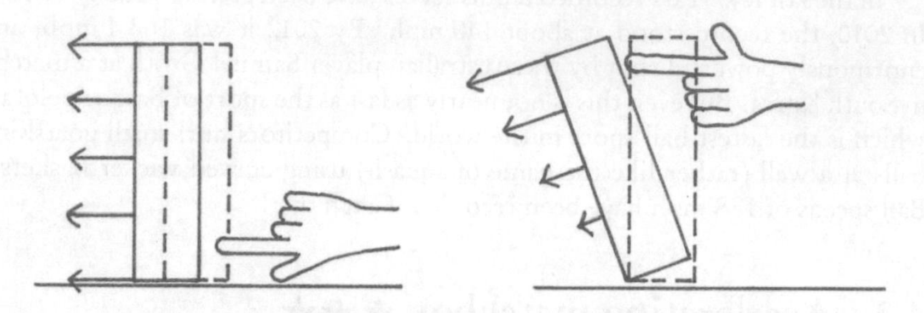

- Case 2: Acceleration Approaching Infinity, $a \to \infty$. To achieve this we need an extremely large force, like the force you would get by firing a pellet from an airgun at the matchbox. To prevent toppling, it should feel appropriate to apply this very large force at the centre of mass of the matchbox. Imagine if you fired the airgun at either the very bottom or the very top of the matchbox, so that as it passed through it imparted a sudden impulse. In either case we would expect the matchbox to be sent spinning through the air. Both these cases should feel instinctively wrong, equivalent to slicing a table tennis ball when we don't intend it to spin.

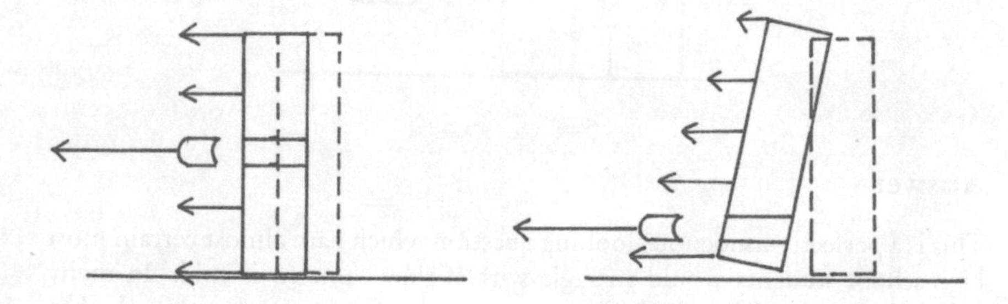

Now let's test our intuition by solving the problem formally. As you will see, it is actually quite straightforward, and follows the same procedure as many of the statics problems we have already considered. Having set many of these problems in the past, however, I'm quite sure that it would initially stump most pre-university students. This is nothing to do with the ability of the average student. Rather it reflects the fact that too few pre-university questions test fundamental understanding of simple concepts, focussing instead on formulaic solutions to standard problems.

We begin, as always, by identifying and drawing the forces on the matchbox.

- The weight, mg, which acts downwards from the centre of mass of the matchbox.

- The reaction force from the ground, R, which acts vertically upwards at the point of contact. This needs to be equal to mg to balance the forces in the vertical direction, so $R = mg$.
- The horizontal driving force, F, applied at a yet to be determined height y, required to cause acceleration a.
- The frictional force[12] with the ground, μR, which acts horizontally at the point of contact with the ground, opposing the driving force F.

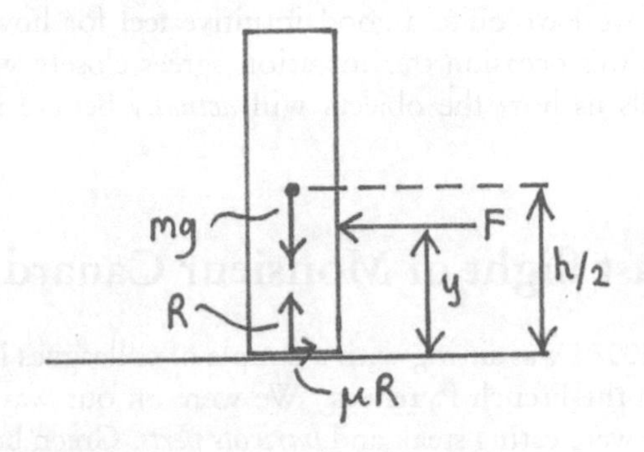

To achieve an acceleration a in the horizontal direction, we require the net horizontal force, F_{net}, to be equal to the mass multiplied by the desired acceleration. That is

$$F_{net} = F - \mu mg = ma$$

We want the box to slide without toppling, so require zero net torque about the centre of mass. The weight and reaction force both pass through the centre of mass, contributing zero torque. So we need to consider only the torques due to the friction force and the driving force. We write

$$\mu mg\frac{h}{2} = \left(\frac{h}{2} - y\right) F$$

Combining these two equations we can eliminate F to get an equation for y.

$$\frac{y}{h} = \frac{1}{2\left(1 + \frac{g}{a}\mu\right)}$$

This is quite a compact answer in the form we expect:[13] $y/h = f\left(\mu, g/a\right)$. Let's return to our intuitive answer to see if it agrees with the solution we have derived.

[12]In many examples in statics we take $|\text{friction}| \leq \mu R$, with $|\text{friction}| = \mu R$ in the case of limiting friction when an object is on the point of slipping. This latter condition also applies to objects in motion. For many real substances the *coefficient of kinetic friction* is lower than the *coefficient of static friction*. In this case we neglect this complexity.

[13]We expect this because the quantities y/h, μ and g/a are all *non-dimensional* (or *dimensionless*). Physical systems are most compactly described by the relationship between non-dimensional quantities, or, as they are often called, non-dimensional *groups* of dimensional variables. In this

- CASE 1: ACCELERATION SLIGHTLY GREATER THAN ZERO, $a = 0 + \delta a$. The derived expression gives $y \rightarrow 0$. That is, pulling or pushing as close to the bottom of the matchbox as possible, which agrees with our intuition.

- CASE 2: ACCELERATION APPROACHING INFINITY, $a \rightarrow \infty$. The derived expression gives $y \rightarrow h/2$. That is, pulling or pushing at the centre of mass, which agrees with our intuition.

As we'd expect, we have quite a good intuitive feel for how objects *should* behave. And on this occasion our intuition agrees closely with our derived result, which tells us how the objects will *actually* behave in this idealised situation.

4.4 The last flight of Monsieur Canard ★★

In mid-January 2007 I was sitting with a couple of colleagues in the restaurant in Pau airport in the French Pyrenees. We were on our way home from an EU meeting. We were eating steak and *haricots verts*. Green beans. But a kind they only serve in France—they are thinner than the sort we eat in England. I remember it quite clearly because it was the third steak with haricots verts we had had in two days. It was either a very strange coincidence or at the design of the EU commission. A reiteration of the Entente Cordiale perhaps. It was cold and clear and sunny and we watched the occasional plane take off as we ate. I noticed that there were birds flocking on the runway. "Extraordinary," I said, "that there are flocking birds on the runway. I would have thought there was some system to keep them away." We then speculated about how the airport authorities could tolerate the birds, and eventually concluded that the birds must simply keep out of the way of noisy aircraft, and not present a problem. Less than a week later, a colleague from Rolls-Royce (one of the people I'd been eating lunch with) emailed me an article.[14]

> Truck driver killed as Air France Régional Fokker 100 hits vehicle during overrun in Pau. French investigators have opened an inquiry after a Fokker 100 operated by Air France subsidiary Régional Compagnie Aérienne Européenne overshot the runway during take-off at Pau Béarn airport yesterday, hitting a works vehicle on a nearby road and killing its driver.

---

case, for example, we did not need to specify the *scale* of the problem (that is, the value of h) to find the point of contact for stable acceleration. Nor did we need to specify the value of g. This is because the behaviour of the system is determined by the three ratios y/h, μ and g/a. These can be thought of as a height ratio (a *geometric similarity condition*), and two force ratios (*dynamic similarity conditions*). The process of reducing problems to the relationship between non-dimensional groups can be quite powerful, and is taught in the first year of many university courses, particularly in engineering science.

[14] Written by Stuart Todd, and published online by www.flightglobal.com, on 26 January 2007.

It was an awful event which could have been a lot worse—miraculously the aircraft's fifty passengers and crew were unhurt. It was investigated by the French civil aviation authority. On the 18th April 2007, the preliminary report was filed.

> The accident resulted from a loss of control caused by the presence of ice contamination on the surface of the wings associated with insufficient consideration of the weather during the stopover, and by the rapid rotation pitch, a reflex reaction to a flight of birds.

As is fairly usual in such reports, the investigators cited a number of contributing factors. I wasn't surprised, however, that birds featured among the causes. I have not flown through Pau airport since, but I expect and hope the problem of flocking birds has now been solved.

Birdstrikes cause an enormous amount of damage to planes every year. According to John Allan of the Central Science Laboratory Birdstrike Avoidance Team (based in Sand Hutton in the UK), the annual repair bill due to incidents of Bird Aircraft Strike Hazard (BASH for short) on commercial aircraft alone is about $1.2 bn.[15] There are also, on average, ten fatalities a year due to birdstrikes. Birds are quite a serious business. The vast majority of strikes occur during takeoff and landing, within a few thousand feet of the ground, when planes are travelling relatively slowly. But it seems that we are never completely safe from our avian friends. The current record for collision height is 37,100 ft, when a plane hit a griffon vulture over the Ivory Coast.[16] You may legitimately be wondering how it is possible to identify species so accurately from the remains of a 600 mph collision. Some poor person has the unfortunate job of collecting *snarge* (I kid you not) from the plane, and subjecting it to DNA analysis.[17]

Let's turn our attention to the case of plane versus duck. We start in the cockpit of a Boeing 747 taking off from Aéroport Paris Charles de Gaulle.

"Monsieur pilote, c'est un canard!" says the co-pilot, alarmed at the winged apparition lining up for a head-on collision. Despite the huge size of the plane, there is a significant thump. The pilot, who has a dry wit of a sort only the French can master, replies, "Oui, un canard mort!" They report the incident and circle back to make an emergency landing so they can inspect the damage. There is, of course, a very large red dent in the nose cone which keeps the plane grounded.

Ruminating over steak and haricots verts in the pilots' lounge after the event, the co-pilot remarks how unlucky they were that the mallard was on a head-on approach. He reasons that because the mallard was flying towards

---

[15] Allan, J.R., Orosz, A.P., 2001, "The costs of birdstrikes to commercial aviation," Bird Strike Committee Proceedings 2001, Bird Strike Committee—USA/Canada, Third Joint Annual Meeting, Calgary.

[16] Laybourne, R.C., 1974, "Collision between a vulture and an aircraft at an altitude of 37,000 feet," The Wilson Bulletin (Wilson Ornithological Society) 86 (4), pp. 461–462, ISSN 0043-5643.

[17] Dove, C.J., Heacker, M., Weigt, L., 2006, "DNA identification of birdstrike remains—progress report," Bird Strike Committee—USA/Canada, Eighth Annual Meeting, St. Louis.

them, its relative speed was increased, resulting in an average impact force that was worse than if the duck had been flying at right angles to the plane. "Tu es stupide!" says the pilot. He reasons that a collision on flight paths at right angles would result in a higher average impact force, because of the shorter duration of the impact, and despite the lower relative speed.

CONSIDER A COLLISION between a plane travelling at 180 mph and a duck travelling at 20 mph. Take the duck to have a length of 50 cm in a head-on collision, and a width of 20 cm in a collision at right angles (we ignore the wings). If the duck has mass 1 kg, estimate the average impact force during both a head-on collision and a collision on a right-angled path.

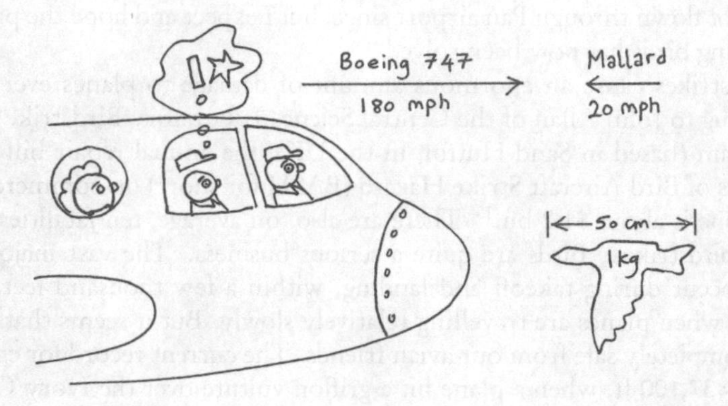

Answer

We should first convert the speeds into SI units. 1 mph is approximately 1.6 km h^{-1}, which is 1600/3600 m s^{-1}. In the case of the head-on collision, the speed of the duck with respect to the plane is 200 mph or $v_r \approx 88.9$ m s^{-1}. It is helpful to consider the problem in the reference frame of the plane, which is the larger object.

To *bring the duck to rest* on the nose of the plane[18] requires a change in the duck's momentum[19] given by $\Delta p = m\Delta v = mv_r$, where m is the mass of the duck. Newton's Second Law tells us that

$$F = \frac{dp}{dt}, \quad \text{so} \quad F_{avg} = \frac{\Delta p}{\Delta t}$$

where F_{avg} is the average force on the duck during the collision time Δt. So we have

$$F_{avg} = \frac{mv_r}{\Delta t}$$

[18] In the frame of reference of the plane.

[19] We often call this change of momentum the required *impulse* I, where $I = \int_{t1}^{t2} F dt = F_{avg}(t_2 - t_1) = F_{avg}\Delta t = \Delta p$.

We need to know how long the impact lasts, Δt. For a high-speed collision of this sort, where one object is flattened against the other, a good estimate is simply to take the length of the object divided by the relative velocity of impact. We have

$$\Delta t = \frac{L}{v_r}$$

where L is the length of the duck. Thus

$$F_{avg} = m\frac{v_r^2}{L}$$

Using $m = 1.0$ kg and $v_r \approx 88.9$ m s^{-1}, and, as given for a head-on collision, $L = 0.5$ m, we have $F_{avg} \approx 15.8$ kN. This force is roughly equivalent to the weight of a 1.6 tonne mass.

Consider now a collision for flight paths at right angles.

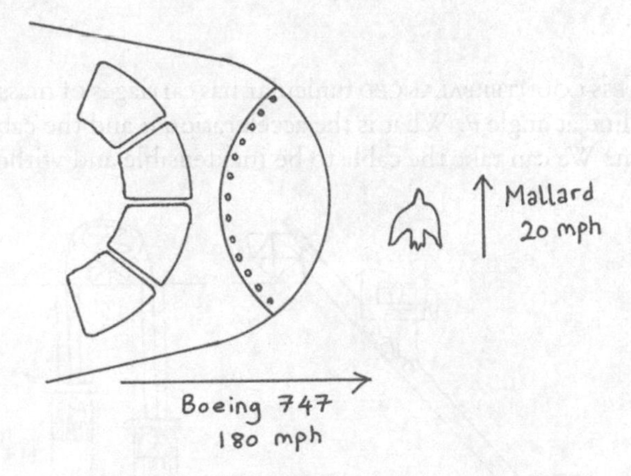

We calculate the relative speed by considering the velocity triangles of the plane and the duck. The speed of the plane is 180 mph, or approximately 80.0 m s^{-1}. The speed of the duck is 20 mph, or approximately 8.9 m s^{-1}. The velocities of the duck and plane are at right angles, so we calculate the relative velocity v_r to be $v_r \approx 80.5$ m s^{-1} at an angle $\theta = \tan^{-1}(8.9/80) \approx 6.3°$ away from the heading of the plane.

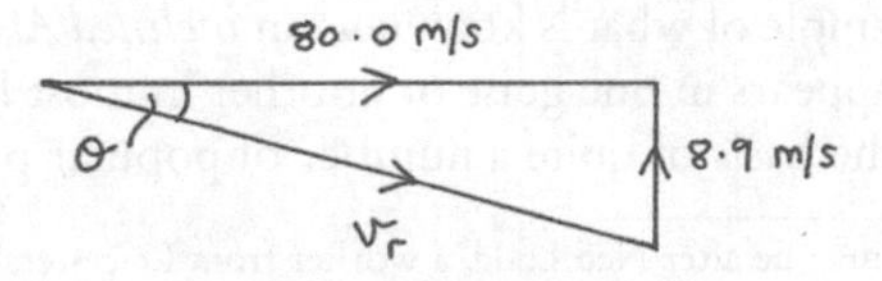

Using $m = 1.0$ kg and $v_r = 80.5$ m s^{-1}, and, as given for a collision at right angles, $L = 0.2$ m, we have $F_{avg} = mv_r^2/L \approx 32.4$ kN. This force is roughly equivalent to the weight of a 3.2 tonne mass.

The pilot is correct. A collision on flight paths at right angles gives a much higher average impact force because the duration of the impact is shorter, despite the lower relative speed. For the numbers given, the peak impact force is more than twice that for a head-on collision.

4.5 Water-powered funicular ★

"In this age of science," said the inventor thoughtfully, "we must surely be able to power a funicular railway with water. Think what an achievement it would be if we needed no energy source other than a nearby stream."

"I doubt it would work", replied the Luddite.[20] "The mass of any useful carriage would be immense."

"With counterbalanced carriages, provided I have an equal load in each carriage, all I need to overcome is resistance", said the inventor. "And with a properly oiled track that will be negligible." The inventor sketched his concept on a napkin.

A FRICTIONLESS COUNTERBALANCED funicular has carriages of mass M and $M + m$ on an incline at angle θ. What is the acceleration a and the cable tension T of the system? We can take the cable to be inextensible and without mass.

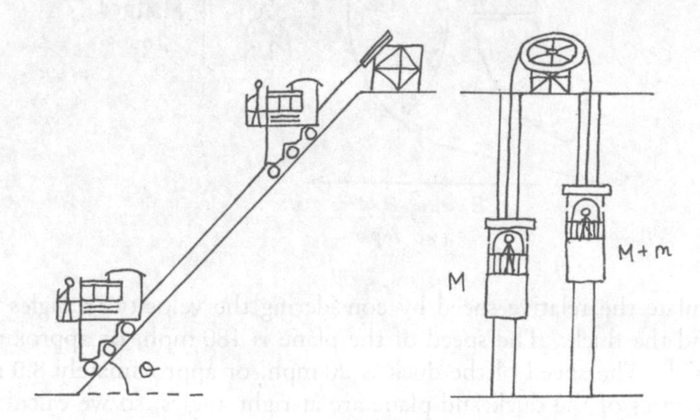

Answer

This is a simple example of what is known as an *inclined Atwood machine*. The Atwood machine appears in one guise or another in most high-school physics syllabuses, and is the basis of quite a number of popular puzzles I have heard

[20]The term Luddite may be after Ned Ludd, a worker from Leicestershire who originated the concept that the rise of manufacturing machinery during the Industrial Revolution would put the honest worker out of a job. Between 1811 and 1817 workers organised uprisings to destroy such machinery, believing that the work of the artisan would die out. The story goes that in 1779 Ned Ludd smashed two stocking frames, the first of many acts of man against machine. The term now refers to a person who is against change, particularly of a technological or industrial nature.

over the years. The last time I posed puzzles in this genre to students they did rather well.

The Atwood machine was developed by Reverend Atwood to demonstrate principles of classical mechanics. It consists of a series of five light brass pulleys on which can be suspended masses on inextensible strings, which are allowed to accelerate when connected in various combinations. Atwood's book[21] describes a number of demonstration experiments that can be performed to determine the accelerations with the aid of timing equipment. Not all the results are immediately intuitive, which is why it has been a classroom favourite for almost 200 years. Sadly, I think most school physics departments are unable to afford and maintain such delicate equipment, so for most it is a *thought experiment*[22] rather than the demonstration it was intended to be. I have done a physics degree, and have never knowingly been face-to-face with a real-life Atwood machine!

Let us consider the problem. We will see that although it is dressed up as something rather different, it is fairly straightforward.

The system can be represented most simply by masses M and $M + m$ suspended over a frictionless pulley on opposing slopes which are both at an angle θ.

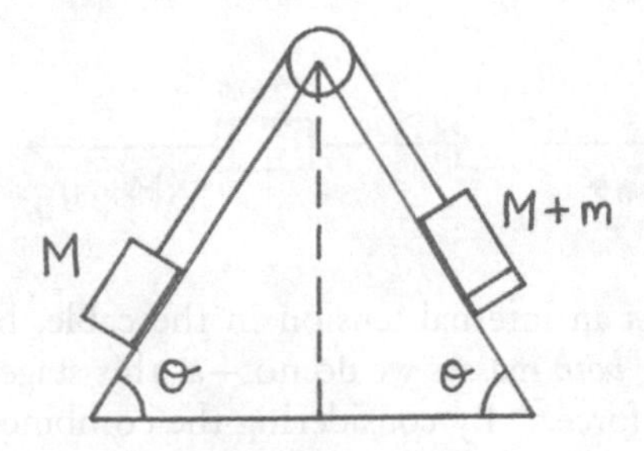

The forces on the masses are:

- The weight terms Mg and $(M + m)\,g$, which act vertically downwards.

- The reaction forces R_1 and R_2, which act perpendicular to the frictionless slope.

- The cable tension T, which acts along the slope.

[21] Atwood, G., 1784, "A treatise on the rectilinear motion and rotation of bodies; with a description of original experiments relative to the subject," printed by J. Archdeacon, Printer to the University of Cambridge.

[22] A thought experiment is a method for considering the logic of an argument. An argument is posed by way of a hypothetical experiment, and that experiment is thought through to its logical consequences.

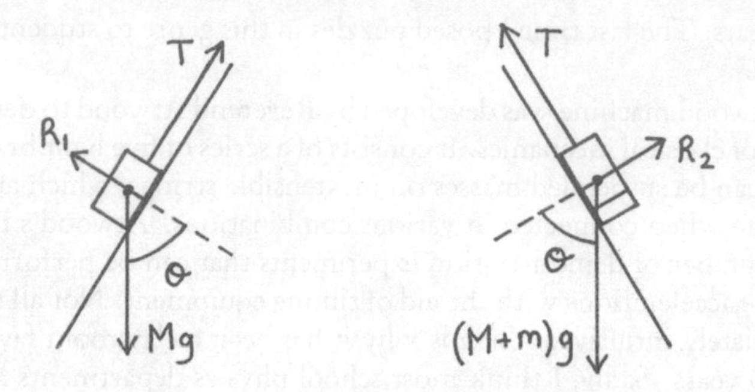

We should assume that the carriages remain in contact with the slope. So, resolving perpendicular to the slope, we require

$$Mg\cos\theta = R_1$$

and

$$(M+m)\,g\cos\theta = R_2$$

I've noted this mainly in the interests of completeness, because we don't need it to solve the problem.[23] We are primarily interested in the forces parallel to the slope. We can represent these forces in a simplified form.

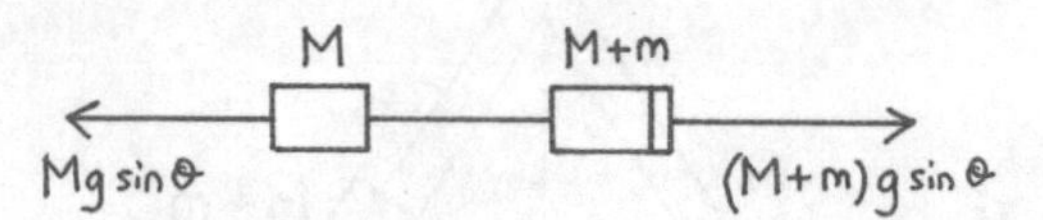

We note that there is an internal tension in the cable, but if we define the system as comprising *both* masses we do not—at this stage—need to explicitly consider the tension force.[24] By considering the combined mass, and the net force on it, we can simplify the system further.

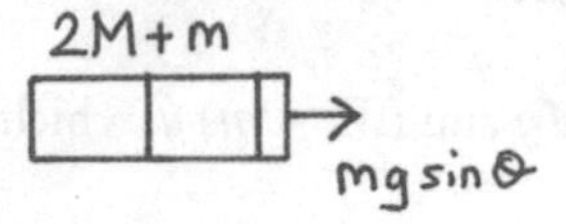

It is now a simple matter to define the acceleration of the system. The combined mass is $2M + m$ and the net force is $mg\sin\theta$. Using Newton's Second Law, we have

[23] Although we would be required to solve for these forces if we were to introduce friction. If you would like to solve a harder problem, pose and solve your own variant with friction included. It is very good practice to set your own questions—it is normally considerably harder than solving equivalent problems.

[24] It is obvious, I hope, that this technique works if and only if all parts of the system have the same magnitude of acceleration. If they do not, then the net external force on the system determines the acceleration of the centre of mass of the system.

$$mg\sin\theta = (2M + m)\,a$$

So the acceleration of the system is given by

$$a = \frac{mg\sin\theta}{2M + m}$$

Mass M accelerates upwards, and mass $M + m$ accelerates downwards. The acceleration is aligned with the slope. The form of the answer is exactly as we might have expected. The acceleration is proportional to the unbalanced force, and inversely proportional to the total mass of the system. No surprises.

What about the tension in the cable? We can consider this by thinking about the tension force required to cause either mass to accelerate. Consider the mass M first. The forces aligned with the slope are the downhill component of weight (in the direction of acceleration, that is), $Mg\sin\theta$, and the tension T. The net force is therefore $T - mg\sin\theta$. This force is responsible for an acceleration a of mass M. Using Newton's Second Law we write

$$T - Mg\sin\theta = Ma = \frac{Mmg\sin\theta}{2M + m}$$

We can rearrange to get the tension T in the cable.

$$T = 2Mg\sin\theta\left(\frac{M + m}{2M + m}\right)$$

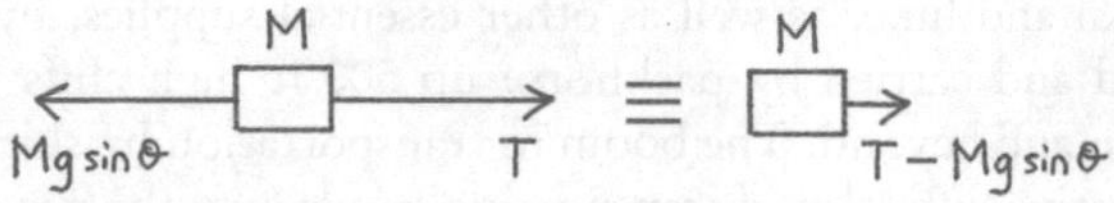

At this stage it's a good idea to check the limits of the equation to see if they make sense. That is, the case of $m \ll M$, and the case of $m \gg M$.

- The Case of $m \ll M$. We find that $a \to 0$. In the limit, the system does not accelerate because there is no unbalanced mass. We find that our equation for tension reduces to $T \to Mg\sin\theta$. That is, the tension force is equal to the component of weight down the slope.
- The Case of $m \gg M$. We find that $a \to g\sin\theta$. In the limit, the system accelerates as though the heavier carriage was in a constrained free fall down the slope (along the inclined plane). The lighter carriage is accelerated upwards at the same rate. In this limit, the lighter carriage is affected by the heavier carriage (in that the latter sets the acceleration of the former), but the heavier carriage is essentially unaffected by the lighter carriage. We see that the tension goes to $T \to 2Mg\sin\theta$. If we think about it a little, it makes sense. The tension now only depends on the lighter mass, M, not the heavier mass $m + M \approx m$. The

tension is responsible not only for opposing the weight of the lighter mass, $Mg\sin\theta$, but also for providing an upward acceleration equal to $a = g\sin\theta$. So the overall tension is twice the weight of the lighter mass.

Further discussion

The inventor mentioned in the introduction was George Marks, and the funicular is the Lynton and Lynmouth Cliff Railway. It was opened in 1890, and has been operating continuously ever since. A remarkable feat, and an exceptional example of both engineering and *sustainable development* (a fashionable phrase in this decade). I think George Marks would have had a word or two to say to those who feel that the concept of sustainability is a new one. In fact, a quick review of the engineering of the Industrial Revolution[25] (1760–1840) and the Technological Revolution[26] (1860–1910) suggests that the scientists of those times were acutely aware of the need to use energy sparingly. It is only in the comparatively recent times of abundant and inexpensive energy (in the *developed world,* at least) that we have become profligate.

For more than 120 years, the Lynton and Lynmouth funicular has been quietly ferrying people and goods back and forth with no energy input other than water from a nearby river. Lynton and Lynmouth are on the North coast of Devon, right in the middle of what is now the Exmoor National Park. And therein lies part of the explanation for the investment in the funicular. By the late 1800s Lynmouth had become a major port, receiving continual supplies of coal and lime, as well as other essential supplies, by sea. Goods were offloaded and carried by packhorse up 500 ft high cliffs to the town, then to Lynton and beyond. The boom in transportation by ship was because of the comparative difficulty of transporting goods over the rugged terrain of Exmoor. The funicular bypassed the long trip by road between the two towns, allowing passengers and goods to be delivered between them in a few effortless minutes.

The railway rises 500 ft on a 900 ft long track with a gradient of approximately 1:1.5. Each of the carriages has a tank which is filled at the top station with approximately three tonnes of water. Once the carriages are loaded with goods or passengers (40 at a time per car), water is discharged from the lower carriage until the cars start to move under the action of gravity. The speed is controlled by releasing a series of brakes on the cars. It appears that throughout its long operation there has never been an accident on the line. So long as the West Lyn River continues to flow (and if you have ever been to Devon you will know this is quite a likely prospect), it seems that the Lynton and Lynmouth Cliff Railway will continue to run.

[25] The development of machines for large-scale manufacturing, driven first by water power, and then by steam, and the transition from wood-burning furnaces to larger coal-burning ones.

[26] The period in which factories started to be electrified, and the first production lines were developed for mass manufacture.

4.6 Sherlock Holmes and the Bella Fiore emerald ★★★

Sherlock Holmes and Dr Watson are attending Agate Manor, where the possible sale of the Bella Fiore emerald to Lord Agate, by an unknown client of Christie's, has been interrupted by the unannounced arrival of a high-calibre rifle bullet. Holmes has sent Watson on ahead to do a preliminary investigation. When Holmes arrives on the scene, Watson appears to be examining the carpet pile in the far corner of the room.

"Well, my dear Watson, what have you uncovered?" asks Holmes.

"The bullet came in from the west window. Just there, where the broken pane of glass is. The emerald was sitting in a gem stand on the table. It got blasted clean off, and the bullet landed here, right in the lap of Lady Agate. Her portrait, that is. Fortunately, Lord Agate and the man from Christie's were unharmed. The would-be jewel thief appears to have missed his intended target and vanished empty-handed."

"His intended target?" says Holmes, raising an eyebrow.

"Why, Lord Agate, Holmes."

"Let us suppose nothing, Watson, but let us collect everything. Data is what we need. So what of this, Watson?" Holmes holds a large brass rifle cartridge case up to the light.

"A shell case...but I haven't seen it before!"

"That, my dear Watson, is because you look, but you do not see. It was sitting in plain view on the grass by the ha-ha. Which is, I'm sure you will agree, the very first place one would look if one were in the business of trying to find shell cases."

"I see," says Watson. "And what sort of shell case is it?"

"I thought you learnt about that sort of thing in the army, Watson!" says Holmes, handing over the shell case.

Watson examines the article in question. "Drawn brass. Point four-five calibre. Black powder charge. Stamped with the word "Kynoch". Almost certainly a Boxer-Henry cartridge." He thinks for a moment. "In all likelihood an 1877 Martini Henry Rifle. Of the type used in Egypt. Bullet weight 480 grains.[27]"

"And the one that punctured Lady Agate?" says Holmes.

"The same weight."

"And the emerald, Watson?"

"Missing, I'm afraid. I've searched the entire room."

"The weight, my dear Watson, the weight! I assumed you knew more about the science of ballistics."

---

[27] The *grain* is a measurement of mass historically based on the mass of a single seed of cereal. For centuries this measure was used for trade. Today the grain is still used in specialist applications—to measure the mass of gun propellants and bullets, for example—and in some pharmaceutical prescriptions. A grain is now taken to be equivalent to 64.8 mg.

"156 carats.[28]"

"Precisely the weight of the bullet, Watson, which means you are searching in entirely the wrong location. Did you observe a gardener within the last hour?"

"A gardener, Holmes?"

"Any intelligent child who has played billiards would not at this juncture be distracted by a methodical search of the carpet, Watson. The hypothetical child would be looking through the window on your right, where if you look now you will almost certainly observe disturbances consistent with a jewel-thief pretending to be a topiarist while searching for something. It seems that they not only hit their intended target, but had time to tidy up the box hedging before leaving."

"Extraordinary, Holmes", says Watson, examining the view from the window. "I don't know how you do it. I really don't."

"It is elementary ballistics, Watson. If a moving object collides elastically with a stationary object of equal mass, the objects always scatter at right angles to each other. You see an approximation of this behaviour every time you play billiards, but you choose not to perceive."

Let us now solve the question that Holmes and Watson were dealing with.

IN AN ELASTIC collision between objects of equal mass, one of them initially stationary, show that the objects always scatter at right angles to each other.

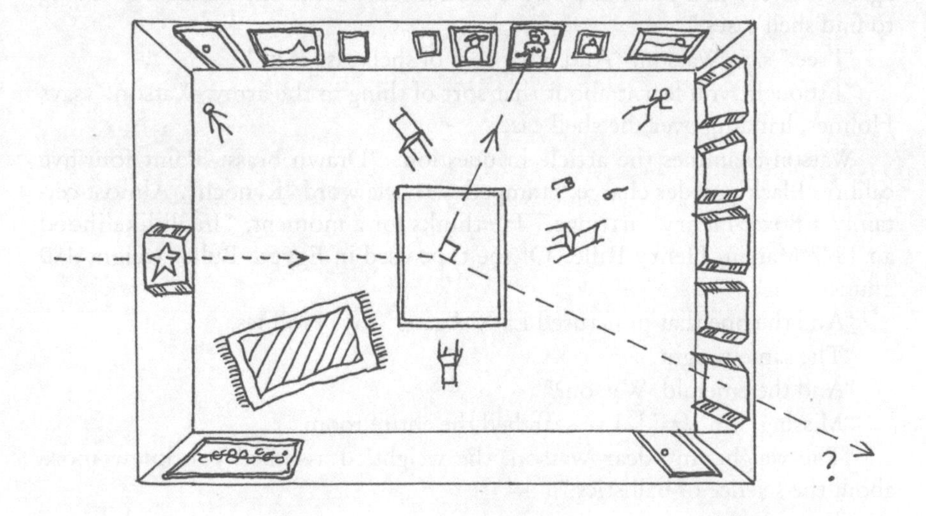

[28] A *carat* is a unit of mass used by jewellers and traders to measure *gemstones* (particularly the four precious gemstones—diamond, ruby, sapphire and emerald) and pearls. Since the metric carat was adopted in 1907, a carat has represented 200 mg (0.0071 oz).

Answer

The answer to this question is partly an excuse to introduce the concept of the centre-of-momentum frame of reference as a method for solving problems. I've given a much shorter, more elegant answer at the end.

Take two objects, 1 and 2, with masses m_1 and m_2 and speeds V_1 and V_2 (where $V_2 = 0$). To simplify the analysis we step into the centre-of-momentum frame of reference. We define this frame as moving with speed V_{CM} relative to the *lab frame* (the one we are observing from). The speeds of the objects relative to the centre-of-momentum frame are $v_1 = V_1 - V_{CM}$ and $v_2 = |-V_{CM}|$. The speed of the centre-of-momentum frame, V_{CM}, is defined by the condition for zero momentum. We have

$$m_1 (V_1 - V_{CM}) - m_2 V_{CM} = 0$$

For $m_1 = m_2 = m$ this gives

$$V_{CM} = \frac{1}{2} V_1 \Rightarrow v_1 = \frac{1}{2} V_1 \text{ and } v_2 = \frac{1}{2} V_1$$

In this simple case, with equal masses and with one object initially at rest, we should not be surprised by the symmetry of the problem. It is easy to extend the analysis to a more general case $m_1 \neq m_2$ if we wish. It is helpful to sketch a diagram of the velocities, which are positive in the directions shown.

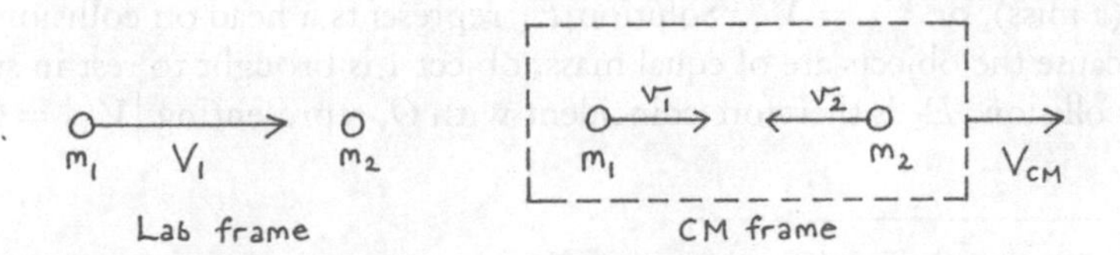

For an elastic collision (with coefficient of restitution $e = 1.0$) between equal masses with equal initial speed, it is easy to show (by conservation of momentum and energy) that the speeds in the centre-of-momentum frame following a collision, v_1' and v_2', are unchanged. For this simple case we have $v_1' = v_2' = v_1 = v_2$. We define the deflection angle as θ. Consider now some special cases.

- $\theta = 0$, the objects miss each other.
- $\theta = \pi/2$, the objects suffer a glancing blow and are deviated perpendicular to their original paths.
- $\theta = \pi$, the particles rebound along their original paths.

We can show that all these solutions are possible, as are any in-between. By symmetry, the solutions for $\boldsymbol{v}_1'$ and $\boldsymbol{v}_2'$ are related by a rotation of π. That is, $\boldsymbol{v}_1' = -\boldsymbol{v}_2'$.

We now consider the velocities of the objects following the collision in both the centre-of-momentum frame ($\boldsymbol{v}_1'$ and $\boldsymbol{v}_2'$) and the lab frame ($\boldsymbol{V}_1'$ and $\boldsymbol{V}_2'$).

- OBJECT 1 FOLLOWING THE COLLISION. In the centre-of-momentum frame the locus of velocities $\boldsymbol{v}_1'$ is a circle of radius $V_1/2$, centred on a zero-velocity condition, marked O in the diagram below. We mark some *arbitrary* particular solutions for $\boldsymbol{v}_1'$, namely A_1, B_1 etc., to E_1. The purpose of these points is simply to allow us to identify the corresponding points in the lab frame, and to allow us to identify corresponding points for object 2. We now arbitrarily pick solution B_1 to discuss. The deflection angle is $0 \leq \theta \leq 2\pi$. The velocity in the lab frame $\boldsymbol{V}_1'$ is the vector sum $\boldsymbol{V}_1' = \boldsymbol{v}_1' + \boldsymbol{V}_{CM}$, giving a locus of points (with respect to zero velocity in the lab frame) which is a circle centred on $\boldsymbol{V}_{CM}$. The centre is marked as O. The solutions A_1, B_1 etc., to E_1 in the lab frame are shown. We see that in the lab frame A_1 represents zero deflection (a miss), or $\boldsymbol{V}_1' = \boldsymbol{V}_1$. Solution E_1 represents a head-on collision. Because the objects are of equal mass, object 1 is brought to rest in such a collision. E_1 is therefore coincident with O, representing $|\boldsymbol{V}_1'| = 0$.

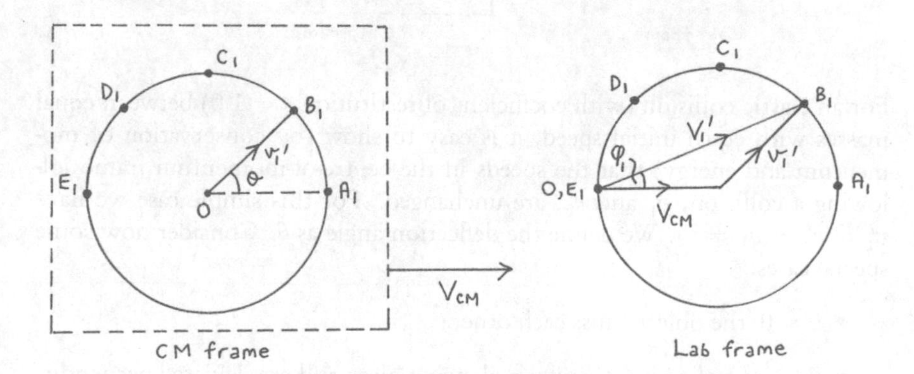

- OBJECT 2 FOLLOWING THE COLLISION. With respect to the centre-of-momentum frame, the locus of velocities $\boldsymbol{v}_2'$ is a circle of radius $V_1/2$, centred on a zero-velocity condition, marked O in the diagram. We mark arbitrary particular solutions A_2, B_2 etc., to E_2, corresponding to solutions A_1, B_1 etc., to E_1 for object 1. We have noted that solutions for $\boldsymbol{v}_1'$ and $\boldsymbol{v}_2'$ are related by a rotation of π. The velocity in the lab frame

$\boldsymbol{V}'_2$ is the vector sum $\boldsymbol{V}'_2 = \boldsymbol{v}'_2 + \boldsymbol{V}_{CM}$. We mark the angle of $\boldsymbol{V}'_2$ (with respect to the direction of $\boldsymbol{V}_1$) as ϕ_2. The locus of points representing $\boldsymbol{V}'_2$ in the lab frame (with respect to zero velocity) is a circle centred on $\boldsymbol{V}_{CM}$. The centre is marked as O. We see that in the lab frame A_2 represents zero deflection (a miss), or $\boldsymbol{V}'_2 = \boldsymbol{V}_2 = \boldsymbol{0}$. A_2 is therefore coincident with O. Solution E_2 represents a head-on collision. The objects are of equal mass, so in the collision object 2 acquires the initial velocity of the incident object 1, which is brought to rest in the collision. We see that E_1 represents $\boldsymbol{V}'_2 = \boldsymbol{V}_1$.

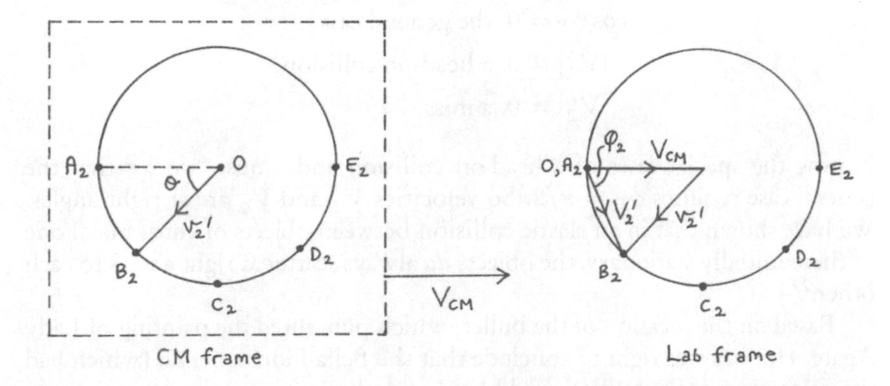

Consider now the relationship between the velocities $\boldsymbol{V}'_1$ and $\boldsymbol{V}'_2$ in the lab frame following the collision. Velocity $\boldsymbol{V}'_1$ is at an angle ϕ_1 to $\boldsymbol{V}_1$, and velocity $\boldsymbol{V}'_2$ is at an angle ϕ_2 to $\boldsymbol{V}_1$. The angle between $\boldsymbol{V}'_1$ and $\boldsymbol{V}'_2$ is $\phi_T = \phi_1 + \phi_2$. The dot product is defined by $\boldsymbol{V}'_1.\boldsymbol{V}'_2 = |\boldsymbol{V}'_1||\boldsymbol{V}'_2| \cos\phi_T$. If the vectors are perpendicular, the dot product is zero, $\boldsymbol{V}'_1.\boldsymbol{V}'_2 = 0$. The vectors are defined by $\boldsymbol{V}'_1 = \boldsymbol{v}'_1 + \boldsymbol{V}_{CM}$ and $\boldsymbol{V}'_2 = \boldsymbol{v}'_2 + \boldsymbol{V}_{CM}$.

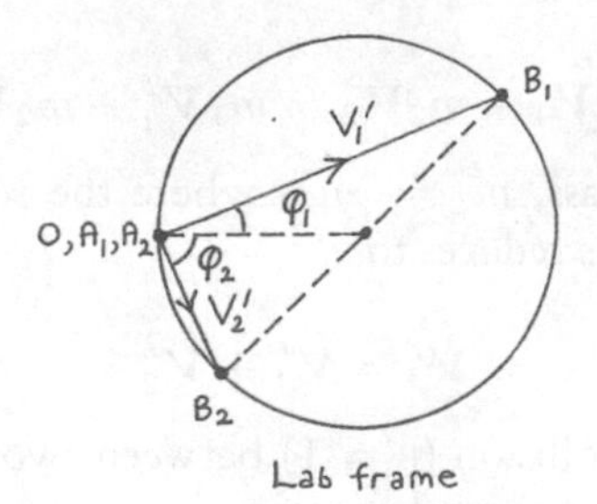

Recalling that $\boldsymbol{v}'_1 = -\boldsymbol{v}'_2$, we see that $\boldsymbol{V}'_1 = -\boldsymbol{v}'_2 + \boldsymbol{V}_{CM}$. So

$$\begin{aligned}\boldsymbol{V}'_1.\boldsymbol{V}'_2 &= (\boldsymbol{V}_{CM} - \boldsymbol{v}'_2).(\boldsymbol{V}_{CM} + \boldsymbol{v}'_2)\\ &= \boldsymbol{V}_{CM}.\boldsymbol{V}_{CM} + \boldsymbol{V}_{CM}.\boldsymbol{v}'_2 - \boldsymbol{v}'_2.\boldsymbol{V}_{CM} - \boldsymbol{v}'_2.\boldsymbol{v}'_2\end{aligned}$$

Simplifying, we get

$$\boldsymbol{V}_1'.\boldsymbol{V}_2' = |\boldsymbol{V}_{CM}|^2 - |\boldsymbol{v}_2'|^2$$

But since $|\boldsymbol{V}_{CM}| = |\boldsymbol{v}_2'| = V_1/2$, we have

$$\boldsymbol{V}_1'.\boldsymbol{V}_2' = 0$$

This implies one of three conditions.

$$\begin{aligned}\cos\phi_T &= 0, \text{ the general case.}\\ |\boldsymbol{V}_1'| &= 0, \text{ a head-on collision.}\\ |\boldsymbol{V}_2'| &= 0, \text{ a miss.}\end{aligned}$$

Noting the special cases of a head-on collision and a miss, we see that the general case requires $\phi_T = \pi/2$. So velocities $\boldsymbol{V}_1'$ and $\boldsymbol{V}_2'$ are at right angles. We have shown that in an elastic collision between objects of equal mass, one of them initially stationary, the objects *do* always scatter at right angles to each other.[29]

Based on the location of the bullet, which punctured the painting of Lady Agate, Holmes was right to conclude that the Bella Fiore emerald (which had the same mass as the bullet) would have passed on a perpendicular trajectory through the window he identified. The effect is slightly easier to spot when playing pool or snooker (in which collisions are almost elastic).

An elegant answer

The general vector equation for conservation of linear momentum in a collision between two objects is

$$m_1\boldsymbol{V}_1 + m_2\boldsymbol{V}_2 = m_1\boldsymbol{V}_1' + m_2\boldsymbol{V}_2'$$

For objects of equal mass, $m_1 = m_2$, where the second object is initially stationary, $\boldsymbol{V}_2 = 0$. This reduces to

$$\boldsymbol{V}_1 = \boldsymbol{V}_1' + \boldsymbol{V}_2'$$

For an entirely elastic collision ($e = 1$) between two objects, the kinetic energy is conserved, and the general equation is

$$\frac{1}{2}m_1\boldsymbol{V}_1^2 + \frac{1}{2}m_2\boldsymbol{V}_2^2 = \frac{1}{2}m_1\boldsymbol{V}_1'^2 + \frac{1}{2}m_2\boldsymbol{V}_2'^2$$

For objects of equal mass, where the second object is initially stationary, this reduces to

[29]Other than for the case of a head-on collision, for which we note that the direction of $\boldsymbol{V}_1'$ is undefined, because $|\boldsymbol{V}_1'| = 0$.

$$\boldsymbol{V}_1^2 = \boldsymbol{V}_1'^2 + \boldsymbol{V}_2'^2$$

Using Pythagoras' theorem, the only way to satisfy both vector equations simultaneously is if vectors $\boldsymbol{V}_1$, $\boldsymbol{V}_1'$ and $\boldsymbol{V}_2'$ form a right-angled triangle, with vectors $\boldsymbol{V}_1'$ and $\boldsymbol{V}_2'$ at 90° to each other.

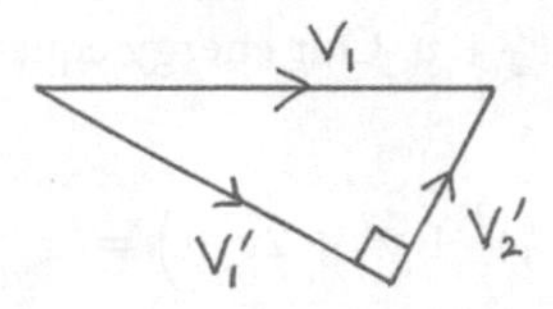

You will agree, I hope, that this is a very much neater way to solve this problem.

4.7 Equivalent statements for linear collisions ★★★

In this question we consider a well-known result that is surprisingly algebraic to prove.

TWO BALLS WITH masses m_1 and m_2 and initial speeds v_1 and v_2 have a head-on (one-dimensional) collision with resulting speeds v_1' and v_2' respectively. We can define an elastic collision as one in which no kinetic energy is lost, or one in which the coefficient of restitution is equal to unity:

$$e = -\left|\boldsymbol{v}_1' - \boldsymbol{v}_2'\right| / \left|\boldsymbol{v}_1 - \boldsymbol{v}_2\right| = 1$$

Show the equivalence of these statements in all inertial (non-accelerating) frames of reference.

Answer

We start with the statement that kinetic energy is conserved in the collision.

$$\frac{1}{2}m_1v_1^2 + \frac{1}{2}m_2v_2^2 = \frac{1}{2}m_1\left(v_1'\right)^2 + \frac{1}{2}m_2\left(v_2'\right)^2$$

Letting $r = m_2/m_1$, and simplifying, we write

$$v_1^2 - \left(v_1'\right)^2 = -r\left(v_2^2 - \left(v_2'\right)^2\right) \qquad (1)$$

The one-dimensional conservation of momentum equation for the system is

$$m_1v_1 + m_2v_2 = m_1v_1' + m_2v_2'$$

Which can be written

$$v_1 - v_1' = -r\left(v_2 - v_2'\right) \qquad (2)$$

First we show the equivalence of these equations in all inertial frames. Moving from the current inertial frame to a new inertial frame travelling at v with respect to the current one, we have initial velocities $v_1 + v$ and $v_2 + v$, and final velocities $v_1' + v$ and $v_2' + v$. Our energy equation (1) becomes

$$\left(v_1^2 + v^2 + 2v_1v\right) - \left(\left(v_1'\right)^2 + v^2 + 2v_1'v\right) = \\ -r\left[\left(v_2^2 + v^2 + 2v_2v\right) - \left(\left(v_2'\right)^2 + v^2 + 2v_2'v\right)\right]$$

Simplifying, we get

$$v_1^2 - \left(v_1'\right)^2 + \left\{2v\left(v_1 - v_1'\right)\right\} = -r\left(v_2^2 - \left(v_2'\right)^2\right) - r\left\{2v\left(v_2 - v_2'\right)\right\} \qquad (3)$$

We see that equation (3) is equal to the sum of equation (1) and $2v$ times equation (2), the latter shown in curly brackets. So we've demonstrated the equivalence of the equations in all inertial frames of reference.

Using this result, to simplify the solution of the momentum and energy equations, we step into a frame of reference in which v_2 is zero. Equations (1) and (2) become

$$v_1^2 - \left(v_1'\right)^2 = r\left(v_2'\right)^2 \qquad (4)$$

$$v_1 - v_1' = rv_2' \qquad (5)$$

Equations (4) and (5) have two unknowns, v_1' and v_2'. We rearrange (5) in terms of v_1', and substitute into (4) to give

$$\left(v_2'\right)^2\left(r^2 + r\right) - v_2'\left(2rv_1\right) = 0$$

This is a quadratic in v_2', which we can solve to give

$$v_2' = \frac{2rv_1 \pm \sqrt{4r^2v_1^2}}{2\left(r^2 + r\right)} = \frac{v_1 \pm v_1}{r + 1}$$

This gives two solutions.

$$\begin{aligned} v_2' &= 0 \\ v_2' &= v_1\frac{2m_1}{m_1 + m_2} \end{aligned}$$

The first is the solution for no collision. Substituting the second solution into (5) we get

$$v_1' \quad = \quad v_1 \frac{m_1 - m_2}{m_1 + m_2}$$

Now that we have the solutions for v_1' and v_2' for linear collisions, we turn our attention to the definition of the coefficient of restitution. Taking

$$e = -\frac{|\boldsymbol{v}_1' - \boldsymbol{v}_2'|}{|\boldsymbol{v}_1 - \boldsymbol{v}_2|}$$

and reducing to a one-dimensional equation, taking $v_2 = 0$, and substituting using our expressions v_1' and v_2', we get

$$e = -\left(\frac{v_1' - v_2'}{v_1 - v_2}\right) = -\left(\left(\frac{m_1 - m_2}{m_1 + m_2}\right) - \left(\frac{2m_1}{m_1 + m_2}\right)\right) = 1$$

We have shown that zero kinetic energy loss ($\Delta KE = 0$) implies a coefficient of restitution of unity ($e = 1$) in all inertial frames. To complete the proof, we also need to show the converse—that is, that $e = 1$ implies $\Delta KE = 0$. I've left that as an exercise for you to try.[30]

[30] Hint—you can use some parts of the proof above in reverse order.

Chapter 5

Circular motion

In this chapter we consider *circular motion,* or the dynamics of rotation. In a way it seems a little odd to separate this from other questions in dynamics. Circular motion is often taught as a self-contained topic in high-school courses, however, and is fertile ground for conceptual puzzles that require almost no mathematics. It is a popular topic in high-school physics, and can be combined with energy conservation principles (the behaviour of a yo-yo, or roller coasters, for example) or with problems involving limiting friction (such as cars going round bends). Let's review the basic equations.

- CENTRIPETAL ACCELERATION. The equation for the centripetal acceleration, a, of an object travelling at uniform speed, v, in a circle of radius R, is

$$a = \frac{v^2}{R}$$

The acceleration is directed towards the centre of curvature. The speed, v, and angular velocity, ω, are related by $v = \omega R$, giving

$$a = \omega^2 R$$

- CENTRIPETAL FORCE. In circular motion, if an object of mass m is accelerated towards the centre of a circle with acceleration $a = mv^2/R$, a force F must be applied to satisfy Newton's Second Law, giving $F = ma = mv^2/R$. We call this force the *centripetal force.* The origin of the force depends on the system, and is best illustrated by examples. If a ball is rotated on the end of a string, the force is provided by string tension. When a roller coaster moves round a tight bend, the centripetal force is (normally) provided by the reaction force with the track. When a car moves round a corner on a level road, the centripetal force is provided by the frictional force between the wheels and the road. When a plane performs a level turn in a banking manoeuvre, the centripetal force is provided by a *component* of the lift force generated by the wings. In all these examples we can see that centripetal force is not an additional force. Rather it is simply a force, or component of a force, that's naturally present in the system that is causing the centripetal acceleration we can see.

5.1 Friction at the superbike races ★

In 2011 I was at a conference in Vancouver, and had the opportunity to cross the border (from Canada into the US) to visit my brother in Redmond, near Seattle. At the time he was working for Microsoft. I hadn't seen him for years, nor met his wife and two daughters, who were by that time about three and four years old. The northwest corner of the US has a similar climate to England, so the showery weather meant we were indoors most of the time. Like most young children, my nieces seemed fascinated with online games. What was perhaps less usual was that the games were all maths puzzles. I guess that's what happens if one parent is a mathematician and the other is an economist.

Eventually the rain cleared and I suggested we set out to climb the highest mountain in Washington. I like to make a habit of getting to the top of the highest local point. I should have realised the expedition was doomed from the outset when I saw that my brother was wearing sandals. I reminded him that the top thousand meters of the mountain would be under deep snow. My brother produced two plastic bags which, he informed me, he was going to wear as the other part of his experimental footwear.[1]

It was a long way to the mountain. More than long enough for me to recall that my brother had always been a slightly erratic driver, alternating between two quite different styles of driving. The first was incredibly leisurely, in which he would take the car in a gentle weaving motion between the near-side lane markers and the off-side rumble strips (which he used as a guide to tell him when to turn back onto the road). He used that style when he was solving equations in his head. The second style was more like that of a racing driver. It involved revving the engine hard, short shifting between gears when moving away from lights, and taking a precise racing line round corners. He used that style when he wasn't solving equations. Both styles required his full concentration, so I had plenty of time to contemplate which I preferred.

[1]In case you are interested, he planned to wear the plastic bags over his socks but under the sandals, as a sort of semi-external waterproof spat. The theory was that heat is lost primarily through the ingress and egress of cold water and air, not by conduction, and that by effectively sealing the possible leaks, the combination would provide protection even against deep snow. It worked perfectly for about five minutes after we reached the snow line, after which time my brother's feet were so numb he could no longer feel them. We returned to the car. It's probably worth noting that the mountain was Mt Rainier, which has a heavily glaciated summit 4,392 m above sea level. So there was more than one reason for the expedition to be doomed.

After a particularly fast turn onto the freeway, during which I could feel the rear tyres start to slip, I decided to ask the obvious question.

"Do you get pulled over frequently?"

"Oh no, not really that frequently," he replied, rather idly.

"So what do you tell them when you do get pulled over?"

My brother's tone implied he thought that my question barely warranted a reply.

"I say that I answer only to the laws of friction."

Having been pulled over by a patrol car once myself in the US (I was a passenger), I can assure you it isn't the cordial experience that you have if you deal with police constables in England. The bristling guns somehow spoil the lightness of the occasion. I doubt that even the most insouciant offender would be inspired to deliver a flippant explanation for their recidivism. With my brother, however, you really can't be sure.

Quite a number of well-known puzzles deal with the case of limiting friction on roads. I know of numerous examples that involve either cars or motorbikes performing circular manoeuvres on flat or *banked* roads. This particular problem is one of the simplest possible examples, and is a fairly standard pre-university physics question. If it's too easy, solve the problem on a banked circular track for ★★.

A SUPERBIKE IS racing around a flat circular track at radius r. The coefficient of friction between the tyre and the track is μ. What is the optimum angle θ, which the rider makes with the vertical, for the bike to go as fast as possible?

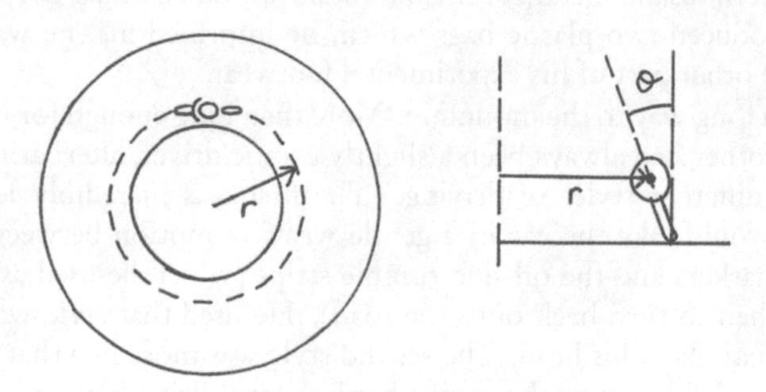

Answer

When the bike is going as fast as possible it is at the point of limiting friction. Friction normally obeys the law $F \leq \mu R$, but at the point of limiting friction $F = \mu R$. There are three external forces acting on the bike.

- The weight, mg, which acts downwards from the centre of mass.
- The reaction force from the ground, R, which acts vertically upwards at the point of contact. This has to be equal to mg because we know

that there is zero acceleration in the vertical direction. According to Newton's Second Law the forces must balance.

- The frictional force at the ground, which is equal to μR for the case of limiting friction.

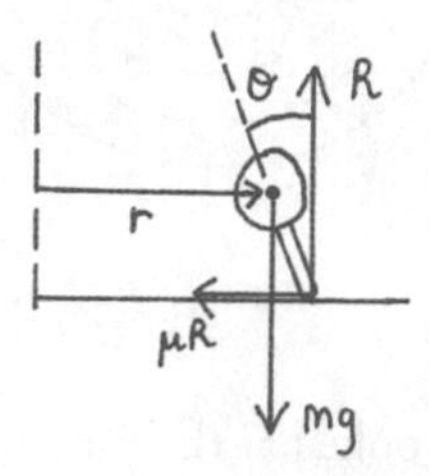

Because we have circular motion in the horizontal plane, we have acceleration towards the centre of rotation. For the case of limiting friction, the acceleration is driven by a force μR. For the bike to be rotationally stable (maintaining constant θ) there should be no net torque on the bike. Consider moments about the centre of mass.

$$RL\sin\theta = \mu RL\cos\theta$$

Here L is the distance between the centre of mass and the point of contact with the ground. Simplifying, we get

$$\mu = \tan\theta$$

That is,

$$\theta = \arctan\mu$$

For a value of $\mu = 1$ we get $\theta = 45°$. Next time you watch superbike races you will see that bikes achieve significantly greater angles with the vertical than this, implying an extremely high value of μ with the ground. Sticky rubber is everything.

5.2 Pole position at the superbike races ★★

At the superbike races, three identical bikes 1, 2 and 3 will do a hundred laps round a tight circular track in lanes at different radii, $r_1 < r_2 < r_3$. Which bike will win?

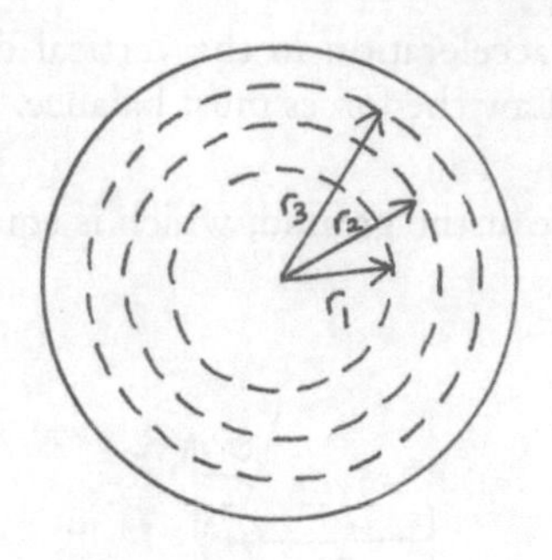

Answer

In the previous question we looked at the forces on the bike in the case of limiting friction. In this question we consider the same forces, but also consider the maximum speed a bike can have.

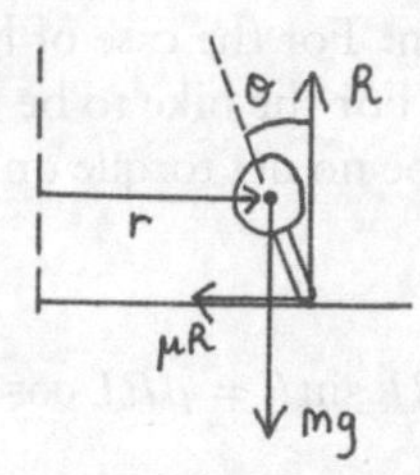

For the case of limiting friction, the centripetal acceleration $a = v^2/r$ is driven by the force μR towards the centre of rotation. Using Newton's Second Law, we have

$$F = ma = \frac{mv^2}{r} = \mu R$$

Using $R = mg$ and simplifying, we get

$$v = \sqrt{\mu g r}$$

The distances the bikes need to travel are proportional to the circumferences of their respective circular paths, $c = 2\pi r$. The lap time is the ratio of the distance and the velocity,

$$t_{lap} = \frac{c}{v} = \frac{2\pi r}{\sqrt{\mu g r}}$$

So

$$t_{lap} \propto \sqrt{r}$$

Bike number 1 has the smallest r and is the winner. A *racing line* is achieved through a combination of minimising the radius and the overall track distance.

5.3 Roller coaster ★★★

I really like this question. I invented it a couple of years ago, and have enjoyed testing students with it since then. I intended it to be a pure discussion, with no pen or paper to solve equations, although I do normally ask students to draw forces on the cars of the roller coaster at various points around the track. I find that most students need help with at least part of the question, and encouragement to examine their answers thoroughly. The question is as follows.

AN UNPOWERED ROLLER coaster with seven cars (marked 1 to 7 in the diagram) is pushed very slowly to the edge of a drop on the frictionless track shown. The track is composed of circular arcs and straight lines. There are seven identical people in the cars. Points of interest $p0$ to $p11$ are marked. The roller coaster has caged wheels, so cannot leave the track. Answer the following questions:

1. At what point(s) is the force between the seats and the passengers greatest?

2. At what point(s) is the force between the shoulder restraints and the passengers greatest?

3. At what point(s) is the force between the cars and the passengers smallest?

4. Do passengers 1 to 7 experience the same forces at the points identified in the answers to parts 1) to 3)? If not, identify the passengers who experience the greatest and smallest force for parts 1) to 3).

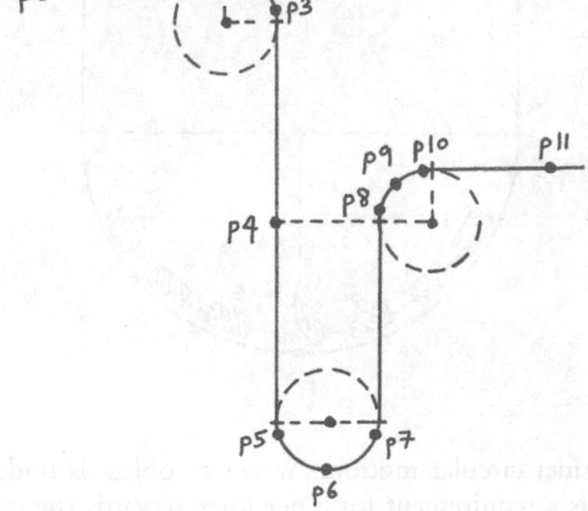

Answer

We will address each of the questions in turn, but first we should make some preliminary remarks about the overall method of solving the problem. We are interested in two primary things: energy conservation, and the equations of circular motion. We do not need to write the equations (this is not a quantitative question); we just need to understand the principles.

Let's first consider energy conservation. As the roller coaster loses height, it loses potential energy and gains kinetic energy. When it gains height again it loses kinetic energy and gains potential energy. The speed of the roller coaster at any given time is dictated by the difference in potential energy between the starting point $p0$ and the point of interest. This difference in energy is proportional to the height difference between $p0$ and the point of interest. So the roller coaster has the highest speed at $p6$. Here we use the term *height difference* rather loosely. The roller coaster assumes different shapes as it moves around the track. What we mean here is the *height difference between the centres of gravity*. Consider two examples.

- When the roller coaster is passing $p0$, the centre of gravity is also at the same height as $p0$ (neglecting the height of the cars above the track). This is the same for all cars 1 to 7.
- In contrast, when the roller coaster is passing $p6$, the centre of gravity is always *above* $p6$. The centre of gravity is above $p6$ by different amounts for cars 1 to 4: it is lowest when car 4 is passing $p6$ and highest when car 1 (or car 7) is passing $p6$. The pairs $(1, 7)$, $(2, 6)$ and $(3, 5)$ have the same solutions. So car 4 passes $p6$ with a higher speed than any other car, and cars 1 and 7 pass $p6$ with a lower speed than any other car.

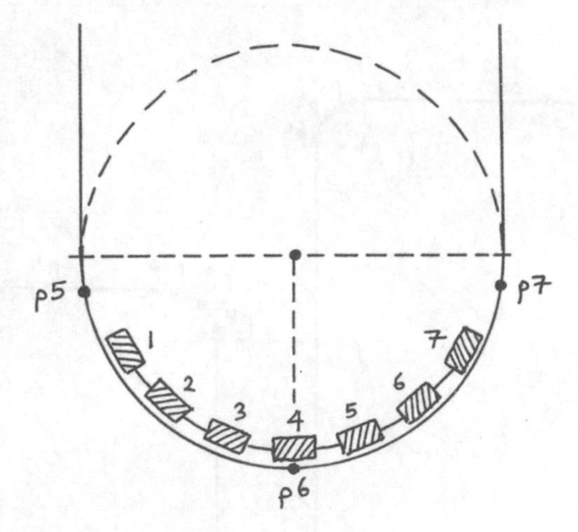

Let's now consider circular motion. When an object is undergoing circular motion, there is a requirement for a net force towards the centre of motion (the centre of the circle). The magnitude of the force is proportional to the acceleration, which is proportional to the square of the speed of the object.

We should now be in a position to answer the questions fairly simply. We consider each of the questions 1) to 3) in turn. The fourth question is answered as we go along.

1) At what point(s) is the force between the seats and the passengers greatest?

The force is greatest at $p6$. Here the force on a mass m must not only oppose the force due to gravity, mg, but must also provide the acceleration towards the centre of the circle. The acceleration towards the centre of the circle is greatest at this point because the speed here is greater than at any other point. The force on a mass m would be given by

$$F_6 = mg + \frac{mv_6^2}{r}$$

where we note that v_6 can take four different values, depending on the car number. The person in car 4 experiences the greatest force, because car 4 is travelling faster than the other cars when it passes $p6$.

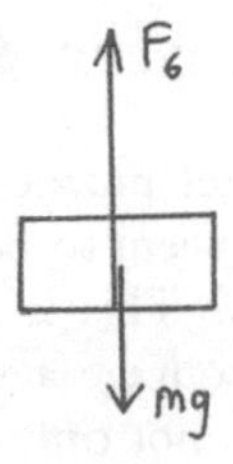

2) At what point(s) is the force between the shoulder restraints and the passengers greatest?

When passengers go round the outside of a bend, rather than the inside, they experience a force from the shoulder restraint rather than from the seat. A force from the shoulder restraint is possible *in principle* at points $p1$, $p2$ and $p3$[2] and points $p8$, $p9$ and $p10$. The lowest of these points is $p8$, so the roller coaster is travelling fastest at this point ($v_8 > v_1, v_2, v_3, v_9$ and v_{10}). In addition, we note that the force due to gravity acts approximately perpendicular to the force between the track and the car[3] at $p8$. We write

[2] As drawn, it is unlikely that any passengers will experience a force from the shoulder restraint at $p1$, because the gravitational force should be enough to provide the acceleration towards the centre of the circle. If any passenger *were* to feel such a force it is most likely to be the passenger in car 1, who is travelling faster than any other passenger when he passes $p1$. Similar comments apply to $p2$, but more passengers may experience a force from the shoulder restraints. At $p3$ it is likely that most passengers would experience a force from the shoulder restraints, with the person in car 1 experiencing the greatest force.

[3] The force between the shoulder restraint and the person behaves in exactly the same way as the force between the track and the car.

$$F_8 \approx \frac{mv_8^2}{r}$$

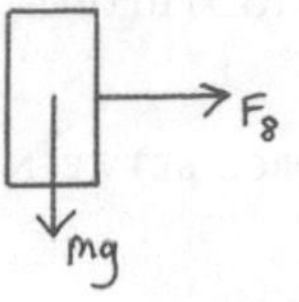

where we note that v_8 can take *seven* different values, depending on the car number. The person in car 7 experiences the greatest force, because car 7 is travelling faster than all the other cars when it passes $p8$. Using a similar rationale, the person in car 1 experiences the smallest force.

3) AT WHAT POINT(S) IS THE FORCE BETWEEN THE CARS AND THE PASSENGERS SMALLEST?

When the roller coaster passes $p4$ it is in free fall and there is zero force between the cars and the passengers. This is true for all passengers.

5.4 Derailed roller coaster ★★

I have always enjoyed roller coaster puzzles, and have invented quite a few over the years. Roller coasters do seem to be one of the most popular topics for pre-university physics puzzles. This is probably because they allow for almost infinite variation, and encourage a variety of discussions. They are also instantly familiar to students, not only because of the physics involved, but also because students can identify with the physical situation, which is important when dealing with questions about dynamics. Of course, although most students *feel* they are familiar with the material, a good question really tests deep understanding. Roller coaster questions tend to separate students into two types: those who have remembered a solution to a similar question; and those with a firm grasp of the tools needed to solve a problem like this, and who are willing to apply them to a new problem. This particular roller coaster question is one that I have especially enjoyed.

A FRICTIONLESS ROLLER coaster car is released from rest at a height h and rolls of its own accord down a 45° slope and through the loop a, b, c, d, e. The loop is of constant radius r and has the slope as its tangent. The car then continues on a level track, also at a tangent to the loop. Answer the following questions.

1. Where is the magnitude of the force between the car and the track greatest? Derive an expression for the force at this point.

2. Where is the force smallest? Derive an expression for the force at this point.

3. Define the minimum h for the car to derail if it has normal wheels.[4]

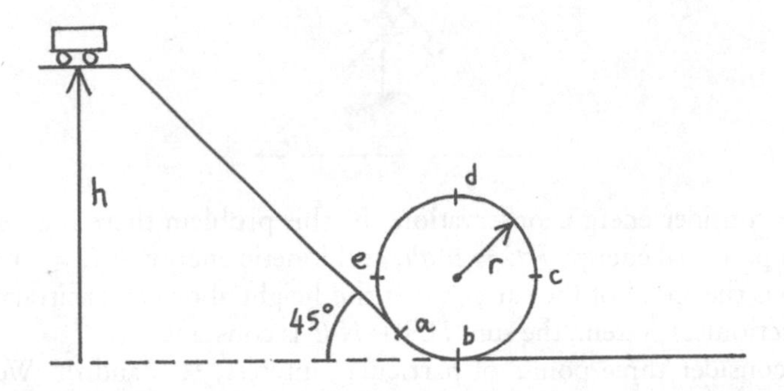

Answer

A good approach for questions of this type is to consider energy conservation—the conversion of potential energy to kinetic energy, and vice versa. The speed at a given point will depend on the difference in height between the initial state (when the car has zero speed) and the point of interest.[5] We should also see that if the car is undergoing circular motion at a particular speed, a force is needed to provide the necessary acceleration. The force must come either from the gravitational force on the car, or from a contact force with the track. It is this latter force that we are being asked to consider.

Before I advance too far with the discussion in this question, I ask students to draw the forces on the car when it is halfway down the 45° slope. This is to check their familiarity with basic force decomposition–if you can't decompose forces it's difficult to progress much further. I have seen a number of students struggle with this step.

Consider the car as it accelerates down the track. There is a downward force mg due to gravity, where m is the mass of the car and g is the force per unit mass due to gravity. There is also a reaction force F, which acts perpendicular to the frictionless track. The car is not accelerating in the direction perpendicular to the track, so the net force in that direction is zero. We write $F = mg\cos\theta$. In the direction aligned with the track there is a net force $mg\sin\theta$ which gives rise to an acceleration $a = g\sin\theta$. This force diagram isn't needed to answer the question as it is set. But it is sometimes useful to reference a situation which should be entirely familiar before moving on to something unfamiliar.

[4]Like a conventional train, without cages to prevent derailment.

[5]In some questions students might also be asked to take account of the effect of friction, but this makes the analysis considerably harder.

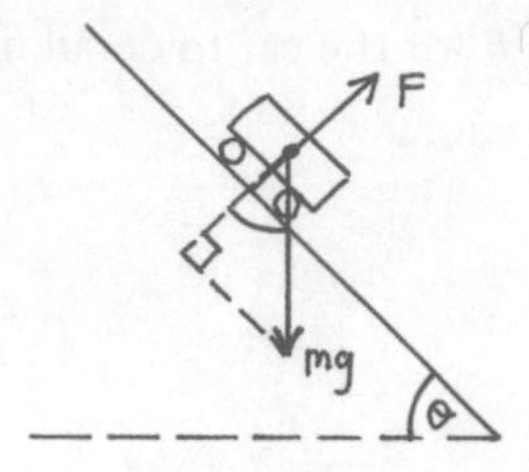

We now consider energy conservation. In this problem there is an exchange between potential energy, $PE = mgh$, and kinetic energy, $KE = (1/2)\, mv^2$, where v is the speed of the car and h is the height above an arbitrary *datum*. For a frictionless system, the sum $PE + KE$ is constant.

We consider three points of particular interest, b, c and d. We should expect the greatest speed at point b and the smallest speed at point d. There is also a change in sign of the reaction force between the car and the track between points b and d. We therefore expect these points to be important to our answer. Point c is of interest simply because the reaction force from the track is normal to the gravitational force. We will discuss this further in a moment. Point e is similar to c, and a is of no particular interest.

We note that b, c and d are at heights 0, r and $2r$ above ground level. The loss of potential energy between the starting point and points b, c and d is given by $mg\,(h-0)$, $mg\,(h-r)$ and $mg\,(h-2r)$ respectively. Setting $\Delta PE = \Delta KE$, we obtain the speeds of the car as

$$v_b = \sqrt{2gh},\ v_c = \sqrt{2g\,(h-r)} \text{ and } v_d = \sqrt{2g\,(h-2r)}$$

As expected, point b has the greatest speed, and point d the smallest speed.

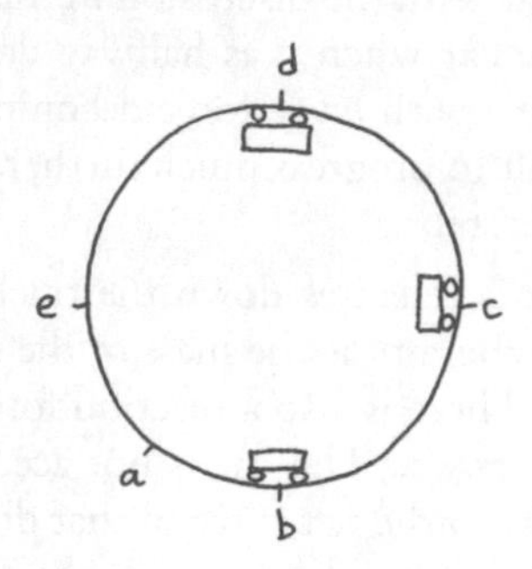

Consider now the angular (or centripetal) acceleration of the car due to motion at a speed v around a circular arc of radius r. The angular acceleration is given by $a = v^2/r$, and is directed towards the centre of the motion, which in this case is the centre of the circular track. The centripetal acceleration requires a centripetal force. In this problem, the force is provided by the gravitational force, or the reaction force with the track, or both. Using the equation for acceleration, the required force is $F = mv^2/r$, and must be directed towards the centre of the circle.

We now consider the points b, c and d in turn.

- POINT b. The gravitational force mg acts downwards, and the track reaction force F_b acts upwards. The required centripetal force is the component of the vector sum of mg and F_b that acts in the vertical direction. We write $F_b - mg = mv_b^2/r$.
- POINT c. The gravitational force mg acts downwards, and the track reaction force F_c acts to the left. The required centripetal force is the component of the vector sum of mg and F_c that acts to the left. We write $F_c = mv_c^2/r$.
- POINT d. The gravitational force mg acts downwards, and the track reaction force F_d acts downwards. The required centripetal force is the component of the vector sum of mg and F_d that acts downwards. We write $F_d + mg = mv_d^2/r$.

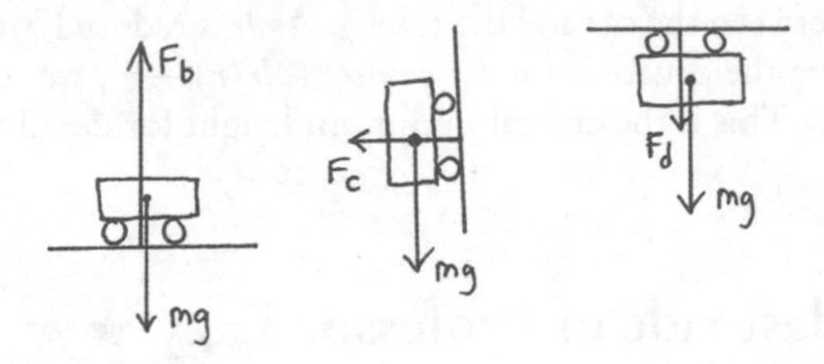

Rearranging, and substituting for v_b, v_c and v_d, we obtain the following equations for the track reaction forces.

$$\begin{aligned} F_b &= mg + \frac{mv_b^2}{r} = mg\left(\frac{2h}{r} + 1\right) \\ F_c &= \frac{mv_c^2}{r} = mg\left(\frac{2h}{r} - 2\right) \\ F_d &= \frac{mv_d^2}{r} - mg = mg\left(\frac{2h}{r} - 5\right) \end{aligned}$$

As expected, we see that $F_b > F_c > F_d$. We are now in a position to answer the questions as originally posed.

1) WHERE IS THE MAGNITUDE OF THE FORCE BETWEEN THE CAR AND THE TRACK GREATEST?

The force is greatest at b. The value is $F_b = mg\,[(2h/r) + 1]$. The force is greatest here because: i) this is the point at which the car has the greatest speed, and therefore the greatest acceleration, so it requires a greater track reaction force; ii) the gravitational force acts away from the centre of rotation, so not only does it not contribute to the centripetal acceleration, it also imposes a requirement for an increased track reaction force to support the weight of the car.

2) Where is the force smallest?

The force is smallest at d. The value is $F_d = mg\left[(2h/r) - 5\right]$. The force is smallest here because: i) this is the point at which the car has the smallest speed, and therefore the smallest acceleration, so it requires only a small track reaction force; ii) the gravitational force acts towards the centre of rotation and contributes to the force required for centripetal acceleration, reducing the required track reaction force. If the car's speed is low enough, the gravitational force may be greater than required for centripetal acceleration, and the track reaction force will be zero (we cover this in the next part of the question).

3) Define the minimum h for the car to derail if it has normal wheels.

The car will derail if the track reaction force goes to zero (that is, there is no contact force between the car and the track). As h is reduced, this will happen first at d. Taking the equation for $F_d = mg\left[(2h/r) - 5\right]$, we see that $F_d = 0$ when $h = 5r/2$. This is the critical minimum height for derailment.

5.5 The last ride of Professor Lazy ★★

"Hemisphere Hill in icy conditions," said Professor Lazy, "is the ultimate place to test my frictionless sled."

"But you have no brakes!" cried his colleagues. "It will surely end disastrously."

"My dear fellows!" said Lazy. "Hemisphere Hill is completely smooth; why would I have need of brakes? It will be a triumph."

And so they donned their crampons and went to the very top of Hemisphere Hill. From there, on a clear day, it is said, one can survey the whole of Puzzleshire.

"I really must protest," said Professor Lucid. "When you hit a critical angle, you will leave the slope and go sailing through the air."

"Nonsense, Lucid!" said Lazy, and gritted his teeth for the descent.

As his colleagues watched him slide slowly over the edge of the hemisphere they shouted and waved their arms in horror, but there was nothing they could do. Professor Lazy was already committed to his final trajectory.

If you were to sit on a frictionless sled, and allow it to slide in a straight line downwards from the crown of a hemisphere, is there a critical angle at which you leave the slope? If so, what is that angle?

Answer

This lovely little question, normally posed as a bead-on-a-ball[6] problem or something similar, is an absolute classic puzzle. It has appeared in numerous places over the years. Let us first consider the general principles.

As Professor Lazy starts to move down the slope, some of his potential energy is converted to kinetic energy, and he gains speed. An energy balance enables us to calculate his speed v_θ as a function of the angle θ from the crest of the hemisphere. For a given speed v_θ, if the sled follows the curve of the hemisphere, a centripetal force is needed–acting towards the centre of the circular motion–to keep the sled attached to the slope. There are only two forces on the sled: the force due to gravity, mg, and a reaction force, R_θ, that is normal to the slope (pointing away from the centre of the circle). Remember that there is no force due to friction. The centripetal acceleration is provided by a component of the force due to gravity. As the sled speeds up the requirement for centripetal force increases. However, the component of the force due to gravity pointing in the correct direction diminishes as the angle increases. And at a critical angle, θ_c, it is no longer enough to sustain the required centripetal acceleration for circular motion. At θ_c, $R_\theta = 0$, and the sled leaves the slope.

Let's consider these principles in more detail. First consider conservation of energy. When the sled has descended to an angle θ (measured from the crest of the hemisphere), from geometry we see that the sled has descended a distance $r\,(1 - \cos\theta)$. The potential energy that has been given up is $mgr\,(1 - \cos\theta)$. We define the kinetic energy at this point as $(1/2)\,mv_\theta^2$. Equating these two terms we have

$$v_\theta^2 = 2gr\,(1 - \cos\theta)$$

[6] A bead rests on a frictionless ball, and is allowed to slide downwards from the crown.

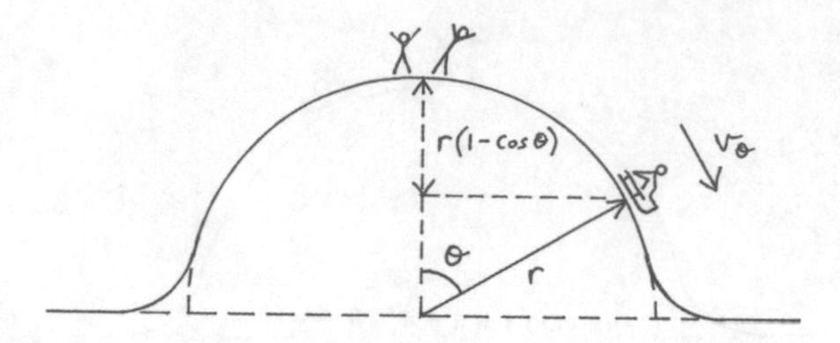

Let's now consider the two forces on the sled: the force due to gravity, mg, and the reaction force from the hemisphere, R_θ. We draw the forces in two general positions, to highlight their angular dependence. In the first case we draw the sled near the top of the hemisphere. In the second case we draw the forces at the sled's point of departure from the hemisphere, when $R_\theta = 0$. We see that if the centripetal acceleration is $a = v_\theta^2/r$, the required force acting towards the centre of motion is mv_θ^2/r. The forces R_θ and mg are therefore related by

$$mg\cos\theta - m\frac{v_\theta^2}{r} = R_\theta$$

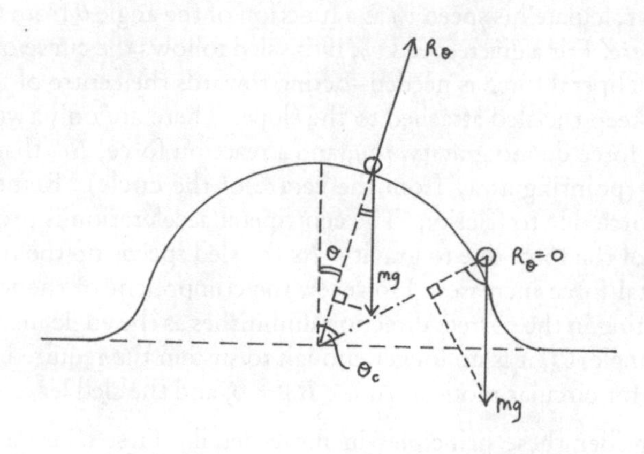

The point of departure from the hemisphere is defined as the point where $R_\theta = 0$. If we combine the two equations above at this critical point we have

$$\cos\theta = 2\,(1 - \cos\theta)$$

Simplifying, we get

$$\theta = \cos^{-1}\left(\frac{2}{3}\right)$$

Professor Lazy leaves the slope at $\theta \approx 48.2°$, and continues in free flight. Professor Lucid was right to be worried, because this was probably Professor Lazy's last ride on his frictionless sled.

5.6 Wall of Death: car ★★★

The Wall of Death is an old amusement park stunt in which cars and motorcycles race around the side of a cylindrical track with a vertical wall. The first track was built in Coney Island in New York in 1911, but the stunt quickly spread to England and India, as well as many other places. The cars or motorcycles start in the centre of the cylinder, gaining speed on a section of inclined wall, before transitioning to the vertical wall when their speed is high enough. The audience stands at the top of the drum, looking down on the riders, and presumably wondering how they stick to the side of the wall. In the US and England the stunt has almost died out, but there are still many tens of walls in India. In the most extreme shows, acrobatics are performed on the sides of cars and motorcycles as the vehicles race around the track. There are many videos of such performances, and they are rather breathtaking. If the stunt is not exciting enough for you, there's always the Globe of Death, in which motorcycles race around the inside of a sphere made of steel mesh. The record for the most motorcycles circling within a Globe of Death is ten, in a performance by a Chinese acrobatic group in 2010.

CONSIDER A CAR driving in a horizontal circle around a Wall of Death with diameter 12 m. Assume that the centre of gravity of the car is 1 m from the wall, and that the separation of wheels on the same axle is 2 m. Calculate the minimum speed when the coefficient of static friction is:

1. $\mu = 1$.
2. $\mu = 1/2$.
3. $\mu = 3/2$.

Draw the forces on the car for solutions 1) to 3).

Answer

First we should draw a diagram of the car. We mark the points of contact with the vertical wall (at the wheels) as A and B, and mark the centre of mass of the car as C. We are told that the distance between the centre of mass and the wall is $x = 1$ m. The vertical distances between C and A and between B and

C are both $y = 1$ m. Let's consider the cases $\mu = 1$, $\mu = 1/2$ and $\mu = 3/2$ in turn.

1) THE CASE OF $\mu = 1$.

Let us now consider the forces on the car in the most general situation. There are five forces.

- The weight of the car, mg, which acts vertically downwards from C.
- Reaction forces R_1 and R_2, which act normal to the wall at points A and B respectively.
- Friction forces F_1 and F_2, which act vertically upwards at points A and B respectively.

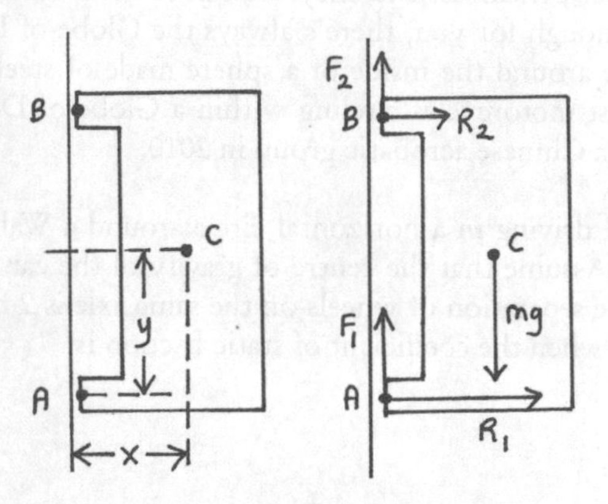

Let's now consider how the forces are related. We resolve forces in the horizontal and vertical directions, and sum moments about the centre of mass. Considering first the forces in the vertical direction, and noting that the acceleration in this direction is zero, we have

$$mg = F_1 + F_2$$

Considering the forces acting towards the centre of the cylinder (the direction indicated instantaneously by R_1 and R_2), and noting that the acceleration towards the centre is v^2/r, where v is the tangential speed and r the radius of the circle that the centre of mass of the car follows (here $r = (12/2) - 1 = 5$ m), we have

$$R_1 + R_2 = \frac{mv^2}{r}$$

Before we take torques about the centre of mass,[7] we consider the case of limiting friction in the vertical direction. In general, the frictional forces are

[7] We do this to ensure that rotational stability is achieved.

limited to being less than or equal to the coefficient of friction multiplied by the local reaction force. That is,

$$F_1 \leq \mu R_1 \text{ and } F_2 \leq \mu R_2$$

In the case of limiting friction, we write

$$F_1 = \mu R_1 \text{ and } F_2 = \mu R_2$$

We can rewrite the vertical equilibrium condition as

$$mg = \mu (R_1 + R_2) = \mu \frac{mv^2}{r}$$

The minimum speed v for the case of limiting friction is

$$v = \sqrt{\frac{rg}{\mu}}$$

Taking $r = 5$ m, $g \approx 10$ m s^{-2} and $\mu = 1$, we have $v = \sqrt{50} \approx 7.1$ m s^{-1} (or 15.9 mph). This seems eminently achievable, and consistent with the fairly low speeds that cars seem to be travelling at in videos of Wall of Death stunts. Of course, riders will drive faster than the speed corresponding to the point of limiting friction, to ensure they stick to the wall. Our speed is in the correct range, however.

We have not taken torques about the centre of mass, to ensure that rotational stability can be achieved at this speed. We have *assumed* that this rotational stability condition can also be satisfied at the point of limiting friction. Let's now check this assumption.

The centre of mass is horizontally $x = 1$ m from the points of contact with the wall (A and B). Reversing this statement, the upward forces at the points of contact (which keep the car in vertical equilibrium) generate a moment about the centre of mass, which tends to tip the car away from the wall. This moment is always equal to the weight of the car multiplied by the distance of action, which is $x = 1$ m. If the car is in equilibrium, this moment must be restored by unequal reaction forces at A and B. If, for a given situation, we can solve for positive R_1 *and* R_2, we know that the condition for rotational equilibrium is satisfied. If we can get "solutions" to the equations with negative R_2 (a negative reaction at B), we know that the rotational equilibrium condition cannot be physically satisfied.

Taking moments about the centre of mass, we get

$$(F_1 + F_2)\, x = (R_1 - R_2)\, y$$

At the point of limiting friction, taking $F_1 = \mu R_1$ and $F_2 = \mu R_2$, we have

$$R_1 \left(1 - \mu \frac{x}{y}\right) = R_2 \left(1 + \mu \frac{x}{y}\right)$$

This is the condition for rotational equilibrium at the point of limiting friction. For $\mu = 1$, and $x = y = 1$ m, the left-hand side of the condition is zero. The equation can only be satisfied for $R_2 = 0$. In the limit, this is an acceptable condition. All the weight is carried on the lower wheel at point A. The reaction and frictional forces are equal, and create exactly opposing moments about the centre of mass. The solution is very simple.

$$F_1 = R_1 = mg \text{ and } R_2 = 0$$

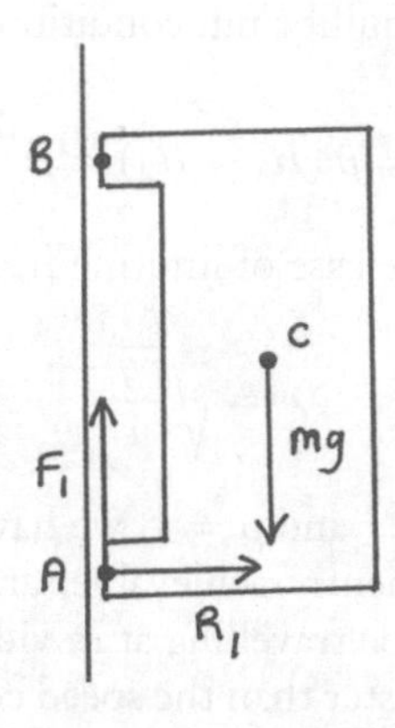

2) THE CASE OF $\mu = 1/2$.

Recalling the minimum speed condition for the case of limiting friction, $v = \sqrt{rg/\mu}$, and taking $r = 5$ m, $g \approx 10$ m s^{-2} and $\mu = 1/2$, we have $v = \sqrt{100} = 10$ m s^{-1} (or 22.4 mph). As expected, we need a slightly higher speed when the friction is reduced. The speed increase is fairly modest, because of the v^2 dependence of the reaction force—the limiting friction force is proportional to this reaction force.

We also need to check the rotational stability condition, $R_1\,(1 - \mu x/y) = R_2\,(1 + \mu x/y)$. For $\mu = 1/2$, and $x = y = 1$ m, we have $R_1\,(1/2) = R_2\,(3/2)$. That is, $R_1 = 3R_2$. We could, perhaps, have expected a result of this sort. To achieve the required frictional force with the reduced coefficient of friction we need to increase our reaction force. The torque due to frictional forces is the same as for the case of $\mu = 1$, so the reaction forces have to redistribute to achieve no net moment about the centre of mass. This is achieved by having a positive reaction force at point B. Solving the equations gives reaction forces

$$R_1 = \frac{3}{2}mg \text{ and } R_2 = \frac{1}{2}mg$$

and friction forces

$$F_1 = \frac{3}{4}mg \text{ and } F_2 = \frac{1}{4}mg$$

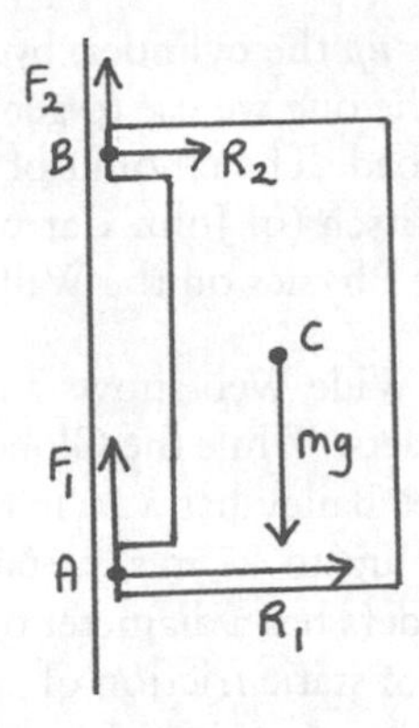

3) THE CASE OF $\mu = 3/2$.

Although it's probably slightly unrealistic, the case of $\mu = 3/2$ is an interesting one. Consider, again, the minimum speed condition for the case of limiting friction, $v = \sqrt{rg/\mu}$. Taking $r = 5$ m, $g \approx 10$ m s^{-2} and $\mu = 3/2$, we have $v = \sqrt{100/3} = 5.8$ m s^{-1} (or 13.0 mph). With super-sticky rubber, we need only a very low speed to meet the condition for vertical equilibrium.

Now we check the rotational stability condition at the point of limiting friction, $R_1\,(1 - \mu x/y) = R_2\,(1 + \mu x/y)$. For $\mu = 3/2$, and $x = y = 1$ m, we have $R_1\,(-1/2) = R_2\,(5/2)$. That is, $-R_1 = 5R_2$. This result needs a little interpretation. To meet the rotational stability condition it seems we need a *negative* reaction force at B. This, of course, is impossible. It is so easy to meet the condition for friction that we achieve it at a lower speed than we need to achieve rotational stability. At this speed the car would topple off the wall. In this situation, the minimum speed is not governed by the condition for vertical equilibrium, but by the condition for rotational stability. We saw in the example for $\mu = 1$ that this condition is met when $v = 7.1$ m s^{-1}, with the conditions

$$F_1 = R_1 = mg \text{ and } R_2 = 0$$

We note that when this rotational stability condition is satisfied, friction is *not* limiting. Our solution requires $F_1 = R_1$, which is considerably lower than the limiting condition $F_1 \leq (3/2)\,R_1$. We simply need to go fast enough that we don't topple over—there is no danger of slipping.

Further discussion

In the above analysis we said that there is a relationship between the reaction (normal) and the friction (vertical) forces, governed by $F_1 \leq \mu R_1$ and $F_2 \leq \mu R_2$. We did not say how the actual frictional force was achieved, however. If we held the steering wheel in the centre, presumably the car would spiral down in a dangerous manner. Probably to our death! We generate the frictional force by attempting to turn the car *up* the cylinder, by turning the steering wheel. It is the same mechanism as the one we use to generate a sideways force when we are driving on a normal road. The amount of force depends on how hard we turn the wheel.[8] Klaus Fritsch (of John Carroll University), takes up this theme in his 1998 letter More Physics on the Wall of Death.[9]

> A survey of the Word Wide Web shows a range of radii, speeds, and angles for different acts. While the Globe of Death of Ringling Brothers and Barnum & Bailey has a 16 m (52 ft) diameter and is negotiated at speeds of up to 27 m s^{-1} (60 mph), the Globe of Death of the Fusion Riders has a diameter of only 4.1 m (13.5 ft). Assuming a coefficient of static friction of $\mu = 0.7$, the minimum speed here becomes 7.6 m s^{-1} (17 mph), a *g*-load of 1.7 and angle of 35° [to the horizontal, for motorbikes] for travel in a horizontal equatorial circle.
>
> Photos of vehicles on the Wall of Death available on yet another website show a cylindrical wooden structure with a radius of the order of 12 m (38 ft). These photos indicate angles of perhaps 30° and show some riders with upraised arms, indicating negligible *g*-loads. One photo shows a go-kart negotiating the wall of death. It is interesting to observe that while the bike riders lean, but keep the front fork aligned with the direction of travel, the go-kart driver must steer to the right (up the wall) at all times to keep from spiralling down the wall, since the cart cannot be leaned.

5.7 Wall of Death: motorcycle ★★ or ★★★★

Quite a few explanations in high-school physics text books are found wanting. Sometimes the physics has been simplified by omission, but sometimes it is simply incorrect, or erroneously applied. My favourite example (because I teach the subject to undergraduates) is using Bernoulli's equation when explaining the lift generated by aeroplane wings. The argument goes roughly as follows:

1. Wings are curved, so the top surface has a longer streamwise surface length than the bottom surface.

2. Because the top surface is longer, a particle must travel faster over the top of the wing to meet up with the corresponding particle that has gone under the wing.

[8] The detailed physical mechanism at the point of contact is complex and relies on deformation of the tyres.

[9] Fritsch, K., 1998, "More physics on the Wall of Death," The Physics Teacher, 36, p. 390, http://dx.doi.org/10.1119/1.879902.

3. Bernoulli's equation tells us that the sum of static pressure and dynamic pressure is constant: $p + (1/2)\,\rho v^2 = C$. So higher speed leads to lower pressure.

4. Thus, the pressure on the top surface of the wing is lower than that on the bottom surface, and the wings generate lift.

It's my favourite example because it is so absurdly and patently illogical. It is plausible-sounding only until we remember to ask the extremely obvious question "why should particles that have gone over the top and bottom of the wing meet up again?" The answer is that there is no reason they should at all. And in fact, experimentally, they don't! Yet strangely, at least in the days when I was at school, this explanation was given in most high-school physics textbooks. A variant of it even appears in an undergraduate-level thermodynamics book I own—one that is still in print, and which is otherwise a very good book. It would be fascinating to know where the argument originated, and why it has prevailed so long. It is particularly curious because there are other perfectly simple, but correct, explanations that rely on maths which is barely more complicated.[10]

Purists argue that such examples are crimes against our students. I'm sure that in some cases there are good reasons to give simplified arguments. In any case, there is plenty of time to refine the arguments at a later stage, when students have access to more mathematical tools. I take the rather pragmatic view that hearing a few inaccurate arguments is an opportunity to learn that our teachers or lecturers are not always right. We need to be critical ourselves. The purists would probably be horrified.

In 1998, an article entitled Don't the Texts Have it Wrong? appeared in the The Physics Teacher. It concerned a similar error that had been repeated in numerous high-school physics text books. The author wrote:[11]

> Here is a problem that, in my opinion, all the texts have wrong. A common amusement-park show consists of a cyclist riding his motorcycle around the inside of a vertical cylinder as shown in the diagram.
>
> Textbooks pose the question: what is the minimum speed required for the cyclist to stay on the wall? Traditionally, we equate N to mv^2/r, F to mg, and using $F = \mu N$ obtain the formula $v = \sqrt{gr/\mu}$. Inserting values from a current text ($r = 15$ m, $\mu = 1.1$) we get $v = 12$ m s^{-1} as the minimum speed required.
>
> This analysis cannot be correct. Forces F and mg constitute a couple that would cause the rider to rotate in a clockwise fashion.

[10]One example is an elegant argument based on streamline curvature, which requires us to derive only the *radial-equilibrium equation*, and which is easily accessible to the average high-school student.

[11]Webb, J., 1998, "Don't the texts have it wrong?", The Physics Teacher, 36, p. 184, http://dx.doi.org/10.1119/1.879999.

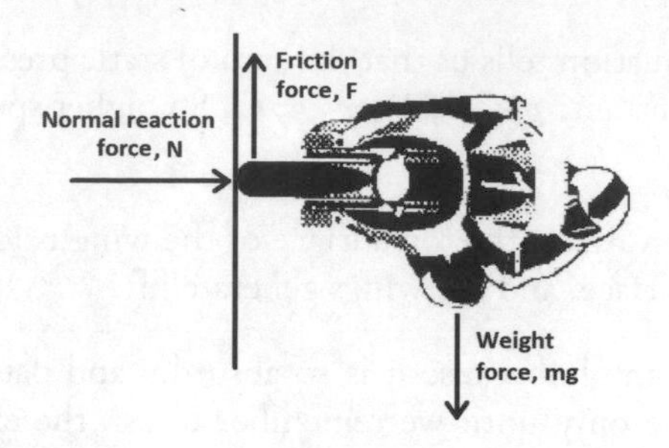

The article then attempted to deal with the couple arising from the two vertical forces, arriving (using curious logic) at a speed of 54 m s^{-1} (120 mph) as the required speed to stay on the track. Let us now consider the problem from first principles, but taking a more realistic value for the diameter of the track.

CONSIDER A MOTORCYCLE driving in a horizontal circle around a Wall of Death with diameter 12 m. Assume that on a level road the mass of the motorcycle-rider system would be concentrated at a point 1 m above the road, but treat the problem as being a *point mass* at the same location.

- For ★★, what is the minimum speed when the coefficient of static friction is $\mu = 1/2$? Draw a force diagram for the motorcycle for this condition, and calculate the angle of the motorcycle to the horizontal.

- For a ★★★★ extension, draw a force diagram for the motorcycle in the situation where the speed is doubled, and calculate the angle of the motorcycle to the horizontal. You may need to use approximations.

Answer

We start by drawing a diagram of the motorcycle. The first thing to note is that it is necessary to *lean* the motorcycle. This allows the moments about the centre of mass to cancel, which is a requirement for rotational stability. The point of contact with the vertical wall is marked A, and the centre of mass is marked C. We are told that the distance between C and A is 1 m. The angle the motorcycle makes with the horizontal is θ. We consider the case $\mu = 1/2$. Our aim is to find the minimum speed, and the angle θ. Let us first draw the forces on the motorcycle. There are three forces.

- The weight, mg, which acts vertically downwards from C.

- A reaction force, R, which acts normal to the wall at the point of contact, A.

- A friction force, F, which acts vertically upwards at the point of contact, A.

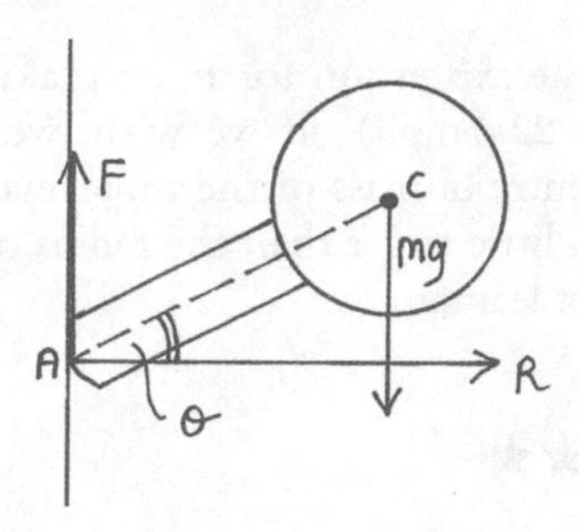

Let us consider the relationship between the forces. Resolving forces in the vertical direction, and noting that the acceleration in this direction is zero, we have

$$mg = F$$

We next consider the forces acting normal to the wall. In this direction there is an acceleration of magnitude v^2/r, where v is the tangential speed and r is the radius of the circle made by the point-mass that represents the motorcycle-rider system. The radius of the circle is given by $r = 12/2 - \cos\theta = 6 - \cos\theta$. In this direction we have

$$R = \frac{mv^2}{r}$$

In general, the frictional force obeys $F \leq \mu R$. For the case of limiting friction, we write $F = \mu R$. For this special case, we get

$$F = \mu R = \mu\frac{mv^2}{r} = \frac{\mu mv^2}{6 - \cos\theta} = mg$$

So

$$v = \sqrt{\frac{g}{\mu}\,(6 - \cos\theta)}$$

Taking moments about the centre of mass to ensure that rotational stability is achieved, we get

$$\begin{aligned} F\cos\theta &= R\sin\theta \\ \mu R\cos\theta &= R\sin\theta \\ \mu &= \tan\theta \end{aligned}$$

Taking $\mu = 1/2$, we get $\theta = 26.6^\circ$. This angle is fairly typical of the angle riders make when driving round a Wall of Death. It allows the moment that is due to the normal reaction force at the point of contact to cancel the moment about the centre of mass that is due to the friction force. At an angle $\theta = 0^\circ$ the motorcycle would topple off the wall.

Substituting θ into the expression for v, and taking $g \approx 9.81\ \text{m s}^{-2}$, we get $v \approx 10.0\ \text{m s}^{-1}$ (or 22.4 mph). If we wish, we can also find the value of r, the radius of the centre of mass of the motorcycle-rider system. We get $r = 6 - \cos\theta \approx 5.1$ m, a little larger than the radius of $r = 5$ m we would get if the motorcycle was not leaning.

Extension for ★★★★

Finally we consider the situation at twice the speed, or $v \approx 20\ \text{m s}^{-1}$. At twice the speed, the normal reaction force is increased by a factor of four. We are no longer at the point of limiting friction. The force due to friction is still equal to the weight, and this balances forces in the vertical direction. At the point of rotational equilibrium the motorcycle is inclined at a lower angle θ than at the point of limiting friction. The *distance of action*[12] of R about the centre of mass is reduced.

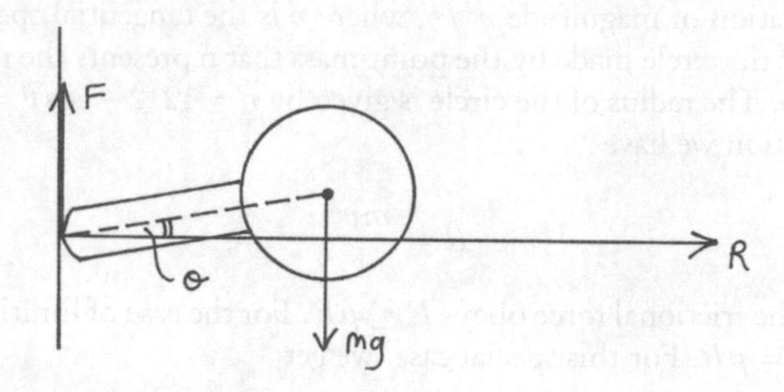

Resolving forces in the vertical direction we get $F = mg$. Resolving forces in the horizontal direction and noting that acceleration towards the centre of the circle is v^2/r, we have $R = mv^2/r$. Taking moments, the requirement for rotational equilibrium is

$$\begin{aligned} F\cos\theta &= R\sin\theta \\ mg\cos\theta &= \frac{mv^2}{r}\sin\theta \end{aligned}$$

Recalling that $r = 6 - \cos\theta$, we write

$$\tan\theta = \frac{g}{v^2}\left(6 - \cos\theta\right)$$

We need to solve this for $v = 20\ \text{m s}^{-1}$. Letting $\cos\theta = c$, and using the trigonometric identity $\tan^2\theta = \left(1 - c^2\right)/c^2$, we have

[12] The perpendicular distance between the line of action of R and the centre of mass.

$$\begin{aligned} \frac{1-c^2}{c^2} &= (6-c)^2\left(\frac{g^2}{v^4}\right) \\ 1-c^2 &= c^2\left(36-12c+c^2\right)\left(\frac{g^2}{v^4}\right) \\ 0 &= c^4-12c^3+\left[36+\left(\frac{v^4}{g^2}\right)\right]c^2-\left(\frac{v^4}{g^2}\right) \end{aligned}$$

This quartic equation has four roots which can be found algebraically, but doing so takes many lines of algebra. It is easier to solve the equation iteratively. We seek the real solutions. We expect a low angle θ, making $c = \cos\theta$ quite close to unity. For completeness, I've given all four roots of the quartic,[13] but an iterative solution close to unity would yield just the first solution. This is the only physically meaningful solution, given that we expect $c \in \mathbb{R}^+$.

$$c \approx \begin{cases} 0.9926 \\ -0.9857 \\ 5.997 & +40.855i \\ 5.997 & -40.855i \end{cases}$$

Taking $c \approx 0.9926$ gives $\theta \approx 6.98°$, a very low angle to the horizontal.

Further discussion

If we perform a more detailed analysis of the Wall of Death, we find a rather interesting difference between the situation of the motorcycle and the car. It is associated with the gyroscopic effect. Because the motorcycle is inclined as it moves around the cylinder,[14] both the angular momentum vector of the motorcycle rotating about its own axis,[15] and the angular momentum vectors of the wheels, are continually changing. To change these angular momentum vectors requires a torque. This torque must be supplied by changing the location of the motorcycle's centre of mass. The mechanism for doing this is identical to the one described above, but the additional gyroscopic effect changes the stability point of the system. According to my own back-of-the-envelope calculations, the effect is far from negligible, changing the required torque by perhaps 5% in a typical Wall of Death situation.

[13] As an exercise you could check that these roots satisfy the equation.

[14] It is inclined at an angle θ in the analysis we have just performed.

[15] The motorcycle rotates about its own centre of mass at the same rate as it rotates about the cylinder.

Chapter 6

Simple harmonic motion

In this chapter we look at simple harmonic motion (SHM), the simplest type of oscillatory motion. A pendulum, a mass on a spring, a body floating partially submerged in water,[1] and an inductor-capacitor circuit (LC circuit), are all examples of systems which for small-amplitude oscillations approximate SHM. What unites these systems is that when they are displaced from their rest position, a restoring force is set up which acts towards the equilibrium position. And this restoring force, F, is proportional to x, the displacement from equilibrium. This defines the condition for SHM. That is, the system must obey

$$F = -kx$$

where we call k the *spring constant*, or—more generally—the *stiffness*, of the system. By invoking Newton's Second Law, $F = ma = m\left(d^2x/dt^2\right)$, we arrive at the equation of motion.

$$m\frac{d^2x}{dt^2} = -kx$$

We now solve this equation[2] to arrive at the *general solution* for SHM. First we note that

$$\frac{d^2x}{dt^2} = \frac{d}{dt}\frac{dx}{dt} = \frac{dx}{dt}\frac{d}{dx}\frac{dx}{dt} = v\frac{dv}{dx}$$

Hence

$$v\frac{dv}{dx} = -\frac{k}{m}x$$

[1] In a situation where the body's cross-sectional area at the plane created by the water's surface doesn't change as the body oscillates.

[2] We use an approach that involves first-order differential equations, in which we can use separation of variables, and which does not require a *trial solution*. This is perhaps not the standard method—you should consult a textbook for the standard method using trial solutions.

Integrating, we get

$$\int_{v_0}^{v} v dv = -\frac{k}{m}\int_{x_0}^{x} x dx$$

where $x_0 = x\,(t = 0)$, and $v_0 = v\,(t = 0)$. We get

$$\begin{aligned}
\left[\frac{v^2}{2}\right]_{v_0}^{v} &= -\frac{k}{m}\left[\frac{x^2}{2}\right]_{x_0}^{x} \\
v^2 &= \frac{k}{m}\left(x_0^2 - x^2\right) + v_0^2 \\
&= \frac{k}{m}\left(x_0^2 + \frac{m}{k}v_0^2 - x^2\right)
\end{aligned}$$

Setting $x_0^2 + (m/k)\,v_0^2 = a^2$, we see that $v^2 = (k/m)\left(a^2 - x^2\right)$, or $v = dx/dt = \pm\,(k/m)^{1/2}\left(a^2 - x^2\right)^{1/2}$. Integrating, we get

$$\begin{aligned}
\int_{x_0}^{x} \frac{dx}{\left(a^2 - x^2\right)^{1/2}} &= \pm\,(k/m)^{1/2}\int_{0}^{t} dt \\
\left[\sin^{-1}(x/a)\right]_{x_0}^{x} &= \pm\,(k/m)^{1/2}\,[t]_0^t \\
\sin^{-1}\left(\frac{x}{a}\right) &= \pm\,(k/m)^{1/2}\,t + \sin^{-1}\left(\frac{x_0}{a}\right) \\
x &= a\sin\left(\pm\,(k/m)^{1/2}\,t + \sin^{-1}\left(\frac{x_0}{a}\right)\right) \\
&= a\sin\left(\pm\,(k/m)^{1/2}\,t + \phi\right)
\end{aligned}$$

where the *phase constant* $\phi = \sin^{-1}(x_0/a)$. This is the general solution for SHM. Here a is the *amplitude* of the oscillation, and $\omega = (k/m)^{1/2}$ is the *angular frequency* of the oscillation in radians per second. As shown in the diagram below, there are four special cases: $t = 0$ occuring at A, B, C and D. Let's briefly consider each of these in turn.

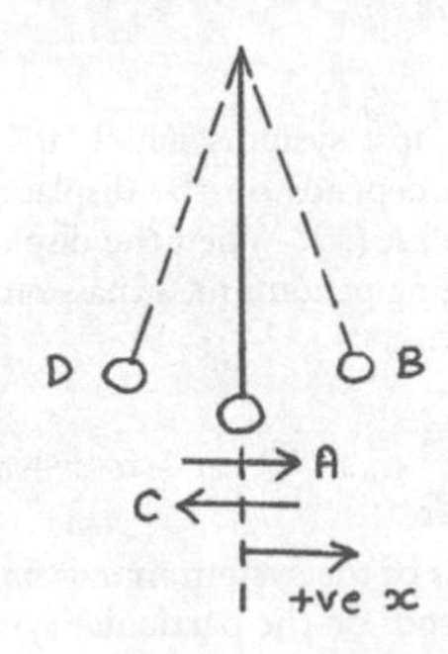

- CASE A. $x_0 = 0$, with motion towards positive x. For $x_0 = 0$, $\phi = \sin^{-1}(0) = 0$, giving $x = a\sin(+\omega t + 0) = a\sin(\omega t)$.
- CASE B. $x_0 = +a$, with zero speed, and at maximum x. For $x_0 = a$, $\phi = \sin^{-1}(1) = \pi/2$, giving $x = a\sin(+\omega t + (\pi/2)) = a\cos(\omega t)$.
- CASE C. $x_0 = 0$, with motion towards negative x. For $x_0 = 0$, $\phi = \sin^{-1}(0) = 0$, giving $x = a\sin(-\omega t + 0) = -a\sin(\omega t)$.
- CASE D. $x_0 = -a$, with zero speed, and at minimum x. For $x_0 = a$, $\phi = \sin^{-1}(-1) = -\pi/2$, giving $x = a\sin(+\omega t + (-\pi/2)) = -a\cos(\omega t)$.

For situations other than the special cases A, B, C and D, we compute the phase constant ϕ to satisfy the starting conditions. We won't consider problems with non-standard starting conditions in this book.

To solve problems in SHM we will need to remember the following important points. This brief summary shouldn't be used instead of a textbook, but rather as an aide-memoire for those who have already studied the subject more formally.

- SYSTEM EQUATION FOR SHM. The system equation for systems that exhibit SHM has the form $F = -kx$. When the object is displaced, a force F is set up which is: i) proportional to x, the displacement from equilibrium;[3] and ii) directed towards the equilibrium position—that is, a restoring force. The constant of proportionality k is often called the stiffness of the system, or, where there is a spring involved, the spring constant.
- GENERAL SOLUTION FOR SHM. Using Newton's Second Law, we can turn the system equation into the equation of motion for SHM:

  $$m\left(d^2x/dt^2\right) = -kx$$

  We can solve this to give the general solution for SHM:

  $$x = a\sin(\pm\omega t + \phi)$$

  This has four special solutions: zero-speed maximum-displacement solutions, $x = \pm a\cos(\omega t)$; and zero-displacement maximum-speed solutions, $x = \pm a\sin(\omega t)$.

---

[3]This is approximately true for most systems which are displaced by a *very small* amount. Consider a restoring force F which depends on the displacement x. In general, F could be a complicated non-linear function of x, $F(x)$. When the displacements become large enough, the function is complex for even a swinging pendulum, a mass on a spring etc. We can expand $F(x)$ about $x = 0$ using a Taylor series. We get

$$F(x) = \sum_{n=0}^{\infty} a_n x^n = a_0 + a_1 x + a_2 x^2 + ...$$

$F(0) = 0$, so $a_0 = 0$. So, regardless of the system, if x is *small enough* we expect $F(x) \approx a_1 x$ (what we mean by *small enough* depends on the particular system we are considering).

- RESTORING FORCE. The restoring force $F = -kx$ depends on the type of system and can take many forms. For example: i) a component of string tension in the case of a simple pendulum, provided that the maximum angle of oscillation is small; ii) the force due to the extension or compression of a spring, in the case of a mass attached to a *linear*, or *Hookean*, spring; and iii) for floating objects undergoing small displacements with no change in cross-sectional area, a buoyancy force that is equal to the difference in weight of fluid that's displaced between equilibrium and non-equilibrium positions. There are many other examples of SHM in which the equivalent force may be less easy to identify with a physical quantity.

- EFFECT OF GRAVITY AND ACCELERATION. In some systems, the restoring force is proportional to gravity. In the case of the simple pendulum, for example, this is because the string tension is proportional to gravity. If we move to a planet on which the gravity is greater than on the surface of the Earth, for example, the restoring force would increase, and the period of the pendulum would decrease. We can construct systems in which the gravitational force is the same, but where the restoring force is changed *as if* the gravity had changed. For these systems we sometimes use the term *apparent gravity*, to distinguish from a system in which the actual gravitational force has changed. We refer here to *accelerating frames of reference*. If you are in an elevator which is accelerating upwards, the apparent gravity is increased—string tension in a simple pendulum would increase to provide the necessary upward acceleration. If you jump out of a plane the apparent gravity decreases (for a few moments at least)—string tension would decrease because some or all of the force due to gravity would act to accelerate the object. We see that when the apparent gravity changes, for some systems the restoring force changes too.

- PERIOD OF MOTION. For a simple pendulum, the angular frequency of the motion is given by $\omega = (k/m)^{1/2}$ rad s^{-1}. The period of the motion, T, is the time the system takes to complete one full cycle from $\omega t = 0$ to $\omega t = 2\pi$. We write $\omega T = 2\pi$, which gives $T = 2\pi\,(m/k)^{1/2}$.

- ENERGY. Energy is conserved in a simple harmonic oscillator. There are two forms of energy: kinetic energy ($\mathrm{KE} = \frac{1}{2}m\dot{x}^2$) and potential energy[4] ($\mathrm{PE} = \int_0^x F\mathrm{d}x = \int_0^x kx\mathrm{d}x = \frac{1}{2}kx^2$). The total energy E is equal to the sum of kinetic and potential energy at every point in the cycle. We write

$$E = \mathrm{KE} + \mathrm{PE} = \frac{1}{2}m\dot{x}^2 + \frac{1}{2}kx^2$$

[4]Here we consider only the simplest case of spring energy. In a more general case we could have a combination of stored potential energy in a spring, and gravitational potential energy. Consider a spring arranged to oscillate in the vertical direction, for example.

When the displacement is at a maximum, all the system energy is in the form of potential energy, so speed is zero. When the displacement is zero, there is zero potential energy with respect to our equilibrium position and the system energy is equal to the kinetic energy. The speed is maximised. There is an exchange between kinetic energy and potential energy as the system oscillates.

6.1 Oscillating sphere ★★

You will have seen boats and other objects bobbing about in harbours. As with all things nautical, there are special terms to describe their motion. Sailors talk about the *sway* (side-to-side movement), *surge* (forward and backward movement) and *heave* (up and down movement) of a boat. You will probably also have noticed that the period of motion depends on the size of boat. If you have ever stood next to a large ferry, for example, you will have seen it heaving slowly up and down for a while after another boat passes, with a period of a number of seconds. In contrast, a seagull alighting on a fishing float sets in motion an oscillation with a period of less than a second. As with most oscillatory motions of mechanical systems, the period is a function of the mass and the restoring force.[5]

A SOLID SPHERE of radius r floats with exactly half its volume above the surface of a dense fluid. It is displaced downwards by a small distance $y \ll r$ and released. Write an equation for the period of the motion. Ignore the motion of the fluid.

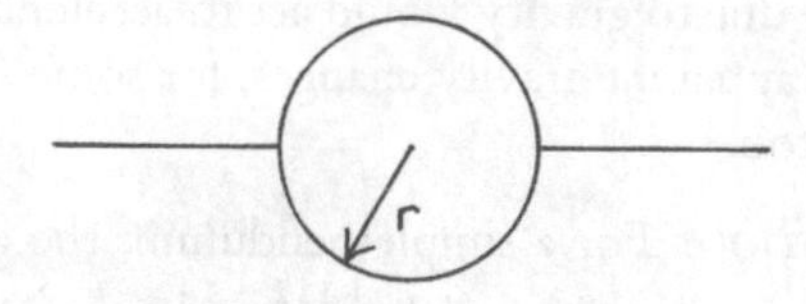

Answer

I have enjoyed asking students this question over the years. It is simple to phrase, and therefore easy to understand. Most students seem to make a reasonable attempt at it without much help. Many spot intuitively that the motion is some form of SHM when the displacement is small enough ($y \ll r$). I do not introduce this idea at the outset, so it is quite pleasing when students spot the connection. I like the question because at first it looks like there is some information missing, which encourages people to think before putting their heads down and getting lost in equations.

[5] In some special cases, like a simple pendulum, the restoring force is proportional to the mass, giving a time period that is independent of the mass: $T = 2\pi\sqrt{l/g}$.

So, how do we begin? By trying to understand the information in the question, of course. We know that the solid sphere floats with half its volume above the surface of the dense fluid. So if the density of the fluid is ρ_f, the density of the solid sphere, ρ, must be given by

$$\rho = \frac{1}{2}\rho_f$$

We have neglected the buoyancy force due to the finite density of the second fluid (presumably air) in which the upper half of the solid sphere is immersed. That is, we assume that the density ρ' is small ($\rho' \ll \rho, \rho_f$).

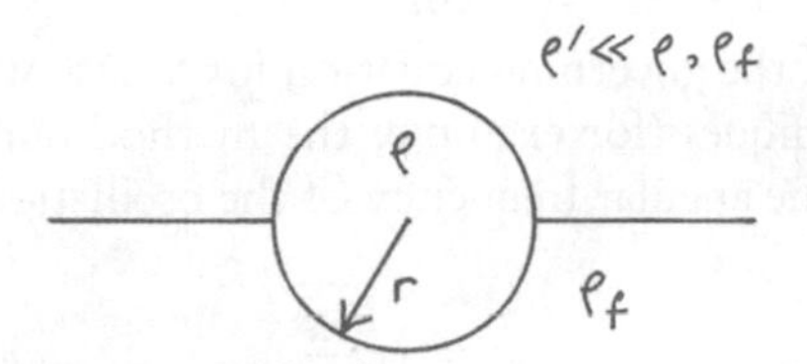

So far so good. Now we need to express, in terms of known quantities, the mass of the solid sphere, and the force acting on the solid sphere. The aim is to derive an expression which is in the form of the equation of motion for SHM. First we consider the mass of the solid sphere.

$$m = \frac{4}{3}\pi r^3 \rho$$

Now we must consider the forces on the solid sphere. There is a downward force due to the weight of the solid sphere, equal to mg. Counteracting this is a buoyancy force, which is equal to the weight of displaced liquid. At equilibrium, the buoyancy force is equal in magnitude to the weight of the solid sphere and acts in the opposite direction. At equilibrium there is no net force on the object.

Now we displace the solid sphere downwards by a small distance y from the equilibrium position and release it. Here *small* means $y \ll r$. There will now be a net upward buoyancy force, F_{net}, equal to the weight of the additional volume of displaced fluid. This force is equal to the *swept volume* of fluid multiplied by the density of fluid multiplied by the force per unit mass due to gravity. For small displacements, the swept volume has approximately the shape of a thin disc, and its volume is $\pi r^2 y$. We write

$$F_{net} = -y\pi r^2 \rho_f g$$

The minus sign arises because the force is in the opposite direction to the displacement vector—a restoring force. We note that the expression is only strictly true for $y \ll r$, when the swept area remains constant.

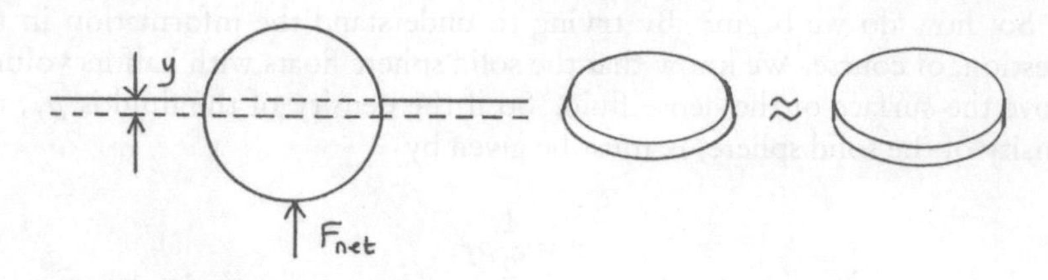

Using Newton's Second Law, substituting for m, and noting that $\rho = \frac{1}{2}\rho_f$, we obtain the acceleration of the solid sphere.

$$a = \ddot{y} = \frac{d^2y}{dt^2} = -\frac{3g}{2r}y$$

This has the form of the governing equation for SHM. We solve the equation using standard techniques (for example, the method outlined at the start of this section) to get the angular frequency of the oscillation.

$$\omega = \sqrt{\frac{3g}{2r}}$$

The period, T, is the time taken to perform one cycle of 2π, and is given by $T\omega = 2\pi$. So

$$T = 2\pi\sqrt{\frac{2r}{3g}}$$

is the expression we were asked to derive. The units are, of course, seconds. It is interesting to note that the period T of the motion is independent of the mass of the object, which may be something of a surprise.[6] The explanation is quite simple. Because the object is floating half in and half out of a dense fluid, the restoring force is effectively defined as being proportional to the mass of the object. There is an analogy with the simple pendulum, another system in which the period is independent of mass. Of course, in any real (non-idealised) system, we would also need to account for motion of the fluid (which would increase the mass that is in motion), and any damping effects arising from viscosity (of both fluids), for example.

Further discussion

Let's calculate the size of solid spheres that would give us periods of, say, 1 s and 10 s. Rearranging the result we get

$$r = \frac{3g}{2}\left(\frac{T}{2\pi}\right)^2$$

For $T = 1$ s we have $r = 0.37$ m, the size of a large fishing float. For $T = 10$ s we have $r = 37$ m, the size of a typical iceberg. We will do the correct

[6] Note that we did not need to specify the density of the fluid, or the density of the solid sphere.

calculation for an ice shelf in a moment, because our current expression is valid only for solid spheres with half the density of the supporting fluid. First we consider quite how large icebergs can be.

Icebergs that calve from the Ross Ice Shelf in Antarctica can be quite fantastic in size. In March 2000, 11,000 square kilometres of ice calved off in a single piece, creating the largest iceberg in recorded history. It was called B-15,[7] and survived for several years until it was eventually broken up by storms. The *freeboard*, or height above water, of these huge *tabular icebergs* is typically only about 30–40 m, giving them an estimated thickness of approximately 300–400 m (freshwater ice has a density of approximately 920 kg m^{-3}; sea water has a density of approximately $1,025$ kg m^{-3}). Using a similar method to that above, we can calculate the period of these huge objects. Let's try it.

We take a tabular iceberg of surface area A and thickness h, where the ice has density ρ_{ice} and the surrounding water has density ρ_{H_20}.

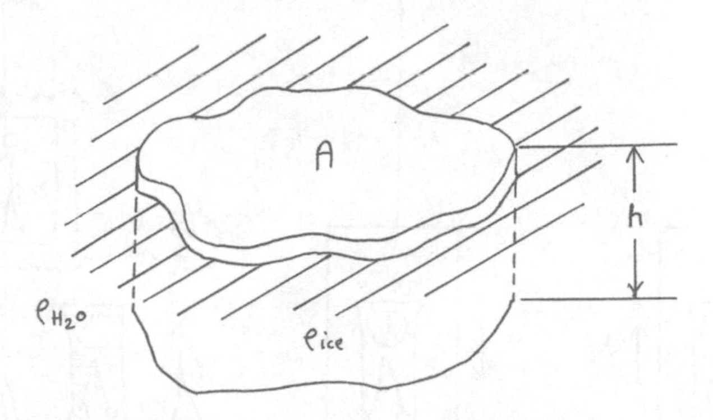

Using the same method as above, for a small displacement y, we write

$$F = -Ay\rho_{H_20}g = Ah\rho_{ice}\frac{d^2y}{dt^2}$$

So

$$\frac{d^2y}{dt^2} = -\frac{\rho_{H_20}g}{\rho_{ice}h}y$$

giving

$$\omega^2 = \frac{\rho_{H_20}g}{\rho_{ice}h} \text{ or } T = \frac{2\pi}{\omega} = 2\pi\left(\frac{h\rho_{ice}}{g\rho_{H_20}}\right)^{1/2}$$

Taking $\rho_{ice} = 920$ kg m^{-3}, $\rho_{H_20} = 1,025$ kg m^{-3}, $h = 300$ m, and $g \approx 9.81$ m s^{-2}, we get $T = 32.9$ s. As we might expect, the result is independent of surface area.

[7] The US National Ice Centre was set up in 1995 to monitor the progress of the world's largest icebergs. Icebergs with a dimension greater than 10 nautical miles (sailors always like to do things differently!) are given an alphanumeric code which indicates where they calved (the letter) and the order in which they calved (the number).

6.2 Professor Stopclock's time-manipulator ★★★

PROFESSOR STOPCLOCK BELIEVES he has found the secret of time. Specifically, he believes he can speed up or slow down time using his famous time-manipulator elevator. This he constructed in a purpose-built tower and well shaft in the basement of Dragging Towers, the seat of the Stopclock family. The time-manipulator elevator has two self-counterbalancing lifts connected so that they travel at the same speed as each other but in opposite directions. This means that their accelerations are also equal in magnitude but opposite in direction.

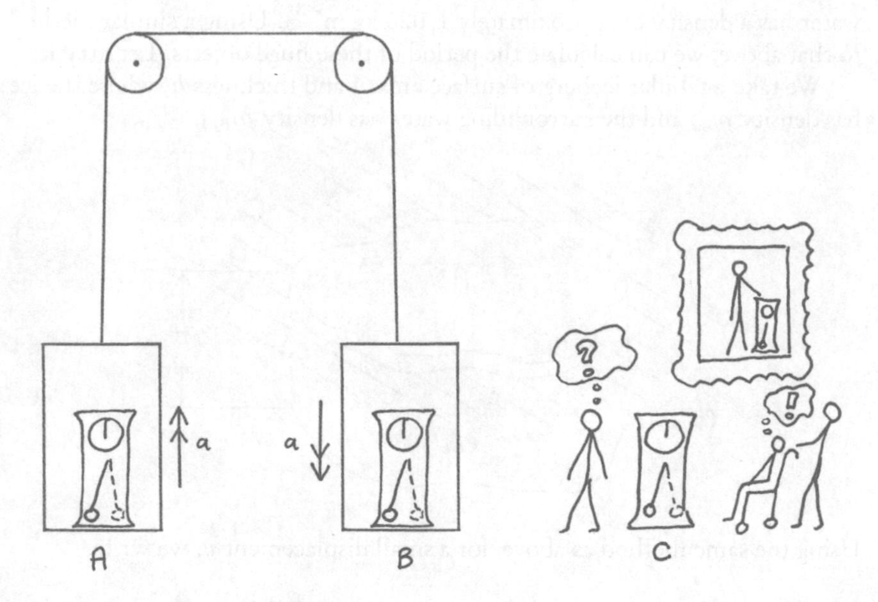

The lifts operate in a continuous cycle, as follows.

- PHASE 1. Lifts A and B are initially at rest and level with each other.
- PHASE 2. Lift A (B) accelerates upwards (downwards) at $a = (3/10)\,g$ for 7.5 seconds.
- PHASE 3. Lift A (B) accelerates upwards (downwards) at $a = -\,(9/10)\,g$ for 2.5 seconds (that is, the lifts slow down until they are at rest).
- PHASE 4. Lift A is *very slowly* lowered at a continuous speed, raising lift B, so that they become level again.

Professor Stopclock says he can demonstrate his marvellous machine with the help of two identical pendulum-driven grandfather clocks, one in lift A and the other in lift B. According to Stopclock, time will pass faster than normal in lift A, and slower than normal in lift B. He invites an elderly guest to test out the time-manipulator and see the effect for himself, while he waits on the landing. Fortunately the infirm gentleman is a physicist, and when they arrive at the lift shaft he explains to Stopclock where his reasoning has gone awry. He requests that a third identical clock C be placed on the landing. They will use this third clock to adjudicate on the true passing of time. The butler synchronizes the grandfather clocks, and the machine is started. The doors are opened at the end of each cycle so that the times on each clock can be noted.

At the end of each cycle of the lifts, what do the clocks read relative to each other, and why?

Answer

The grandfather clocks operate on the principle of simple harmonic motion. In the governing equation for SHM, $\ddot{x} = -\omega^2 x$, the angular frequency ω reduces to $\omega = \sqrt{g/L}$ for the case of a simple pendulum of length L in a uniform gravitational field g.[8]

In an accelerating frame of reference with constant acceleration of magnitude a in the opposite direction to the gravitational force, the restoring force becomes $m\,(g + a)$. The frequency of oscillation increases. For a constant acceleration of magnitude a in the same direction as the gravitational force, the restoring force is $m\,(g - a)$. The frequency of oscillation decreases. In the limiting case of free fall, $|a| = |g|$, the restoring force goes to zero and the pendulum no longer oscillates. The frequency of oscillation is zero.

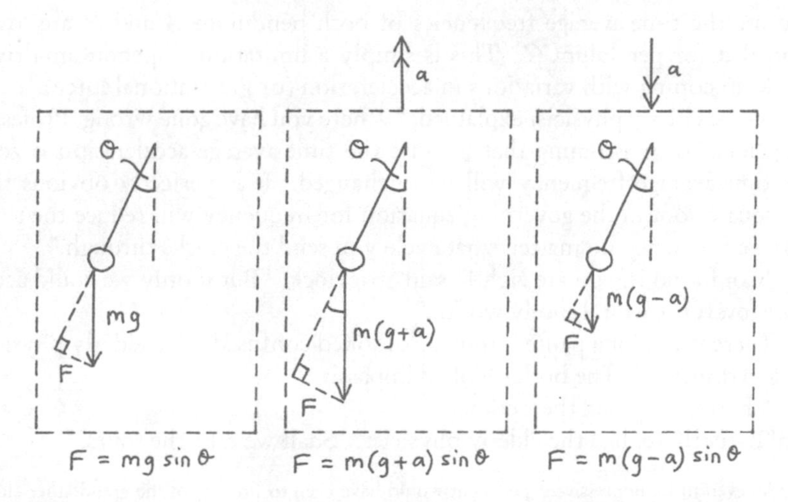

Returning to the governing equation, the simple conclusion is that in all three situations the period is proportional to the square root of the restoring force, which is enhanced by positive (upward) acceleration, and reduced by negative (downward) acceleration. Consider the motion of clocks A, B and C during the two time intervals $\Delta t_1 = 7.5$ s and $\Delta t_2 = 2.5$ s. The total number of

[8] g appears in the equation because the restoring force is a component of the weight, mg.

oscillations, n, is the sum of the frequency in Hz in each time interval, $f = \omega/2\pi = (1/2\pi)\sqrt{g/L}$, multiplied by the length of the time interval, $n = f_1\Delta t_1 + f_2\Delta t_2 = (\omega_1/2\pi)\Delta t_1 + (\omega_2/2\pi)\Delta t_2$. We ignore the part of the cycle (phase 4) in which the clocks are slowly returned at continuous speed. The effect of the acceleration required to bring the clocks to this very slow speed will be negligible. During the 10 s of the cycle in which there is significant acceleration (phases 2 and 3), the clocks behave as follows.

$$\text{Clock A} \qquad n_A = \frac{1}{2\pi}\sqrt{\frac{g}{L}}\left[\sqrt{\left(1+\frac{3}{10}\right)}\Delta t_1 + \sqrt{\left(1-\frac{9}{10}\right)}\Delta t_2\right]$$

$$\text{Clock B} \qquad n_B = \frac{1}{2\pi}\sqrt{\frac{g}{L}}\left[\sqrt{\left(1-\frac{3}{10}\right)}\Delta t_1 + \sqrt{\left(1+\frac{9}{10}\right)}\Delta t_2\right]$$

$$\text{Clock C} \qquad n_C = \frac{1}{2\pi}\sqrt{\frac{g}{L}}\left[\sqrt{1}\Delta t_1 + \sqrt{1}\Delta t_2\right]$$

The ratio of the number of oscillations $n_A : n_B : n_C$ during that 10 s is

$$9.34 : 9.72 : 10.00$$

So for every cycle of the machine, relative to clock C on the landing, clocks A and B run slow by 0.66 s and 0.28 s. This is an entirely real effect which occurs because the time-average frequencies of both pendulum A and B are lower than that for pendulum C. This is simply a limitation of pendulum-driven clocks in coping with variations in acceleration (or gravitational force).[9]

As the elderly physicist explained, "Where you have gone wrong, Professor Stopclock, is in assuming that because the time-average acceleration is zero, the time-average frequency will be unchanged. It is perfectly obvious that the square root in the governing equation for frequency will reduce the time-average frequency no matter what cycle you send the clocks through."

"Confound it, you are right!" said Stopclock. "But if only we could accelerate forever it would surely work..."

There was a long pause. Stopclock looked confused. The elderly physicist looked dismayed. The butler looked impassive.

"Sherry, sir?" said the butler.

"Rather!" replied the elderly physicist. "Shall we take the stairs?"

---

[9]An even more impressive experiment would have been to fire one of the grandfather clocks upwards from the barrel of a large cannon. Ignoring atmospheric drag, while the clock is in free fall (with an acceleration $a = g$) the pendulum does not swing at all, and *apparent time* (as measured on the clock) is at a standstill. This is true both as the clock goes up and as it comes down. It is only during the (almost instantaneous) firing of the cannon and the (even closer to instantaneous) crash with the ground that apparent time slows down. These periods are very short and can almost be ignored. If a grandfather clock were fired on a huge ballistic trajectory and reloaded very quickly every time it hit the ground, ignoring atmospheric drag, Professor Stopclock would have a very effective way of bringing apparent time (as measured on the clock) almost to a standstill.

6.3 Dr Springlove's Oscillator ★

It is a sad fact that, throughout human history, we have conceived of the most abhorrent ways of harming and killing each other, often using science to this end. We have used science to develop ever more sophisticated weapons to perpetrate violence. Indeed, many technological breakthroughs were driven by the need to have more advanced weaponry than one's opponents.

Consider the following developments: the spear and the bow and arrow with stone tips (about 20,000 BC); the domestication of the horse (about 4000 BC); advances in metallurgy for strong and light swords (5000–1000 BC); the invention of gunpowder and the first firebombs (about 1000 AD); high-velocity firearms (about the year 1200 onwards); explosive rockets (1800s); submarines (1775); high-impact shells (1800); the machine gun (1884); tanks (1914); the nuclear bomb (1945); the maser and laser (1960); the Taser or electric stun gun (1974); and the airborne laser (2008). These, as well as many technological developments in aviation, space flight and shipbuilding technology, were driven by the need to compete on the battlefield, in the skies or on the high seas. Of course, the resulting scientific knowledge has also been used for great good as well as harm. The developments of the nuclear age and the arms race led to nuclear reactors, which currently supply a large proportion of the world's energy, and to developments in fusion reactors, which may yet offer a solution to our growing and unsustainable desire for more and more energy.

Even the humble pendulum has been turned into a torture device. From the end of the Middle Ages (around the year 1500) until as recently as the nineteenth century, the Spanish Inquisition used it to make people talk, and as a way to execute them if they did not. The weight on the end of a long (rigid-armed) pendulum was replaced by an axe. This was swung back and forth over the victim, while a special mechanism lowered it by a small amount on each pass. In the preface of an 1826 book by Jean Antoine Llorente,[10] the Secretary of the Inquisition, there is the following disturbing passage.

> One of these prisoners had been condemned, and was to have suffered on the following day. His punishment was to be death by the Pendulum. The method of thus destroying the victim is as follows: the condemned is fastened in a groove, upon a table, on his back; suspended above him is a Pendulum, the edge of which is sharp, and it is so constructed as to become longer with every movement. The wretch sees this implement of destruction swinging to and fro above him, and every moment the keen edge approaching nearer and nearer: at length it cuts the skin of his nose, and gradually cuts on, until life is extinct. It may be doubted if the holy office in its mercy ever invented a more hu-

[10] Llorente, J.A., 1826, "The history of the Inquisition of Spain, from the time of its establishment to the reign of Ferdinand VII," printed in London for Geo. B. Whittaker.

> mane and rapid method of exterminating heresy, or ensuring confiscation. This, let it be remembered, was a punishment of the Secret Tribunal, A. D. 1820!

The concept is so awful, and vivid, that it has been taken up many times in literature, most famously in the 1842 short story *The Pit and the Pendulum* by American mystery and suspense writer Edgar Allan Poe. The story has been endlessly adapted since, and has formed the basis of a number of films with the same name (in various languages). I am not a great fan of this genre, but the story is a classic which I suppose we should all read at least once.

Consider a slightly less awful device based on the same principles. We will eliminate the more macabre aspects of the Inquisition's pendulum machines.

DR SPRINGLOVE HAS a curious way of making his victims talk. In the basement of his Chicago apartment is the machine known as the Springlove Oscillator. It consists of a frame to which the victim is strapped. The whole assembly (including the victim) has mass M, and is suspended from the ceiling by a massless spring of stiffness k. Dr Springlove sets the device in motion (by displacing it and releasing it from rest) until his victim speaks. Draw the forces on the mass M, and calculate the period of the motion if the mass is displaced by a very small amount from the equilibrium position and then released from rest.

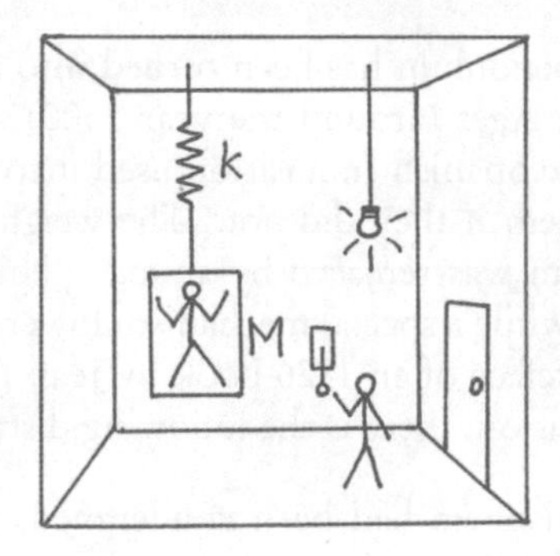

Answer

This is a perfectly straightforward example of SHM with no tricks. If I ask this question, I'm keen to see a very clear explanation of the forces, both in terms of how they arise, and also where they act. In general, even for relatively simple questions of this sort, I expect to give some hints, and I encourage students to be precise in their descriptions of the forces. For instance, many students would instinctively draw the force due to gravity as acting from the bottom of the mass, rather than from the centre of gravity. Equally, many students draw the force due to the spring as acting from the centre of the mass, rather than from the point of attachment. I have no idea why they do this—maybe most high-school courses don't require this level of precision. If you are wondering why it matters, and why I would gently correct someone who drew a force vector incorrectly, the reason is as follows. Other than in the most simple problems, where forces are aligned along a single axis, the torque (or moment) due to a force about the centre of mass of an object can also be very important in determining its motion.

Let's return to the problem. We have a mass M suspended from a massless spring with spring constant k. In the equilibrium position the acceleration is zero, so the net force, $\sum F = F_{net}$, is also zero. The upward (spring) and downward (weight) forces are equal in magnitude and opposite in sign. The force due to gravity acts (equivalently at least) downwards from the centre of mass,[11] and an opposing force in the spring acts upwards from the point of attachment. If we arbitrarily take forces and displacements that act downwards as positive, we write

$$F_{net} = -kY + Mg = 0$$

where Y is the extension from the *natural length* of the spring, and F_{net} is the net force on the mass. The object is in equilibrium and the forces balance.

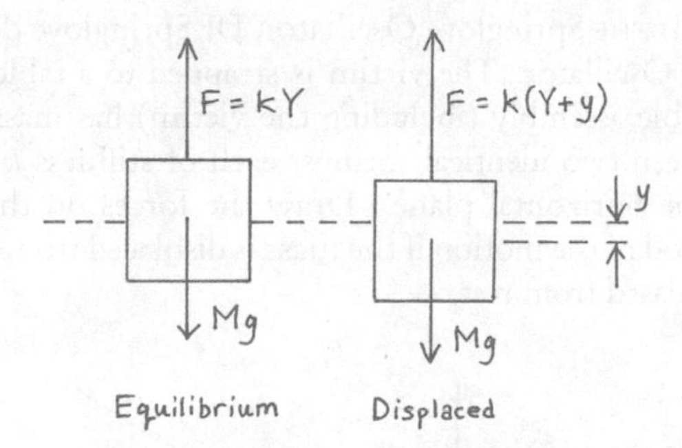

We displace the mass by a small distance y downwards from the equilibrium position (to position $Y + y$ with respect to the natural length of the spring). When we release it, there is an increase in the spring force to $k(Y+y)$. The unbalanced force, or net force, is now given by $F_{net} = -ky$. Applying Newton's Second Law we obtain

$$F_{net} = -ky = M\ddot{y}$$

The general solution to this equation is

$$y = a\sin(\pm\omega t + \phi)$$

where the phase constant $\phi = \sin^{-1}(y_0/a)$, and the angular frequency of the oscillation in radians per second is given by $\omega = (k/M)^{1/2}$. Here a is the

[11] The centre of mass and the centre of gravity are the same in a uniform gravitational field.

amplitude of the oscillation. For a starting displacement $y_0 = y\,(t = 0) = a$, and $v_0 = v\,(t = 0) = 0$, we get $\phi = \sin^{-1}(1) = \pi/2$, giving

$$y = a \sin(\omega t + \pi/2) = a \cos(\omega t)$$

where the angular frequency of oscillation is given by $\omega = (k/M)^{1/2}$. The period of the oscillation, T, which we were asked to find, is defined by $\omega T = 2\pi$. We write $T = 2\pi\,(M/k)^{1/2}$.

The only way this system differs from the simplest possible mass-spring oscillator we can imagine, namely one in the horizontal plane, is that the mass M acts both to lower the equilibrium position (that is, to increase Y) and to increase the period of the oscillation. A mass-spring oscillator in the horizontal plane with the same mass and spring constant has the same period of motion, but no pre-extension of the spring.

6.4 Dr Springlove's Infernal Oscillator ★★

Unsatisfied with the Springlove Oscillator, Dr Springlove develops what he calls his Infernal Oscillator. The victim is strapped to a table on frictionless wheels. The whole assembly (including the victim) has mass M. The table is arranged between two identical springs, each of stiffness k. The assembly is arranged in the horizontal plane. Draw the forces on the mass M, and calculate the period of the motion if the mass is displaced from the equilibrium position then released from rest.

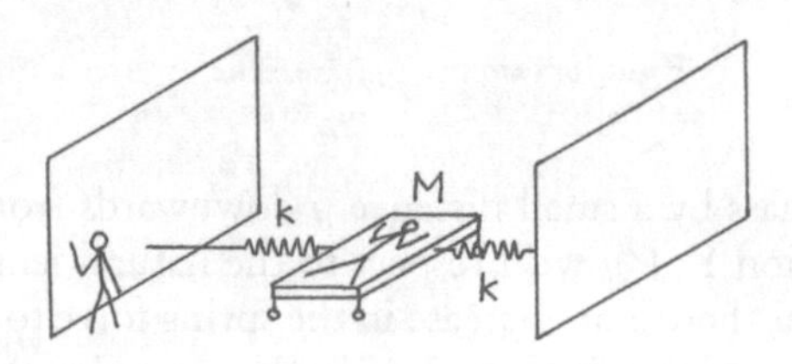

Answer

This is one of the simplest variants of a standard SHM question that one could dream up. In discussing it with students I discovered, to my surprise, that many found it more difficult than they thought it would be. Some made mistakes with the signs of forces, and others had a poor intuitive grasp of what might happen. For instance, some students felt that the forces would cancel in all displaced positions and that there would be no oscillation. They were bright individuals, so I can only assume that they were not used to non-standard problems. All it took was a small twist and everything they knew about SHM went out the window! So it's important to consider a few variants on standard SHM, in the hope that we can apply the principles we know in unfamiliar territory.

Let's address the problem.

The mass M is connected to two springs of stiffness k. We are not told whether the springs are extended, compressed or in their relaxed position. Provided the behaviour is linear, however, it doesn't matter. We proved this in the previous question, albeit for a single vertically oriented spring. But we are asked to draw the forces, so we should consider the possibility of both initial tension and initial compression. In the equilibrium position, the forces for the two cases look like this.

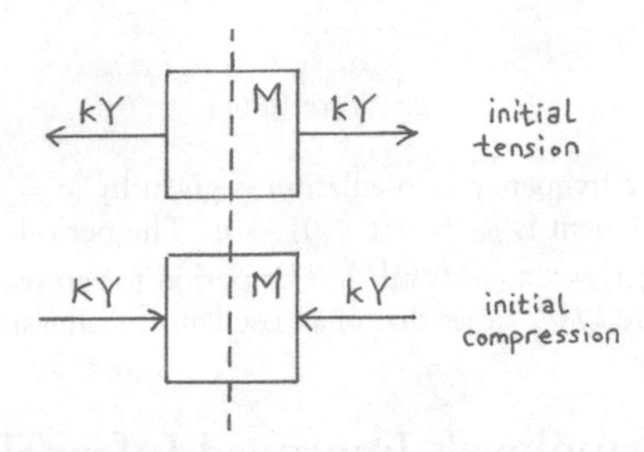

The diagram is drawn for an extension or compression from the natural length of the springs, Y, which is equal on either side of the mass. The net force in both cases is zero at the equilibrium position.

Now consider a small additional displacement y. The forces are changed to $k(Y + y)$ and $k(Y - y)$. That is, the force on one side is increased and the force on the other side is reduced. For a displacement to the right, if the springs are in initial tension, the larger of the two forces is on the left. If the springs are in initial compression, the larger of the two forces is on the right. The forces for the two cases are shown in the diagram. We are concerned with the net force, which, in both cases, opposes the displacement and is equal to $-2ky$.

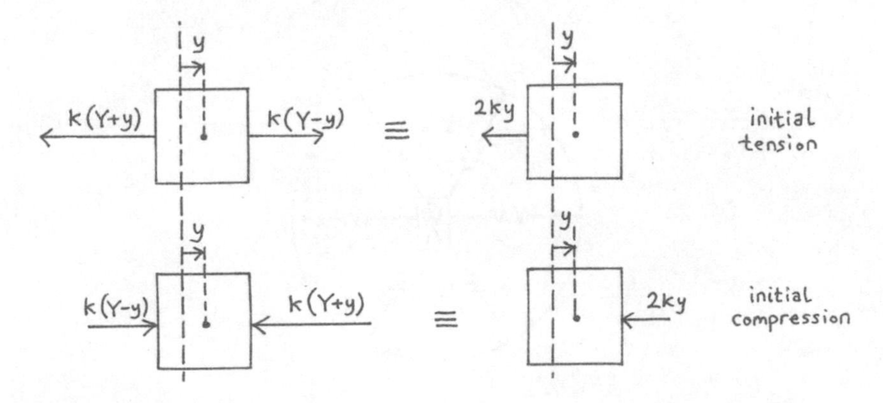

You may consider this rather trivial, but I make no apology for including it. Consider again that some students I have tried this question on were unable to draw a diagram of this type and explain the origin of the forces. Indeed, quite a number were confused about the possibility of an equilibrium position exactly between the two springs, and others argued that in all displaced positions the forces due to the springs would cancel. To get all the forces correct clearly requires a little thought at least.

Return to the fact that the force arising from a displacement y is $-2ky$. The answer now proceeds as for the simpler variant of this question. We apply Newton's Second Law and obtain $F_{net} = -2ky = M\ddot{y}$. This is satisfied by

$$y = a\cos(\omega t)$$

where the angular frequency of oscillation is given by $\omega = (2k/M)^{1/2}$, and the initial displacement is $y_0 = y(t=0) = a$. The period, T, is defined by $\omega T = 2\pi$, giving $T = 2\pi(M/2k)^{1/2}$. The period for an oscillator with two springs is therefore $1/\sqrt{2}$ times that of an oscillator with a single spring.

6.5 Dr Springlove's Improved Infernal Oscillator ★★★

UNSATISFIED EVEN WITH his Infernal Oscillator, Dr Springlove develops what he calls his Improved Infernal Oscillator. This system has six identical springs, each of stiffness k, attached symmetrically around the victim, who is strapped to a table on frictionless wheels. The whole assembly (including the victim) has mass M. The assembly is arranged in the horizontal plane. When the system is in the equilibrium position, the springs have length L. Draw the forces on the table, and calculate the period of the motion if the mass is displaced by a very small distance ($\Delta x \ll L$) from the equilibrium position along the line of action of one of the springs, and then released from rest.

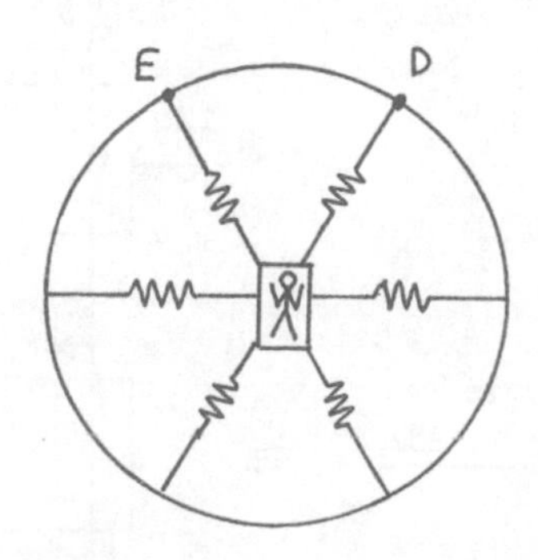

Answer

I have not yet tried this question on many students. Systems of two springs seem to cause a good deal of confusion, so it would probably be optimistic to think that most students could solve a problem like this without considerable help. It is not very different from the other examples of SHM we have considered, however, so I hope in the context of this collection of problems it won't prove too challenging. The following answer, although it's correct, is somewhat sloppy mathematically, in that we take geometric approximations at an early stage. A more formal approach (for ★★★★, at least) would be to derive the full equation of motion, expand it using series approximations, then take a first-order approximation valid for $\Delta x \ll L$. You should try this more formal approach if you fancy a challenge!

From the symmetry of the problem the initial tension or compression of all six springs is the same in the equilibrium position, where the acceleration and therefore the net force must both be zero.

Our first job, then, is to determine the extension of all the springs for a displacement along the line of action of one spring. Consider, arbitrarily, that the mass is displaced by Δx to the right when viewed in the orientation shown in the initial diagram. From a previous question, we already know how to deal with two springs aligned with the direction of displacement Δx. Now we need to consider the other four springs.

Consider the two springs at the top of the diagram, connected between points DA and EA, where the equilibrium position of the mass is represented by point A. The displaced position of the mass is B, where $|AB| = \Delta x$. The triangle ADE is equilateral, with internal angles equal to $\pi/3$. When the mass is in the equilibrium position, A, the springs DA and EA have length L.[12] Construction line AC is perpendicular to line EA. For a very small displacement of the mass Δx, from A to B, the change in angle of the spring from $\angle AED$ to $\angle BED$ is small;[13] that is, $\angle AED \approx \angle BED$. Thus $\angle EA'C \approx \angle EAC = \pi/2$, and $\angle EBF \approx \angle EAF = \pi/3$. From the geometry of $AA'B$, which approximates a right-angled triangle (with $\angle AA'B \approx \pi/2$, and $\angle A'BA \approx \pi/3$), we can show that $|A'B|$, which we defined to be ΔL (the change in length of the spring), is given by $|A'B| \equiv \Delta L \approx \Delta x \cos(\pi/3) = (1/2)\,\Delta x$. For very small displacements Δx, the new length of the spring is given by

$$|EB| \approx L + \Delta L \approx L + \frac{1}{2}\Delta x$$

[12] This length L is, in general, different from the natural length of the springs. We showed in an earlier question that we do not need to consider this in the analysis.

[13] In the diagram the changes in angle are grossly exaggerated.

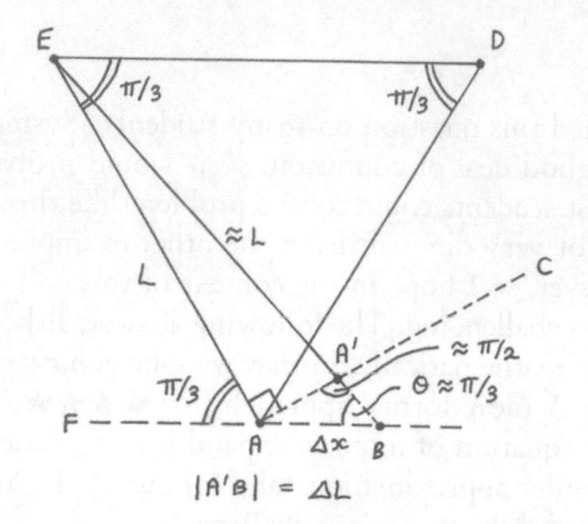

We can show, using almost identical analysis, that when we move the mass from A to B, spring DA is compressed from length L to a new length $|DB|$, given by

$$|DB| \approx L - \Delta L \approx L - \frac{1}{2}\Delta x$$

So for a displacement Δx from A to B, we compress/extend *two* of the springs by Δx, and the remaining four springs by $\Delta x/2$. We can now draw a force diagram for the displaced mass. We will draw the diagram for the case in which the springs are unextended in the equilibrium position, for springs which are linear for both positive and negative displacement (that is, extension or compression). The force diagram is as follows.

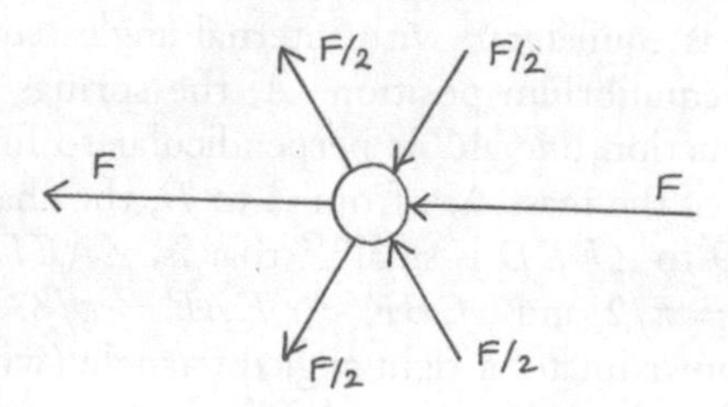

where $F = -k\Delta x$. The forces are symmetrically disposed around the mass. The net force in the horizontal direction can be calculated using basic trigonometry.

$$F_{net} = 3F = -3k\Delta x$$

Letting x now represent a very small, variable displacement, we have

$$F_{net} = -3kx$$

Applying Newton's Second Law we obtain $F_{net} = -3kx = M\ddot{x}$. This is satisfied by

$$x = a\cos(\omega t)$$

where the angular frequency of oscillation is given by $\omega = (3k/M)^{1/2}$, and the initial displacement is $x_0 = x(t=0) = a$. The period, T, is defined by $\omega T = 2\pi$, giving $T = 2\pi(M/3k)^{1/2}$. The period for an oscillator with six springs in this configuration is $1/\sqrt{3}$ times that of an oscillator with a single spring.

Chapter 7

Mad inventions and perpetual motion

Arguably, one of the most important functions of a higher education is to equip a person with the ability to tell when someone is speaking sense or nonsense. An ability to cut through complex and perhaps deliberately obfuscating language or arguments, and to see whether an idea is something, or nothing. To be able to do this in a field which is not our own is a sign that we have learnt how to *think*. But when we wish to use sound physical arguments, we may need to rely a little on our scientific training. The best way to understand physical systems is usually to reduce them to their essentials. Conversely, in the minds of mad inventors there is often surplus complexity. Surplus complexity is a hallmark of mad inventors. It helps fool them into thinking they have something of value, when perhaps they have nothing of any worth at all. Here I am talking in particular about the *perpetual motion machine*.

The perpetual motion machine is a (hypothetical) device that can remain in motion indefinitely without a reduction in its speed. It is, of course, impossible to create such a device. The second law of thermodynamics tells us that the quality of energy in any system slowly degrades over time (the *entropy*, a measure of the reduction in the *quality* of the energy in a system, rises over time). Even the planets, which enthusiasts of perpetual motion have cited as examples of such systems in nature, are slowing down due to the dissipative effects of tidal forces and internal mechanical stresses. There are also the so-called *overunity* machines, devices which (in general) take in some energy, but which put out more energy than they consume. They generate energy indefinitely. These devices disobey the first law of thermodynamics[1] because they create energy from nothing. Energy in any closed system that contains them is not conserved.

There have been countless inventors of perpetual motion machines over

[1]Conservation of energy.

the years. In the drawers full of perpetual motion patents, *epicyclic gears* and complex mechanisms abound. Any perpetual motion machine worth its salt contains enough ball bearings, springs and gears to cause even the most rational mind a bit of confusion. So numerous are the would-be inventors of such machines that the US patent office requires any inventor claiming perpetual motion to build a working model to demonstrate *the motion*. It is a unique requirement in the office's regulations. This makes the life of the US patent examiner considerably easier—no such machine has ever been successfully demonstrated.

In this chapter we critically examine inventions, including some perpetual motion machines. We look at them from first principles to see if they make sense or not. And if not, we'll see if we can find their flaws and crystallize them in simple language. We will try to avoid dismissing the inventions by invoking the laws of thermodynamics. Instead we will see if we can find some simpler principle of physics that each machine violates.

7.1 Stevin's clootcrans ★

Simon Stevin (1548–1620) was a Flemish mathematician and engineer who made significant contributions to trigonometry, mechanics, architecture, musical theory, fortification, and navigation. This was clearly the age of the polymath! In addition to extensions of the work of Euclid and Archimedes in the field of geometry, Stevin is credited with introducing the decimal system to Europe (it had been in use for some time in Chinese, Arab and Indian cultures); devising the general method of solution for quadratic equations; unifying the concept of real numbers (up until that time square roots and irrational numbers, as well as others classes, were treated as distinct); developing both mechanics and statics (most significantly with the method for resolving forces); and—according to some—essentially defining a new field of *hydrostatics*. He was also the first to note that acceleration is independent of mass—something he demonstrated by dropping two lead balls of different mass from a church tower (three years before Galileo performed his famous thought experiment[2]). One of his more curious innovations was a system for rapidly

[2]The story goes that Galileo Galilei (1564–1642), to test Aristotle's theory that greater weights fall faster over a particular distance than lesser weights, dropped two balls (one heavy, one light) from the top of the Leaning Tower of Pisa. According to this story, Galileo noted that they hit the ground at the same time, thus showing that acceleration due to gravity (under frictionless conditions) is independent of mass, and disproving Aristotle's theory. The story is apocryphal, but the real story is much more interesting. In Galileo's final book, his 1638 Discorsi e Dimostrazioni Matematiche, Intorno à Due Nuove Scienze (one translation of which is Galileo, G., 1914, "Dialogues concerning two new sciences," Translated from the Italian and Latin into English by Henry Crew and Alfonso de Salvio, Macmillian, 1914), he describes his scientific understandings by way of conversations between three men (Simplicio, Sagredo and Salviati). Simplicio starts by outlining Aristotle's theory.

> SIMPLICIO: [Aristotle] ...supposes bodies of different weight to move in one and the same medium with different speeds which stand to one another in the same

flooding the lowlands of Holland to slow down an invading army.

His most significant development in mechanics was probably the concept of the chain of balls, or *clootcrans*, arranged on two inclined planes. This was devised as a thought experiment to analyse forces on systems to determine whether motion was possible. The same method was later adapted to analyse machines, and is an important landmark in the development of mechanics. Stevin's starting point was to assert that perpetual motion of the system was impossible, and then to look at the consequences of this assumption. His starting assertion is thought to have emerged from what has become known as Stevin's Principle, or the principle of *virtual work*. The principle is to look at the physical change that has occurred in the system after a finite movement. If we consider rotating the system by the distance between the centres of a pair of balls, we see that it is unchanged after the rotation. In a frictionless system, no work would be required to perform the rotation (whether taken in or given up).[3] We can think of this as a statement of equilibrium for a system.

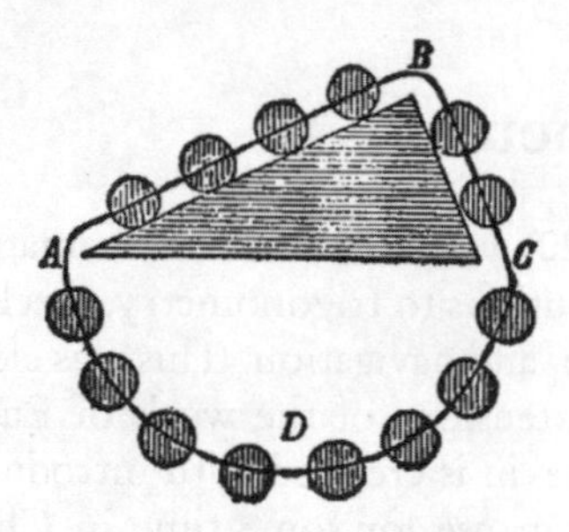

There is enough confusion over this simple principle that numerous perpetual motion machines have been proposed based on energy transfer in systems with

ratio as the weights; so that, for example, a body which is ten times as heavy as another will move ten times as rapidly as the other...

SALVIATI: If then we take two bodies whose natural speeds are different, it is clear that on uniting the two, the more rapid one will be partly retarded by the slower, and the slower will be somewhat hastened by the swifter. Do you not agree with me in this opinion?

SIMPLICIO: You are unquestionably right.

SALVIATI: But if this is true, and if a large stone moves with a speed of, say, eight while a smaller moves with a speed of four, then when they are united, the system will move with a speed less than eight; but the two stones when tied together make a stone larger than that which before moved with a speed of eight. Hence the heavier body moves with less speed than the lighter; an effect which is contrary to your supposition...

And so the contradiction inherent in Aristotle's theory is cleverly exposed. Later in the conversation the resolution follows.

SALVIATI: ...We infer therefore that large and small bodies move with the same speed provided they are of the same specific gravity.

[3]It's worth noting that if such a hypothetical frictionless system were set in continuous motion, if we observed it at very small intervals of time, its speed would vary above and below an average speed (at the frequency of the passing balls). This is due to cyclic exchange of kinetic and potential energy.

rolling weights and inclined planes. Indeed, Percy Verance[4] devotes an entire chapter of his 1916 review of perpetual motion to the topic. He comments that "the inclined plane with rolling weights has been a fertile field of folly among Perpetual Motion seekers." Let's look at the principles involved, and try to prove from first principles that in a system such as Stevin's clootcrans we are always at an equilibrium position with no potential to do work.

CONSIDER A FRICTIONLESS double ramp with angles ϕ_1 and ϕ_2, over which is placed a perfectly flexible (zero-stiffness) frictionless rope with uniform mass per unit length. Prove that the system is in equilibrium. Assume that the rope below the inclined planes hangs in a symmetric curve.[5]

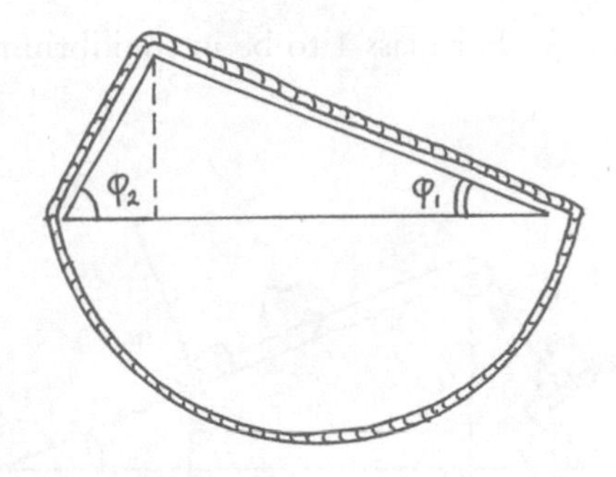

Answer

We are told that the rope hanging as a *catenary* (below the planes) does so symmetrically. We can assume therefore that if the system is in equilibrium, the forces due to the catenary act equally in magnitude on the ropes at the base vertices of the inclined plane, and that their effect can be disregarded when looking for non-equilibrium forces. Here we use another of Stevin's methods, and make imaginary cuts in the rope where the catenary meets the rope on the inclined planes, replacing the hanging rope with two forces of equal magnitude which we then disregard.

We now consider the forces acting on the rope that is in contact with the inclinded planes. Let us refer to the slope with angle ϕ_1 to the horizontal as

[4]Verance, P., 1916, "Perpetual motion: comprising a history of the efforts to attain self-motive mechanism...collection and explanation of the devices whereby it has been sought and why they failed," 20th Century Enlightenment Specialty Co.

[5]It is a very interesting problem to find the shape of this curve, which is referred to as a *catenary*. I have heard that this question has been used in university entrance tests in the past, although I think it's rather challenging. The catenary is the curve that an idealised chain forms hanging under its own weight when supported only at its ends. It is superficially reminiscent of a parabola, but has a different mathematical form. It can be seen in light suspension bridges, which are flexible and have uniform weight per unit length. It also defines the shape of an anchor chain, and, for example, the wire hanging between telegraph poles. In Cartesian coordinates a catenary has the hyperbolic cosine form $y = a\cosh(x/a) = (a/2)\left(e^{x/a} + e^{-x/a}\right)$. The variable a scales the curve for different ratios of length of chain to width of gap (the *span*).

slope 1. The other is slope 2. Let the rope on slope 1 have mass m_1, and the rope on slope 2 have mass m_2. To visualise the mechanics of the system more clearly, we replace the rope lengths with two point masses (which are frictionless with the planes) connected by a light inextensible string over a light frictionless pulley. We then consider the forces on the two masses, taking mass 1 first as an example. The forces are as follows.

- A force $m_1 g$ due to the weight of the mass, which acts vertically downward.
- A normal reaction force, $R_1 = m_1 g \cos \phi_1$, which opposes the component of weight acting normal to the plane.
- A tension force T_1. For mass 1 to be in equilibrium, we require $T_1 = m_1 g \sin \phi_1$.

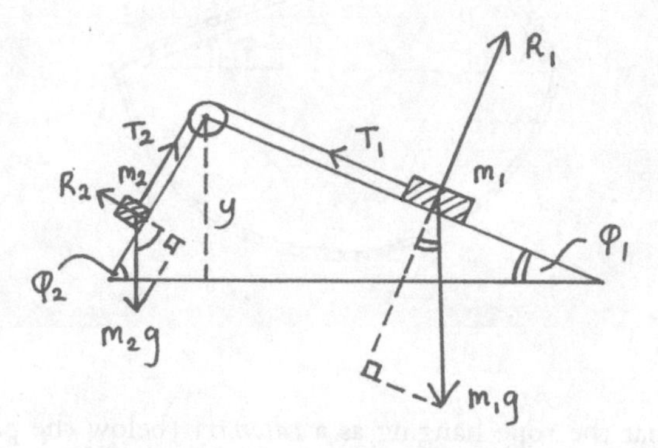

Mass 2 has an identical system of forces acting on it. By replacing subscript 1 with 2 in the preceding lines we have the correct equations. For mass 2 to be in equilibrium we require $T_2 = m_2 g \sin \phi_2$. A light, inextensible rope implies the condition $T_1 = T_2$. For both masses to be in equilibrium with a light, inextensible rope, we therefore require $m_1 g \sin \phi_1 = m_2 g \sin \phi_2$. Rearranging, we get

$$\frac{m_1}{m_2} = \frac{\sin \phi_2}{\sin \phi_1}$$

From geometry, the lengths of rope 1 and 2 are $L_1 = y / \sin \phi_1$ and $L_2 = y / \sin \phi_2$, where y is the height of the inclined planes. The ratio of lengths is therefore given by

$$\frac{L_1}{L_2} = \frac{\sin \phi_2}{\sin \phi_1}$$

If the rope is of uniform mass per unit length, we have $m \propto L$, and the geometric condition and the condition for equilibrium are the same. That is, for any pair of inclined planes, a frictionless rope arranged as described will be in equilibrium.

Verance writes about a few particular contrivances designed to achieve perpetual motion via systems of weights and planes.

> ...persons who had had schooling advantages and were supposed to be versed in the rudiments of mechanics...ought at first sight to have perceived the fallacy and hopelessness of the inventor's dreams. All of these claimed inventions relying on the inclined plane with rolling weights were so nearly alike in the principle involved...

A more elegant answer

When I discussed this with a friend, he proposed a more elegant answer which avoids the need to make assumptions about the shape and behaviour of the catenary. His answer is as follows.

The forces in the system are independent of the relative speeds of its parts. So, if the system is not in equilibrium, it would have non-zero rotational acceleration. It would therefore accelerate clockwise or anticlockwise without limit, due to the symmetry of the problem after rotation. This is absurd. The system must, therefore, be in equilibrium.

7.2 Power-producing speed humps ★★

AN INVENTOR HAS the brainwave that speed humps could be used to generate power. Venture capitalists are thinking about investing in the idea, and bring you into a meeting as a scientific consultant. Your job is to independently assess the idea, and to present arguments for or against the proposed technology. What do you tell them?

Answer

First we need to appreciate that the hump can only extract energy from the moving car, not create it from thin air. The energy the hump gains must be less than or equal to the energy lost by the car. So what do we say to the venture capitalists? Any number of things might spring to mind, but among them we might say:

- Unless the car would in any case be slowing to the speed required to pass over the hump, this isn't free energy that would otherwise be wasted. It is a way of transferring a small amount of energy from the car to a device in the road.

- In most places where there are roads there is also power, especially in cities or large population centres. The place such a device *might* be useful is in very rural locations. We could imagine it being used to power a fridge or an illuminated road sign, for example. Although in rural locations there would not be many cars.

- It feels like the infrastructure might be quite expensive in comparison to the amount of energy transferred. If a small amount of power is needed in a remote location, why not use solar cells, say, or a small generator, or batteries?

To argue the case much further we really need to do some calculations, to estimate the amount of power we could transfer from the car to the speed hump. The mechanism is simple. If the hump is depressed with a force applied by the car, energy is transferred during the displacement. The work done, δW, is equal to the force on the hump, F_h, times the displacement, δy. We write

$$\delta W = F_h \delta y$$

So how much energy could be transferred? To answer this question we need to estimate the force F_h that a car might impose on a hump, and the possible displacement δy. We estimate the mass of a small car to be approximately $m = 1,000$ kg. Taking $g \approx 10$ m s^{-2}, the weight of the car is $mg \approx 1 \times 10^4$ N. The normal reaction force that opposes the weight of the car is split between the front wheels and the back wheels. If the weight is evenly distributed, the road pushes up on each set of wheels (each axle) with a force $F = \frac{1}{2}mg \approx 5 \times 10^3$ N. Let's assume that we design the hump so that it can be depressed with half the reaction force at each set of wheels, or $F_h = 2.5 \times 10^3$ N. The average car wheel is about 0.5 m in diameter. It would struggle to depress a hump much more than 0.2 m high, so let's take $\delta y = 0.2$ m. The work done on the hump by the car as the first set of wheels passes over it is

$$\delta W = F_h \delta y = 2.5 \times 10^3 \times 0.2 = 500 \text{ J}$$

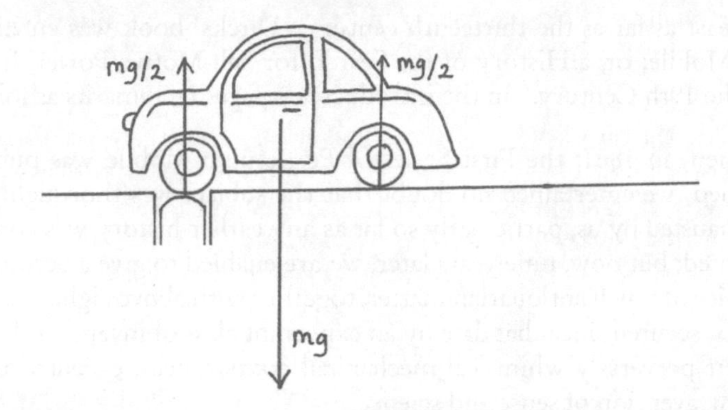

We need to remember that there are two sets of wheels, so the total work is about $\Delta W = 2 \times \delta W = 1,000$ J. How much energy is $1,000$ J? Let's convert it into something that we have more of a physical feel for. An average fridge uses about $P = 100$ W of power. What would be the maximum time interval between passing cars, δt, to keep the fridge powered up? We use the power equation

$$P = \frac{\Delta W}{\delta t} \text{ or } \delta t = \frac{\Delta W}{P} = \frac{1000}{100} = 10 \text{ s}$$

One car every ten seconds! That's six cars per minute, or 360 cars per hour, or a massive 8,640 cars per day. Any village that attracts 8,640 cars per day should be able to find an alternative way of powering a single fridge.

To put this in financial context, the current cost of 1 kWh (3.6×10^6 J) of electricity in the UK is roughly 10 pence. The fridge uses 2.4 kWh of electricity in a day, so costs about 24 pence a day to run. The revenue each car generates is about 0.003 pence, before we take into account the costs of implementing and maintaining the scheme. We can safely tell the venture capitalists that this scheme is an utterly mad idea which should be scrapped.[6]

7.3 The overbalanced wheel ★★

As early as 1870, Henry Dircks wrote a fascinating historical review of the obsessive search for perpetual motion. This seductive but vain pursuit stretches

[6]There are, however, several companies currently manufacturing power-producing speed humps. They make all sorts of wonderful, but generally rather vague, claims about the environmental benefits, using phrases like "green energy harvesting". Around the year 2009, a number of newspapers ran stories about some pilot schemes. One manufacturer claimed a power output of 36 kW from a single hump. Compounding my horror, I then learned of a £150,000 scheme in London to test the humps, as well as a number of high-profile installations at hotels and supermarkets in both the US and the UK. I think we should take the ratio of the power quoted in the advertisements to the power we have calculated for a fairly steady traffic flow: 36 kW/100 W = 360. Oh dear! Perhaps we have made a mistake? Or perhaps not. I have seen only one application of power from a speed hump. It was used to operate a light that confirms you've just been over a speed hump. Clearly we need more science and less marketing.

back at least as far as the thirteenth century. Dircks' book was entitled Perpetuum Mobile; or, a History of the Search for Self-Motive Power, from the 13th to the 19th Century.[7] In the introductory pages he remarks as follows.

> When, in 1861, the First Series of Perpetuum Mobile was published, we entertained no doubt that the subject was thoroughly exhausted by us, particularly so far as any earlier history was concerned; but now, nine years later, we are enabled to give a Second Series of much antiquarian matter, together with above eighty patents, secured since that date by an expectant class of inventors, for their perversely whimsical mechanical contrivances, executed in contravention of sense and science.
>
> The archæology of engineering has no similar phase of enduring and pertinacious pursuit, despite conflicting evidence and multitudinous instances of degrading failure...no amount of disappointment is sufficient to weaken or entirely destroy the indomitable desire to rediscover an alleged lost discovery.

Over 358 pages, Dircks catalogues the chronology of some of the most bizarre and inventive supposed perpetual motion machines imaginable, devices that rely on all manner of basic scientific misconceptions. An updated review by Percy Verance[8] in 1916 expands the range of supposed motions to include all the major types available to the modern perpetualmotionist: devices using wheels and weights; devices using rolling weights and inclined planes; hydraulic and hydromechanical devices; pneumatic siphon and hydropneumatic devices; magnetic devices; devices using capillary attraction and physical affinity; liquid air devices; radium and radioactive-substance devices; devices relying on a misconception of the relation between momentum and energy.

The allure today is as strong as it was in the seventeenth century, and countless patents are filed every year for machines that will never work. There are even perpetual motion societies. I have encountered two perpetual motion obsessives myself. Both of them, as the saying goes, were as mad as hatters. One was introduced to me on the basis that I might do some consultancy for him, to help move his project forward. He was a man of considerable means, and the proposal for my charge-out rate was attractive. I listened patiently for an hour and said almost nothing. Afterwards I had no choice but to break it to him, as gently as possible, that I thought his scheme was deluded, that I couldn't possibly consult for him, and that I would distrust the motives of anyone who took money from him before they could state clearly the mechanism of action of his device, in a fundamental physical way. He was of the fervent

[7]Dircks, H., 1870, "Perpetuum mobile; or, a history of the search for self-motive power, from the 13th to the 19th century (illustrated from various authentic sources, in papers, essays, letters, paragraphs, and numerous patent specifications)," London, E. & F. N. Spon.

[8]Verance, P., 1916, "Perpetual motion: comprising a history of the efforts to attain self-motive mechanism...collection and explanation of the devices whereby it has been sought and why they failed," 20th Century Enlightenment Specialty Co.

type, however, and faith outweighed reason. He went on to find another consultant. The other perpetual motion obsessive was a good friend who I used to sail and climb with. He was both a trained engineer and a medical doctor, and extremely well respected in his professional sphere. But he had a weakness for epicyclic gears that at times became a sort of obsession. It was upsetting to see. I suppose we all have our demons.

We can learn a lot by considering the supposed method of action of perpetual motion devices. Invariably the action relies on some simple misunderstanding of physics. Let us, then, look at the oldest, most classical, example of a perpetual motion machine. It is known as the *overbalanced wheel*, and its first invention is attributed to Villard de Honnecourt, an architect who worked in Paris in the thirteenth century. It has been reinvented endlessly and variously over the past eight hundred years. The first illustration below shows Honnecourt's original plan.[9] The second shows a device from England in 1831 published in The Mechanics' Magazine (and reproduced from Verance's book). The third is an updated drawing (again reproduced from Verance's book) of a sketch originally included in one of Leonardo da Vinci's sketchbooks.

In all three perpetual motion schemes shown below, the supposed method of operation is as follows. As the wheel (or drum) rotates, weighted levers (or weighted balls) are arranged so that they are further from the axis on one side and closer to the axis on the other. Thus, due to the greater torque about the axis in one direction, the wheel rotates continuously. Without giving detailed calculations, discuss the reasoning presented in favour of the machines acheiving perpetual motion, and the expected actual motion.

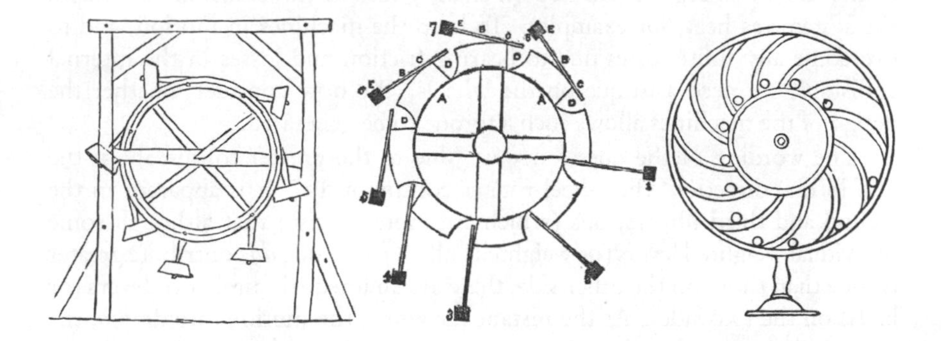

Answer

Let's first consider the motion of a regular wheel rotating on an axle. If we set it in motion, we expect it to rotate for a period during which it gradually slows down and eventually stops. Only an entirely *lossless* and frictionless device (in

[9] The drawing is reproduced from Dircks's book. Curiously, the drawing lacks perspective, and is included primarily for historical interest.

which no energy can leave the system through bearing drag or drag due to air resistance) would rotate forever. This is, of course, impossible to achieve in practice.

Let's also consider a frictionless wheel with a small unbalanced mass. When we set it in motion, the rotational speed oscillates (in time or angle) around an average value. The speed is greatest when the mass is at the bottom, and smallest when the mass it at the top. Potential energy is being converted to kinetic energy and vice versa. If we add the effect of friction, we would not expect the motion to have *monotonically* decreasing rotational speed,[10] but we would expect the wheel to continually lose energy to friction, and eventually to stop.

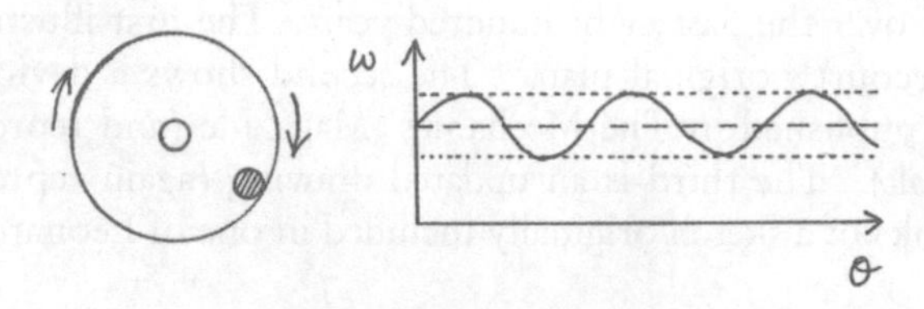

Each of the three perpetual motion machines shown above has a parallel to the wheel with unbalanced mass. They are more complex, however, because each can change its moment of interia due to the action of a weighted lever opening or closing, or a weighted ball rolling into a new position. As a lever or ball is set in motion *with respect to the wheel,* it gains some additional kinetic energy—this suggests that another part of the system is losing energy. If a lever or ball is brought suddenly to rest (by reaching a stop, for example), some of the kinetic energy it had is permenantly lost in the collision—it escapes the system (as heat, for example). To keep the machines in motion, and to overcome any finite losses due to bearing friction and losses in the internal collisions, we need a torque on the wheels. We now consider whether the design of the machines allows such a torque to be generated.

The wording of the question was, "due to the greater torque about the axis in one direction, the wheel rotates continuously." It is apparent in the second and third illustrations (which are more precise) that although some individual weighted levers (or weighted balls) on one side *do* contribute greater torque than those on the other side, there are an unequal number of levers (or balls) on the two sides. At the instant for which the motions are drawn, the second illustration has three levers acting clockwise and five acting counter-clockwise, and the third illustration has five balls acting clockwise and seven acting counter-clockwise. This is at the heart of the fallacy of the argument for rotation of any overbalanced wheel.

[10]This is not a paradox. The *energy* of the system can decrease monotonically, even for non-monotonic speed. For an entirely lossless system, the energy of the system is constant, but the rotational speed is not (although the *mean* rotational speed averaged over a cycle is constant over many cycles).

A careful analysis of the geometry of any of these machines would show that in moving between identical positions (that is, the angular spacing between weighted levers or balls), for a frictionless system the mean torque *in the direction of rotation* is at best zero.[11] That is, the wheel neither accelerates nor decelerates over long periods of time. The instantaneous torque can and will be non-zero, leading to alternating acceleration and deceleration of the system (over the angular spacing of the levers or balls). In a real system with finite friction and other losses, this non-uniform motion is superimposed onto the gradual slowing of the device. It is perfectly possible using fairly elementary mathematics to model the behaviour of a perfect or imperfect device of the forms shown. We'd learn very little more about the underlying principles by doing this, however.

It is also intersting to examine the machines from the point of view of energy exchange. Because the machines are cyclic, any gravitational potential energy that is given up by a mass when it moves downwards must be put in again when it moves upwards. Over a full revolution, the net work that gravity performs on one of the weighted levers or balls is exactly zero. Let's consider this in more detail.

Taking the third machine, we plot the cyclical path $ABCDA$ of one of the balls about the centre of the machine O. Consider the descent ABC and the ascent CDA separately. On the descent, the work done on the ball by the force of gravity is equal to the weight of the ball multiplied by the vertical distance AC. This is true no matter what the path ABC, provided the end limits are the same. The work required (given up by the wheel) to raise the ball against gravity on the journey CDA is exactly equal to the work performed by gravity during the descent. In a system with no internal loss the journey $ABCDA$ would require no net energy. By the same token, the system cannot generate energy.

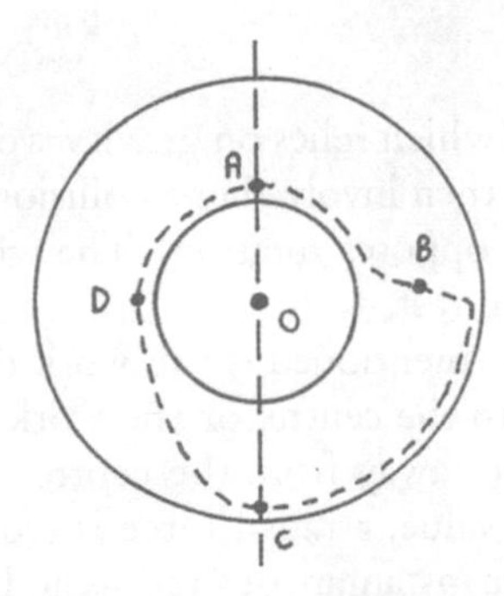

Let's consider whether it is possible for energy to be lost in internal collisions. If energy were lost in a collision, it would require the wheel to slow down, and

[11]Interestingly, even for a system with no bearing friction or aerodynamic drag, if the system is such that internal accelerations and collisions are allowed (which is the case for all three of the mechanisms shown in this question), it is possible to have a mean torque with a value *less than zero* in the direction of rotation. In other words, a torque which slows the machine down.

this would require an external moment. We will now see how this is generated. We know that gravity always acts on a ball of mass m with a force mg. For a very slowly turning wheel, during most of the motion we can assume that the ball is not undergoing significant acceleration.[12] The force on the wheel due to the ball is therefore the weight force mg. If we consider the path CDA, we see that the wheel has to work against this force during the whole of the ball's ascent. The work done is the integral of the force multiplied by the distance moved in the direction of the force, which is the vertical distance between C and A. Consider now the descent, path ABC. During most of the descent the ball is not accelerating significantly, and the force on the wheel is the weight mg. As the ball suddenly accelerates from B' to B'', however, the force on the wheel is reduced. Less work is done on the wheel during the ball's vertical descent between B' and B'' than the work the wheel has to do to raise the ball through the corresponding vertical distance on the other side. Yet, as we have established, the amounts of work done by and against gravity *on the ball* are equal in magnitude. The *missing energy* has turned into kinetic energy of the ball. When the ball collides at B'' most of this energy is lost in the inelastic collision required to put the ball back on the path $B''C$. In this process energy is lost, and there is a net torque on the wheel which causes it to slow.

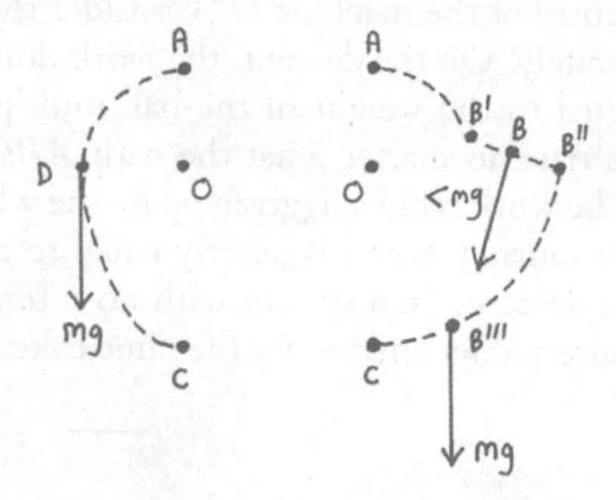

Importantly, this process (which relies on gravity working to increase the kinetic energy of a ball that is then involved in a collision) can only lead to energy reduction and torque that opposes rotation. The wheel rotates in a decaying oscillation until it comes to rest.

One thing we have not mentioned is the work done by the wheel on the balls as they move closer to the centre, or the work done by the balls on the wheel as they move further away from the centre. If we take the rotational speed to have some finite value, a radial force is required to sustain the balls in circular motion (at their instantaneous radius and velocity). As each of the balls goes through a full cycle, we see that the net work required around the cycle is zero.

Finally, we should very briefly consider an entirely different way of viewing the energy conservation of the system, using Stevin's Principle. It is to

[12] To simplify the argument, we ignore the force required for acceleration towards the centre of the circle. This is the same as assuming that the wheel rotates very slowly.

note that at n intervals during the motion, where n is the number of weighted levers (or balls), the system returns to the same geometric configuration. If we were to freeze the motions at those n instants, the configurations would be indistinguishable from the starting condition. Ignoring the possibility of energy loss due to internal collisions, it is an obvious requirement of energy conservation that the system should have the same rotational speed at these instants.

We seem to have a number of simple, rational refutations of the means by which these devices are supposed to achieve perpetual motion. Yet right now, all over the world, otherwise apparently intelligent individuals are spending huge amounts of effort on the pointless quest to build machines very much like these and achieve the fantastic dream of perpetual motion. And they will be doing the same, no doubt, in a century's time.

Nobody, perhaps, is more cutting than Dircks in his introductory essay.

> While the pursuit of Perpetual Motion is not of itself evidence of insanity, it is unquestionably a proof of ignorance, or of mental inability to master elementary knowledge...The shame is...in the persistence in attempting performances without sufficient rules, against scientific reasoning, and in defiance of laws of which he either has never heard, or which he has never sufficiently studied.

The would-be perpetualmotionist would do well to both consult the historical reviews of this topic, and get a basic grounding in science.

7.4 Professor Sinclair's syphon ★★

Professor George Sinclair was a Scottish mathematician, physicist (after a fashion), water engineer, and demonologist.[13] A good deal of his work seems to have been rather questionable. Nevertheless, even for the time he was a man of some prominence. He is credited not only with advancing demonology, but also with making one of the first descents in a diving bell, to explore a wreck from the Spanish Armada that was supposed to have foundered off the Isle of Mull after being defeated by the English fleet. It's believed he also came up with the first example of a perpetual motion machine based on the principle of the syphon. This latter accolade is particularly worrying, because in 1673 he was put in charge of a major scheme to pipe water to Edinburgh.

The proposed operation of Professor Sinclair's perpetual motion syphon was included in a 1669 Latin work[14] (which includes the following diagram) written by the professor himself. It goes as follows.

[13]In fact, according to the Dictionary of National Biography (1885–1900), his best known work was on demonology: Satan's Invisible World Discovered, published in 1685. He wrote this to "prove the existence of devils, spirits, witches, and apparitions."

[14]Sinclair, G., 1669, "Ars nova et magna gravitatis et levitatis," pp. 418–472.

THE GLASS BULB shown below is partially evacuated, so draws water up a tube from a dish. A syphon introduced through the wall of the bulb returns the water at the same rate to the dish, maintaining low pressure in the bulb, and completing the cycle. Perpetual circulation of water is therefore achieved. A small amount of excess energy is available as the water exits the syphon, and this could be harnessed usefully. What is wrong with this scheme?

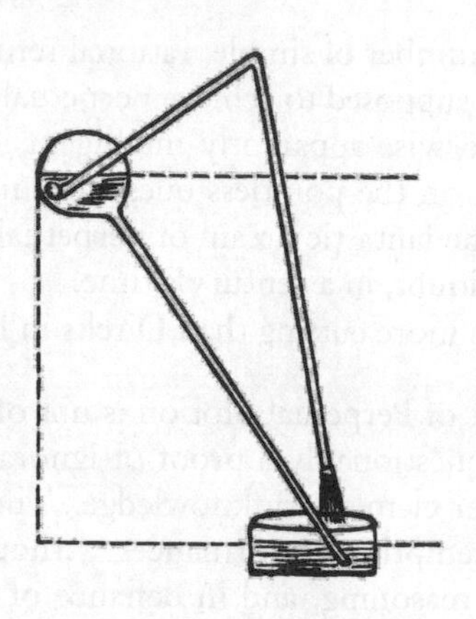

Answer

This seems to me a particularly interesting example of a proposed perpetual motion machine. It is interesting because of its extreme simplicity. I would have imagined that anyone with even practical familiarity with the syphon would quickly see the flaw in its proposed operation. Similarly, where the inventors of most perpetual motion machines can make comforting excuses about the prohibitive cost of actually producing their devices (because of the need for low-friction bearings and so on), no such excuses are available here—it would be easy to make this device and either demonstrate its operation or understand its failing. And there's no doubt that Professor Sinclair had the contacts and the means to accomplish this easily. Yet, as Henry Dircks notes in his 1870 Perpetuum Mobile, Professor Sinclair devotes "eighteen pages to discuss[ing] the merits of the impossible scheme...representing a syphon delivering water or mercury at the short instead of the long leg thus absurdly enough endeavouring to reverse its natural principle of action." It really is most peculiar, and demonstrates the curiously strong lure perpetual motion holds for some obsessives.

It's probably worth noting that in those days research into fluid dynamics was in its early stages. It certainly pre-dated Daniel Bernoulli's[15] famous 1738 work Hydrodynamica, in which the formula we now know simply as

[15]Daniel Bernoulli (1700–1782) was a Swiss mathematician and physicist particularly famous for his work in fluid mechanics. He is perhaps best known for the *Bernoulli Principle*, which describes the conservation of energy in fluids.

Bernoulli's equation[16] was first derived. We must not give Professor Sinclair too much leeway, however, because the syphon was used by the Egyptians three thousand years[17] before Sinclair wrote his work, and has been studied by scientists in all ages thereafter, appearing in many well-known books. In short, the principle of operation and its limitations were far from unknown in 1669.

Let's examine Professor Sinclair's scheme. First we consider the scheme in terms of pressure forces acting on the liquid, and show that it cannot work. Finally we describe the problem in terms of the energy of the liquid as a way to understand how syphons work.

Consider the fact that both the ends of the inverted V-tube and the straight tube are below the levels of liquid in the globe and the dish respectively. The distance the ends of the tubes are below the liquid surfaces cannot affect the pressures at the other ends of the tubes. The *hydrostatic pressure difference* due to liquid in the arms of a tube is related to the difference in height between the inlet and the outlet. Consider each tube in turn.

- For the straight tube, the hydrostatic pressure difference due to the liquid is proportional to Δh_1, the difference in height between the *free surfaces.* The pressure difference is equal to $\rho g \Delta h_1$. Denoting the pressure in the globe as p_A, and the external pressure due to the atmosphere as p_{atm}, we see that to sustain the height of liquid Δh_1 requires $p_{atm} - p_A = \rho g \Delta h_1$. For the liquid to flow into the globe from the lower dish we require $p_{atm} - p_A > \rho g \Delta h_1$.

- For the inverted V-tube, by the same argument we see that the hydrostatic pressure difference between the globe and the lower end (which is at atmospheric pressure) is $\rho g \Delta h_2$. To sustain the height of liquid in the inverted V-tube requires $p_{atm} - p_A = \rho g \Delta h_2$. For liquid to flow from the globe to the free end requires a higher pressure p_A, giving $p_{atm} - p_A < \rho g \Delta h_2$. If the pressure p_A was lower than that required to sustain the column of liquid in the inverted V-tube, the liquid would flow the other way, draining the tube into the globe.

For clockwise flow through the system, we need to satisfy $p_{atm} - p_A > \rho g \Delta h_1$ and $p_{atm} - p_A < \rho g \Delta h_2$, or $\rho g \Delta h_1 < \rho g \Delta h_2$. This is impossible for the system as drawn, for which $\Delta h_2 < \Delta h_1$. So clockwise flow is impossible. If the system was set up as described, it would flow in an anticlockwise direction until all the liquid had drained into the lower dish.

---

[16]This famous *energy equation* describes the exchange of energy between different forms as a fluid moves along a *streamline*, or an imaginary line defining the path of a fluid element. As with discrete objects that exchange potential and kinetic energy, the same happens in systems described by *continuum mechanics* (where we treat a fluid as being infinitely divisible). The forms of energy the fluid can have are: gravitational potential; kinetic (dynamic head); and pressure, due to the action of an external agent such as the atmosphere, or the fluid's own weight (hydrostatic head), or both.

[17]According to evidence from Egyptian reliefs dating from about 1500 BC.

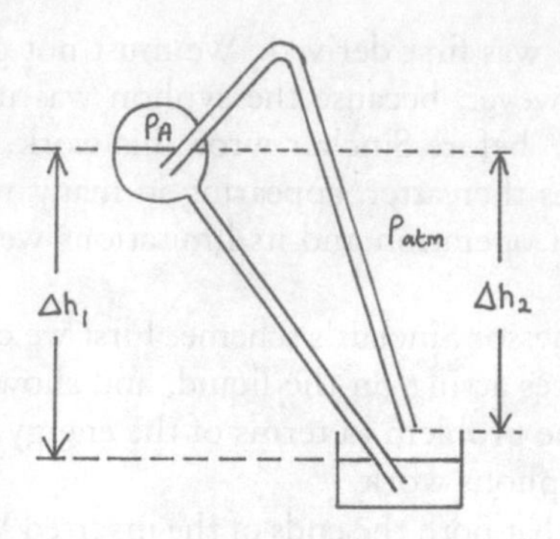

If you're finding it hard to visualise the process it may help to consider the systems A and B below. System A is the same in spirit as Professor Sinclair's scheme. It is a single reservoir with a U-bend of asymmetric length, with an intermediate plenum at the highest point.[18] In terms of the analysis, the intermediate plenum is of no consequence—the inflow and outflow rates would be the same, and the plenum would remain partially filled with air at lower than atmospheric pressure. The free surface of the reservoir is lower than the outlet of the pipe, however, so the U-bend flows anticlockwise, draining all the fluid into the reservoir. System B is a modified system such that, in the nomenclature of the diagram above, $\Delta h_2 > \Delta h_1$. This system flows clockwise, filling the lower dish until the tube in the upper dish is level with the surface of water in that dish. It is not, of course, a perpetual motion machine.

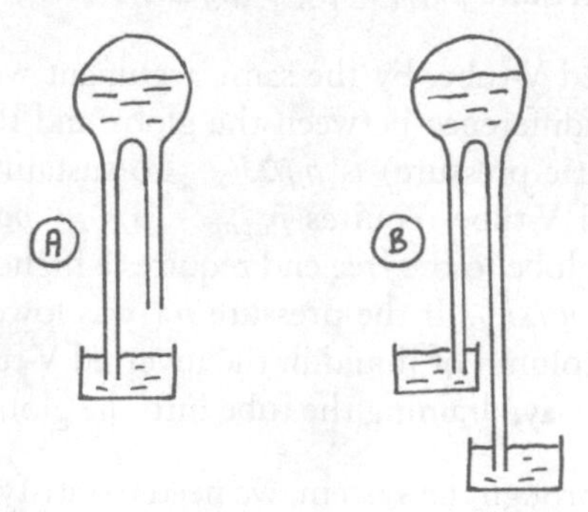

Discussion of energy in syphons

Is is interesting to also consider the action of the syphon in terms of the energy of the fluid. This subject is quite often misunderstood. It must, after all, have

[18] It is actually pretty common in large syphon systems to provide an *air chamber* at the highest point in the pipe, which is generally the point of lowest static pressure. Low pressure encourages dissolved air to come out of solution and accumulate at the high point. A pump is usually provided to remove the air. See, for example, Gibson, A. H., 1961, "Hydraulics and Its Applications," 3rd edition, London, Constable & Company, p. 629.

been Professor Sinclair's confusion on this point that led him to erroneously believe that he had developed a perpetual motion machine.

Water naturally flows to the lowest energy state accessible to it. In the case of two reservoirs, one lower than the other, as in situation A, water will flow from the higher to the lower reservoir. This is common sense. The motion is possible, however, because the fluid is *giving up*[19] energy as it moves to the pipe outlet. We know that the gravitational potential energy PE of a mass m at a height h above an arbitrary datum in a uniform gravitational field g is $PE = mgh$. If, in passing from the upper reservoir to the pipe outlet, the water loses height h_A, the change in energy is $-gh_A$ joules per unit mass.[20] We measure the height difference between the surface of the upper reservoir and the outlet of the pipe. The former defines the energy state of *all*[21] the fluid in the upper reservoir at a given time, while the latter defines the energy state at the last point of resistance before the fluid falls into the lower reservoir.[22]

If we consider the reservoirs in B, exactly the same arguments hold. If a pipe is provided that connects the two reservoirs, it is the height difference between their free surfaces, rather than the drop of the pipe, that defines their relative energy states. The change in energy in moving to the lower reservoir is $-gh_B$ joules per unit mass.

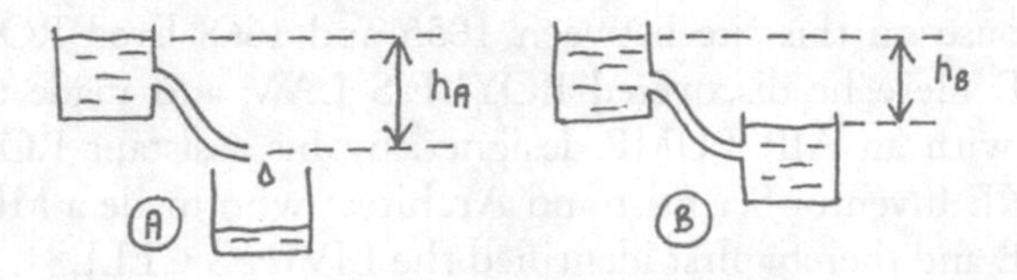

The reservoirs in C differ from those in A only in that the pipe rises above the level of the free surface of the upper reservoir before continuing to the lower reservoir. This is a typical syphon. The change in energy between the inlet and outlet is $-gh_C$ joules per unit mass.[23] The fluid, and the system, loses energy as it flows. For every particle that enters the pipe from the upper

[19] By *giving up*, we mean that potential energy is converted to other forms that are ultimately dissipated. This is usually in the form of a very small increase in temperature in the lower reservoir. But, if it's resisted, it can be in some form of energy that leaves the system (such as through the shaft work of a machine).

[20] This is the reduction in gravitational potential energy, but also the reduction in total energy when the fluid is non-moving in the lower reservoir. The energy is given up against frictional forces, ultimately leading to a very small increase in the temperature of the fluid.

[21] This is a rather subtle point. In a given stagnant reservoir, all fluid is *isoenergetic*. That is, all of the fluid particles have the same energy and therefore the same potential to do work. As we move below the *free surface*, we lose potential energy but increase *hydrostatic head* at an exactly compensating rate. We see that all particles in a reservoir have the same total energy, or the same ability to do work.

[22] To exit the pipe, the fluid requires a finite kinetic energy (in the field of fluid dynamics we call this kinetic energy *dynamic head*). In slow-moving systems this term will be small, and most of the gravitational potential energy loss will be given up to friction. In fast-flowing systems there will be a significant loss of kinetic energy at the pipe's exit.

[23] Interestingly, the energy of a particle *monotonically decreases* as it moves from the higher to the lower reservoirs in C. That is, a particle only loses energy and never gains energy, even when

reservoir, a particle leaves the pipe at the lower reservoir with gh_C joules per unit mass less energy than the particle that entered. The reservoirs in D differ from those in B only in that the pipe rises above the level of the free surface of the upper reservoir before continuing to the lower reservoir. The change in energy between the inlet and outlet is $-gh_D$ joules per unit mass. Once again, the system will flow because the fluid is giving up energy. In all these examples (A to D) the fluid can, and will, move to a lower energy state.

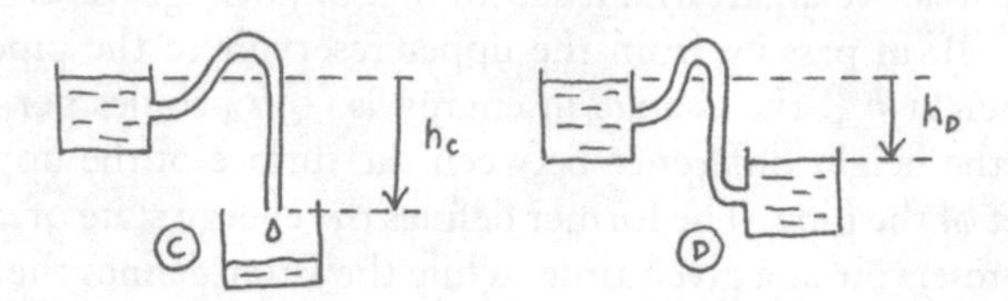

7.5 Boyle's perpetual vase ★

On the outer wall of my college, on the High Street in Oxford, there is a plaque which reads:

> In a house on this site between 1655 and 1668 lived ROBERT BOYLE Here he discovered BOYLE'S LAW and made experiments with an AIR PUMP designed by his assistant ROBERT HOOKE Inventor Scientist and Architect who made a MICROSCOPE and thereby first identified the LIVING CELL

I sometimes park my bike under the plaque, and have wondered more than once what it would have been like to have Robert Hooke as an assistant! Both Boyle and Hooke were fantastic scientists of the old school. Both were polymaths and both were prolific.

Robert Boyle (1627–1691) is regarded by many as the father of modern chemistry, breaking from the *alchemical*[24] tradition. He is best known for Boyle's Law, which relates pressure and volume for a gas at constant temperature, though he made many other developments in chemistry, experimental technique, and scientific methodology. It seems he was also quite a visionary.

it rises above the level of the upper reservoir. This apparent contradiction is explained by the fact that the static pressure falls to a low value, reducing the total energy in the fluid, even when the increase in gravitational potential energy is accounted for. (By static pressure we mean the sum of the atmospheric pressure and the hydrostatic head, which is negative at the top of the pipe.)

[24]Alchemy was the quest for the *philosopher's stone*, a substance that could turn base metals into gold and silver. It was also the *elixir of life*, conferring longevity. Which was useful, because alchemists have been at it for thousands of years. Alchemy started to die out fairly rapidly around Boyle's time (1627–1691), but it was not quite the end. Sir Isaac Newton (1642–1727), for example, devoted considerable effort to alchemy, a generation later than Boyle. The term *magnum opus* (great work) originated in the alchemical tradition, and refers to the quest for the philosopher's stone. Although clearly misguided, alchemy was the precursor to modern chemistry, and led to considerable developments in laboratory equipment, techniques, and experimental methods.

At some point in the 1660s he wrote a list of 24 major scientific problems that he felt deserved attention and which would benefit mankind if they were solved. Among them were:

> The prolongation of life.
> The recovery of youth, or at least some of the marks of it, as new teeth, new hair colour'd as in youth.
> The art of flying.
> The art of continuing long under water.
> The practicable and certain way of finding longitudes.
> The use of pendulums at sea and in journeys, and the application of it to watches.
> Potent druggs to alter or exalt Imagination, waking, memory, and other functions, and appease pain, procure innocent sleep, harmless dreams, etc.

and 17 others, almost all of which have been solved in fairly concrete terms by one or more breakthroughs in science. Considering that he was writing in the 1600s, Boyle's foresight was significant. You can go and see Boyle's list at the Royal Society, where it is exhibited to celebrate 350 years of the institution, of which Boyle was a founding member. Nevertheless, Boyle was around at the start of the Age of Enlightenment, as we are reminded when we see the few items on his list that have yet to be solved scientifically—most notably 'the transmutation of metals.'[25] Robert Boyle, to the end, held out hope that alchemy was possible.

Though perhaps not endowed with the same vision, Robert Hooke (1635–1703) was also a major figure in the early days of the Enlightenment. He was famously hard-working, and simultaneously held the posts of Gresham Professor of Geometry and Surveyor to the City of London after the Great Fire. In this latter capacity it is said that he personally performed more than half the surveys of burned buildings. Hooke was also an architect, he built telescopes to measure the motion of Mars and Jupiter, made contributions to the development of the study of biological evolution (based on his microscope observations), developed the wave theory of light, came up with a particle-based model for air, and developed the fields of surveying and map making. He is most famous for Hooke's Law, which states that the force required to extend or compress a spring (what we now call a *linear-elastic* or *Hookean* spring) is proportional to distance.[26]

[25] I suppose that in its strictest sense this has been achieved by science in certain controlled fission or fusion reactions, which turn one element into another. The purpose of fission and fusion is very different to what Boyle had in mind in his search for the philosopher's stone, however, and the new elements produced are byproducts rather than the goal.

[26] In Hooke's time this was a purely experimental law, but with later mathematics (due to the English mathematician Brook Taylor, 1685–1731) it was shown that it applies to small changes in most systems. Brook Taylor is best known for Taylor's theorem and the Taylor series, which are methods of approximating functions with polynomial series. We can use these techniques to explain the approximately linear, or Hookean, behaviour of many systems subject to small

One of Robert Boyle's more curious inventions was the perpetual vase, or self-flowing flask. The principle is very simple. The weight of liquid in the vase is significantly greater than the weight in the neck, so a pressure difference is exerted across the spout. Liquid is forced around the neck and falls back into the vase, creating a perpetual cycle. Explain the fallacy in this argument.

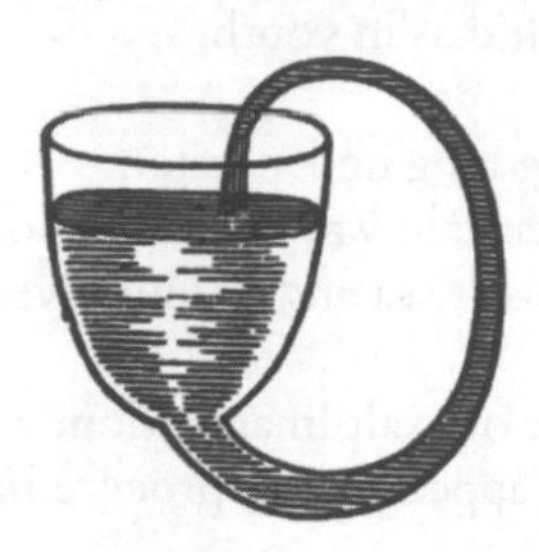

Answer

The argument for perpetual flow is based on the assumption that because the weight of liquid in the vase is greater than the weight of liquid in the spout, a pressure difference will cause the liquid to flow up the spout. It clearly *is* the case that the weight of fluid in the vase is greater than that in the spout. But the pressure[27] in stationary fluid is related only to the depth of the fluid, and not to the shape of the vessel. The amount of fluid, and its total weight, are irrelevant. This contradiction (between misguided expectation using the *weight* argument and actual physical behaviour) is known as the *hydrostatic paradox*.

One way to view the problem in terms of pressure is to introduce an imaginary cut in the tube at its lowest point. For equal pressure to be imposed on either side of the cut we require the same height to the free surface on both

displacements. We argue as follows.

1. Assume that a force F depends only on displacement x (that is, there is no *hysteresis*). In general it could be some complicated function. We define the equilibrium position, $x = 0$, to be the point where $F = 0$.
2. Whatever the form of $F(x)$, provided it is a continuous function (that is, it has no discontinuities), using a Taylor series we can expand $F(x)$ as a polynomial series about the origin $x = 0$. That is, $F(x) = \sum_{k=0}^{\infty} a_k x^k = a_0 + a_1 x + a_2 x^2 + \ldots$ What we mean here is that we can find a set $\{a_k\}$ such that we approximate the original function with the polynomial series. As we move further from the origin we need more and more terms to recreate the function.
3. $F(0) = 0$, so $a_0 = 0$.
4. For small enough x, higher terms in x become very small, so we can approximate $F(x) \approx a_1 x$. For small enough x this holds for any set $\{a_k\}$. This is the definition of Hookean/linear behaviour.

[27] We normally call this the *hydrostatic pressure*.

sides. That is, the levels must be the same. As Aristotle[28] said a very, very long time ago, *water seeks its own level*. Almost two thousand years later, however, there was still confusion about this principle. This is evidenced by a number of proposed perpetual motion machines that used the false principle of fluid being displaced by more weight on one side of a system.

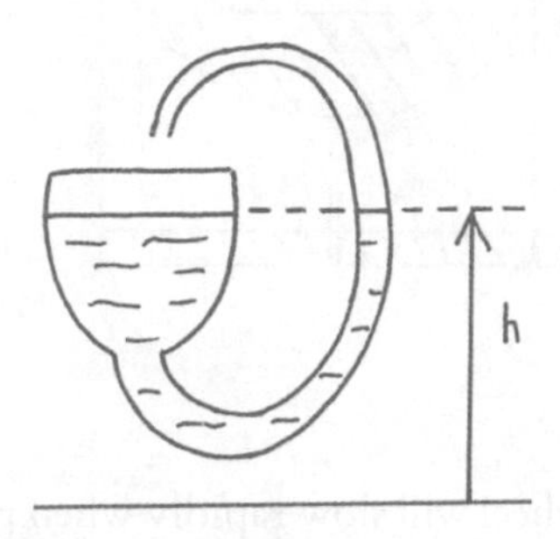

7.6 The curious wheel ★★★

I came up with this puzzle myself but never got round to testing anyone with it. This is a shame because I happen to really like it. The reason it's in this section is that it is an obvious *anti-perpetual motion* machine. A real machine, that is, which slows down, and quickly. Or does it? I'd always thought I would pose a line of reasoning and then ask students to criticise it (or agree with it, if they could give an explanation). I was going to pose it something like this.

THE DRAWING SHOWS a hollow wheel composed of two parallel (and uniform) disks mounted in the vertical plane on a frictionless horizontal pivot C, sealed at the perimeter so that the thin cylindrical cavity can be filled with liquid. When the wheel is empty of water, once set in motion, and in the absence of air resistance, it would turn forever. When the wheel is partially filled with water, as shown in the diagram, once it's set in motion we might expect it to slow down rapidly "due to *internal* friction". We know, however, that this means that the angular momentum of the rotating wheel must be reduced, and this requires an *external* moment. In the absence of bearing friction and air resistance there is no mechanism for an external moment, however, so the wheel must rotate forever, as before. Is this line of reasoning correct?

[28] Aristotle (384–322 BC) was a philosopher and scientist from ancient Greece. His work covers logic, ethics, metaphysics, physics, biology, zoology, politics, systems of government, rhetoric, music and theatre. He must have been extremely busy. His influence was so significant that his name is synonymous with methods in logic (*Aristotelian logic*), the natural sciences (*Aristotelian physics*) and philosophy (*simply, Aristotelianism*).

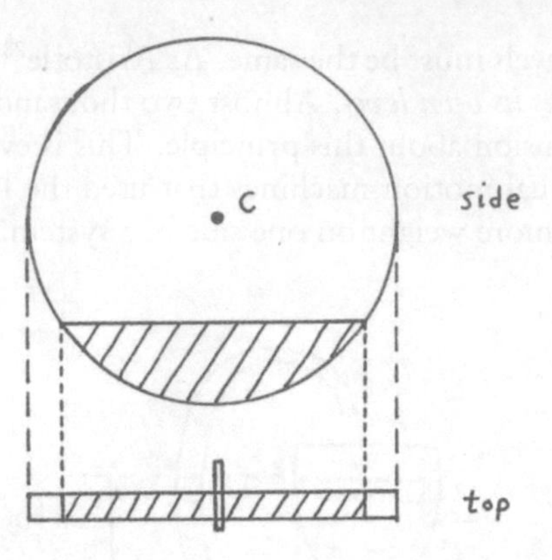

Answer

Our intuition that the wheel will slow rapidly when partially filled with water is clearly correct, which means that the line of reasoning given is wrong. It is true that internal friction between the (approximately) stationary water and the inner rotating walls of the wheel is reducing the energy of the system—kinetic energy is being converted to other forms of energy (mostly heat). To reduce the angular momentum of the wheel we require an external moment that cannot (in the terms of the question) come from either the pivot or air resistance. The external moment comes from the weight of water. Under the action of internal friction the volume of water on the left of the axis C is greater than that on the right of the axis. The centre of gravity of the water on the left, G_A, is also further from the axis than the centre of gravity of the water on the right, G_B. This situation gives rise to an unbalanced external moment on the wheel, which causes it to slow down.

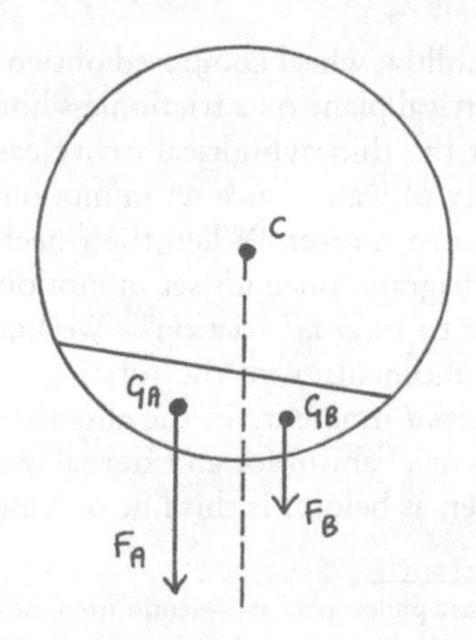

Chapter 8

Kinematics

In this short section we look at puzzles in *kinematics*. This is the study of motion without reference to force and mass, so without reference to Newton's laws, or the *cause* of the motion. Kinematics describes the relationships between distances, velocities and accelerations. It is sometimes called the *geometry of motion*. We need very little specialist knowledge to solve problems in kinematics, but here are a few tips which might help.

- When we are dealing with two- or three-dimensional space, it is often convenient to use *vector notation* to represent the quantities of interest.
- We often need to be able to break a problem down into simpler parts to allow us to think our way clearly and logically to a solution.
- We might need to formulate simple differential equations that link one variable in a problem to another. For example, we might link velocity to displacement, $\boldsymbol{v} = d\boldsymbol{x}/dt$, or acceleration to velocity, $\boldsymbol{a} = d\boldsymbol{v}/dt$. In these differential equations, the variables can be vectors.

But there is nothing special about these techniques—they crop up in many areas of physics.

8.1 Professor Lazy ★★

PROFESSOR LAZY LIVES on a long straight road 4 km due east of his college, and refuses to go into college unless the wind is blowing from the east. When it blows at 10 m s^{-1} he can cover the 4 km distance in 300 s. The first 2 km is down a gentle hill of constant gradient, and he travels at a breakneck speed of 20 m s^{-1} on his frictionless penny-farthing (or, as the professor calls it, his *ordinary*). The second 2 km is on flat ground, and he travels at exactly 10 m s^{-1}. The bicycle is, in fact, not particularly ordinary, because the professor has removed the pedals and its motion is due to wind power alone. Professor Lazy will not contemplate returning home until the wind blows at 20 m s^{-1} from the west. When this happens, how long will it take him to get home? Assume that in each phase of the journey the professor reaches his terminal velocity quickly compared with the overall length of his trip. Assume also that the wind blows parallel to the ground both on the flat and on the slope.

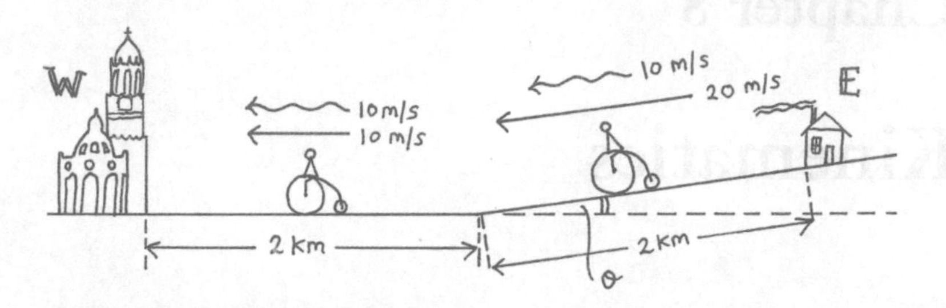

Answer

On the flat the professor travels at the same speed as the wind, because his ordinary is frictionless. We can draw the velocities in the frame of reference of the world, the *lab frame*, or the frame of reference of the bicycle, the *bike frame*. In the lab frame both velocities are 10 m s^{-1} to the west. It takes the professor 200 s to cover the flat leg of his journey. Because the velocities are the same, in the frame of reference of the ordinary, there is no *apparent wind*.

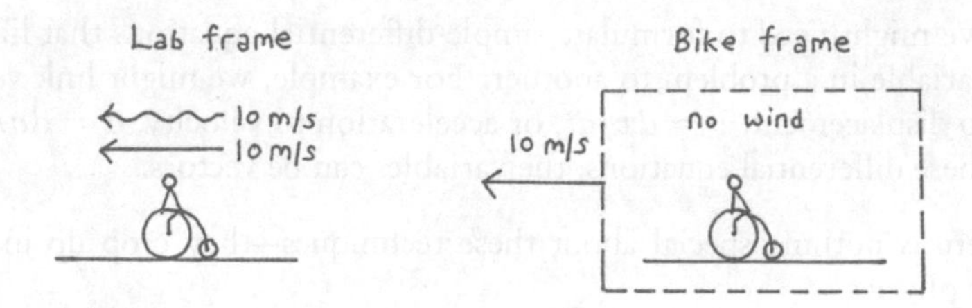

When the professor goes downhill, he makes a cracking 20 m s^{-1}. It takes him only 100 s to cover the downhill leg of his journey. If we look at the velocities in the lab frame, the professor is outgunning the wind by 10 m s^{-1}. This is because a component of his weight acts in the downhill direction. In the frame of reference of the ordinary, there is an apparent wind of 10 m s^{-1} to the east. We know that forces must balance, or the frictionless ordinary would accelerate out of control, so the force due to the apparent wind must match the component of weight acting in the direction of motion. We write $F_w = mg \sin\theta$. We do not explicitly need this equation to solve the problem, but it is good to remind ourselves how to draw force diagrams.

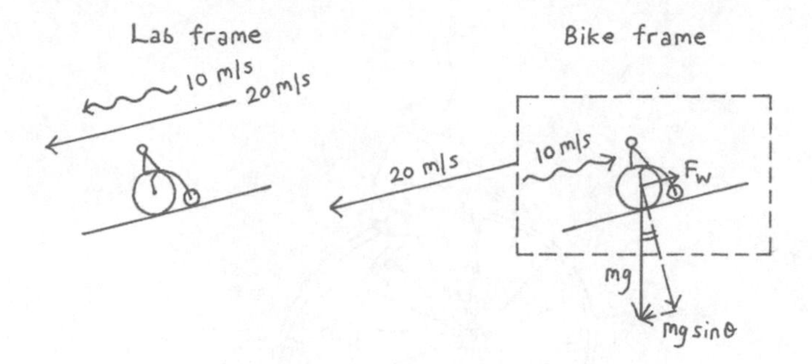

Now consider the return trip. On the flat, the frictionless ordinary travels at the same speed as the wind. Because the wind is blowing at 20 m s^{-1} from the west, the ordinary carries the professor east at the same speed. In the bike frame there is no wind. At 20 m s^{-1} it takes the professor 100 s to cover the first 2 km on the flat.

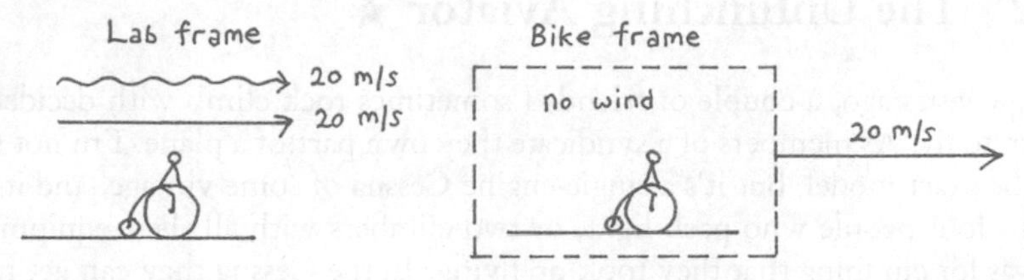

Now the ordinary encounters an uphill leg. For the forces to be balanced—the condition for no acceleration—we know that in the frame of reference of the ordinary there must be a 10 m s^{-1} apparent wind. The true wind is 20 m s^{-1} from the west, so the ordinary must be travelling at 10 m s^{-1} to the east, giving the correct apparent wind. The professor is blown uphill at 10 m s^{-1}, and covers the uphill leg of his journey in 200 s.

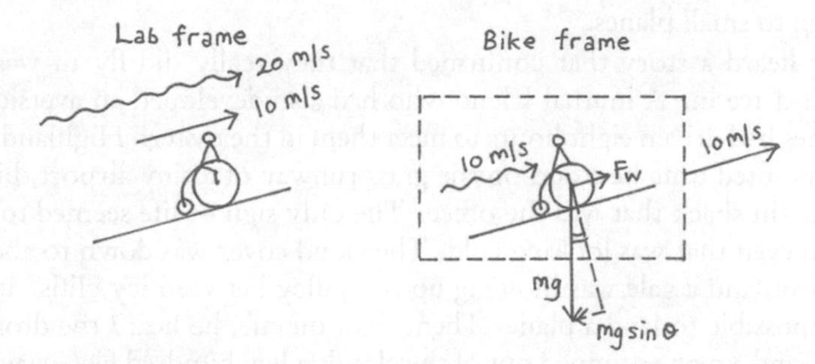

The total time for the professor's journey from college to home is 300 s, which is the same time it takes him to get from home to college. During the journey, the professor idles away the time performing experiments to check that his ordinary is truly frictionless. When he sets off in a stiff wind and his pipe smoke rises in a perfect column (when he's on flat ground), he knows that his frictionless bearings are operating perfectly.

8.2 The Unflinching Aviator ★

A few years ago, a couple of friends I sometimes rock climb with decided to learn to fly. As members of a syndicate they own part of a plane. I'm not sure of the exact model, but it's a single-engine Cessna of some vintage, and it can carry four people who pack light, or two climbers with all their equipment. It was for climbing that they took up flying. In the Cessna they can get from Oxford to North Wales (where some of the best summer rock climbing is) in just over an hour, they tell me, and to the Cairngorms (a range of mountains in the eastern Scottish Highlands, known for ice climbing in the winter months) in just two and a half hours. Having spent countless hours in cars heading to or from cliffs, I could certainly see the advantage. "And the brilliant thing," they added, building up to asking if I wanted to join them for a flight, "is that we fly in weather it would be horrible to drive in." At that point I developed an aversion to small planes.

I later heard a story that confirmed that they really did fly in weather I wouldn't drive in. A mutual friend who had also developed an aversion to small planes had driven eight hours to meet them in the eastern Highlands. At the preappointed time he stood on the grass runway of a tiny airport, hiding behind the tin shack that was the office. The only sign of life seemed to be a sheep, and even that was looking cold. The cloud cover was down to about a hundred feet, and a gale was blowing up the valley between icy cliffs. It was clearly impossible to land a plane. Then, right on cue, he heard the drone of an engine, and a plane dropped out of the clouds a few hundred feet away, and touched down at a terrifying angle on the grass. He helped our friends strap the wings to ties in the patch of hard surfacing, and swore never to get inside a plane that was too small to have cabin service.

Of course, even small planes these days carry some form of GPS, but it can't be relied on absolutely. Pilots navigating in dense cloud rely on *dead reckoning.* This is a method for calculating the course over the ground by vector addition of what are called the *wind vector* and the *air vector*. Pages and pages of the course books for the private pilot's licence are devoted to the principles of vector addition in different situations. But the concept is simple, so I don't need to explain it here. If you know what vectors are and how to add them, and you can develop basic logic, that's good enough.

Quite a number of pre-university physics questions test basic concepts in vector addition. There may or may not be some twist to the problems. For example, a problem might be phrased as an optimisation problem (finding a minimum or a maximum, for example), or it may be testing a student's ability to formulate algebraic equations for simple problems. Sometimes the questions are straightforward—calculating a particular result using vector addition, for example. This question is in that last category.

THE UNFLINCHING AVIATOR flies from West Island to East Island in a straight line, making a track across the ground from west to east. There is a steady and uniform[1] wind with speed v_w, directed with a compass bearing of 30°. At the top speed of the plane, the aviator makes a speed over the ground of $3v_w/2$. The journey takes exactly one hour. If the wind speed is unchanged, what is the minimum time it takes the aviator to make the return journey? If he relies on dead reckoning what compass bearing should he fly at?

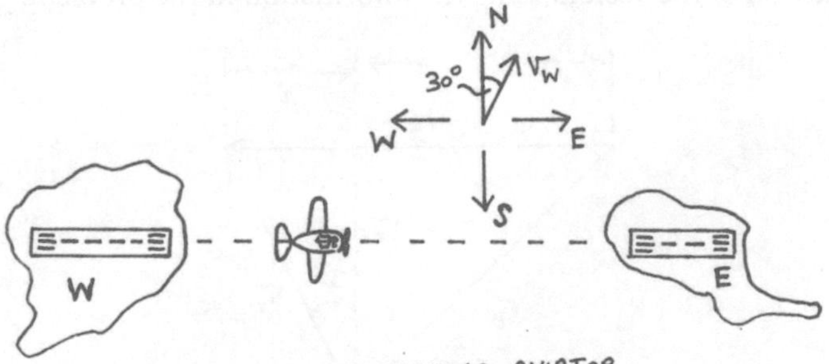

THE UNFLINCHING AVIATOR

Answer

The vector addition in this question is very simple, but the problem of extracting the information from the question is not necessarily so. Let's set out

[1]By *steady* and *uniform* we mean unchanging in both space and time. A velocity vector $\boldsymbol{v}$ can be thought of as a function of all the spatial coordinates, and also time. We might represent it as $\boldsymbol{v}(x, y, z, t)$. *Steady* implies the condition $\partial\boldsymbol{v}/\partial t = \mathbf{0}$, where the curly d ($\partial$) shows that we have taken the *partial derivative* of $\boldsymbol{v}$ with respect to a variable, in this case time. A partial derivative (taught in very few pre-university courses) is the derivative of a function of several variables (in this case x, y, z and t) with respect to a single variable, with the others held constant. *Uniform* implies that the condition $\partial\boldsymbol{v}/\partial x = \partial\boldsymbol{v}/\partial y = \partial\boldsymbol{v}/\partial z = \mathbf{0}$.

the information we are given in the question about: the *ground vector*[2] (or velocity) $\boldsymbol{v}_g$; the *wind vector* $\boldsymbol{v}_w$; and the *air vector* $\boldsymbol{v}_p$. We'll define these terms fully in a moment. For the outward journey:

- The ground vector $\boldsymbol{v}_g$ represents the direction and speed of the plane travelling over the ground. We are told that $\boldsymbol{v}_g$ is directed with a compass bearing of 90°, or due east, and that the magnitude is $|\boldsymbol{v}_g| = 3v_w/2$.
- The wind vector $\boldsymbol{v}_w$ represents the direction and speed of the wind over the ground. We are told that $\boldsymbol{v}_w$ is directed with a compass bearing of 30°, and that the magnitude is $|\boldsymbol{v}_w| = v_w$.
- The air vector $\boldsymbol{v}_p$ represents the direction and speed of the plane (hence the subscript p) *with respect to the air*, which might itself be in motion with respect to the ground. $\boldsymbol{v}_p$ is defined by the equation $\boldsymbol{v}_g = \boldsymbol{v}_p + \boldsymbol{v}_w$. That is, the ground vector is the sum of the air vector and the wind vector. The magnitude and bearing result from the vector addition.

We can add the vectors in one of two ways. We can write each vector out explicitly in terms of x and y components (that is, $\underline{i}$ and $\underline{j}$ components, respectively), and then simply add them (this is all that vector addition is, after all). Alternatively, we can use a simple geometric construction of the vectors[3]—the *vector triangle method*. We'll use the geometric method here, because it enables us to visualise the problem more easily. It should be relatively simple to draw the interaction of the vectors using the information in the problem.

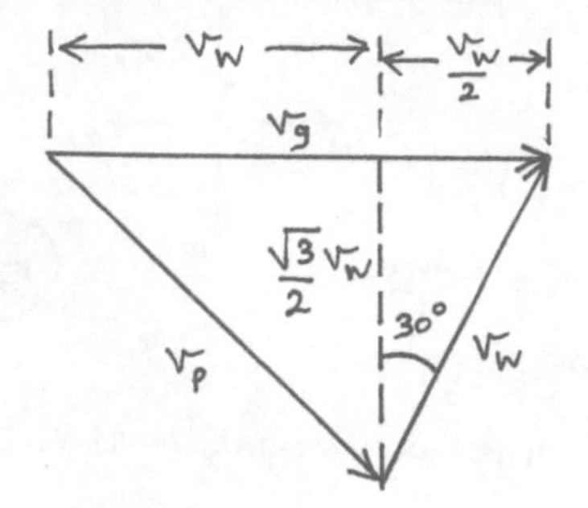

From here we can solve for $\boldsymbol{v}_p$. To begin with we're only interested in the magnitude of the air vector, so we write

$$|\boldsymbol{v}_p| = v_w\sqrt{1 + \left(\frac{\sqrt{3}}{2}\right)^2} = \frac{\sqrt{7}}{2}v_w$$

[2]This slightly unusual terminology is used by pilots and sailors. In a way, this makes sense, because in normal—non-scientific—usage, the words *speed* and *velocity* are used somewhat interchangably. So deliberately using the word *vector* clarifies the intent.

[3]The x-y decomposition method requires us to frame the problem in terms of a particular coordinate system, which we can arbitrarily define for a particular problem. The vector triangle method is known as a *coordinate-free* approach, because it does not require us to choose particular coordinate axes. Both methods achieve the same result.

If we wanted to, we could easily calculate the compass bearing of the air vector, but we don't need it.

We now consider what we know about the return journey. In the second case $\boldsymbol{v}_w$ is known, and is unchanged from the outward journey. We know the *direction* of $\boldsymbol{v}_g$ (on a compass bearing of 270°, or due west) and we know the magnitude of $\boldsymbol{v}_p$, namely $|\boldsymbol{v}_p| = (\sqrt{7}/2)\, v_w$. The vector addition rule, $\boldsymbol{v}_g = \boldsymbol{v}_p + \boldsymbol{v}_w$, still holds. Using this information we can draw the construction for the return journey.

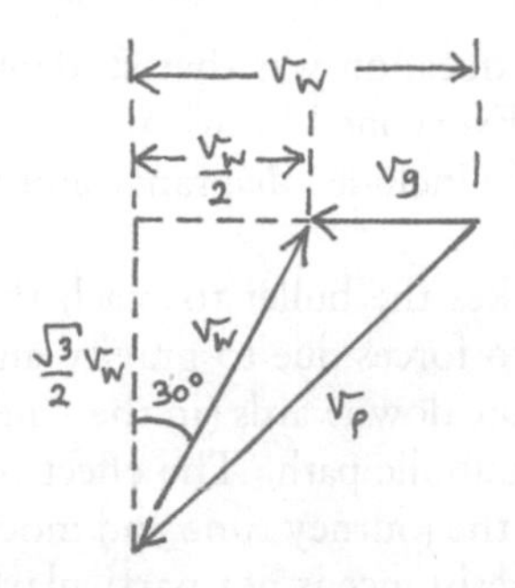

From the construction we calculate $|\boldsymbol{v}_g| = v_w/2$. If the outward speed is $3v_w/2$ and the return speed is $v_w/2$, the return journey takes three times longer (three hours). From the construction we can calculate that the bearing of $\boldsymbol{v}_p$ is

$$270° - \tan^{-1}\left(\frac{\sqrt{3}}{2}\right) \approx 229.1°$$

The Unflinching Aviator can now fly back even in dense cloud, and can be certain that he will land dead on target on West Island three hours later.

8.3 Target shooting ★

I really like this short question. I was asked it during my Oxford physics interview, as were a number of friends who studied with me at the same college. One of my friends is absolutely convinced that it was invented for him. He did target shooting at school and wrote about it on his application form, and he remembers the interviewer saying he had invented the perfect question to ask. Instinct says he misunderstood, or that the interviewer was joking. It seems too clever to be a spur of the moment question. But our physics tutor was exceedingly clever, so it is at least plausible.

An accurate target-shooting rifle has been calibrated to hit the centre of a target at a long-distance indoor rifle range. While facing the target, the marksman rotates the gun by 90° anticlockwise about the line between the gun and the target. He takes aim and fires. In what sector of the target does the bullet land?

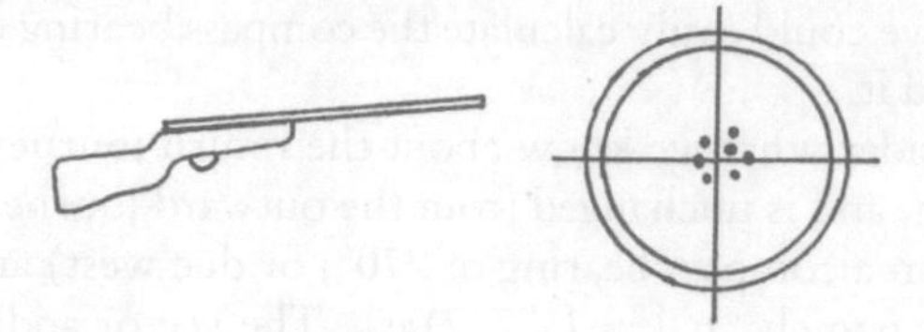

Answer

I think the idea with this question was that students would answer without diagrams or calculations. Everyone I know who was asked it worked out the answer in their head. I have included diagrams simply to make the discussion clearer.

In the finite time it takes the bullet to reach the target (which is in the x-direction), it is subject to forces due to gravity and air resistance. Gravity causes the bullet to accelerate downwards (in the $-y$-direction) and to *fall away* from the target along a parabolic path. The effect of air resistance is to slow the bullet a bit, increasing the journey time and modifying the parabolic path slightly. The effect of air resistance is not particularly interesting here, so we can ignore it.

Let's consider the motion of the bullet when the rifle is held in the conventional, non-rotated, orientation. The *sights*, by definition, are aligned with the centre of the target. We draw a horizontal line between the tip of the rifle and the centre of the target[4] and mark the location on the target S. The path of the bullet is a symmetric parabola (in the absence of air resistance, when we are vertically aligned with the target). We draw this parabola and mark the location of the bullet P. The sights have been aligned with the point of impact, so P and S represent the same point. The initial gradient of the parabola is set by the barrel of the rifle. The barrel is therefore inclined upwards very slightly to acheive the shot. A straight-line extension of the barrel marks point B on the target.

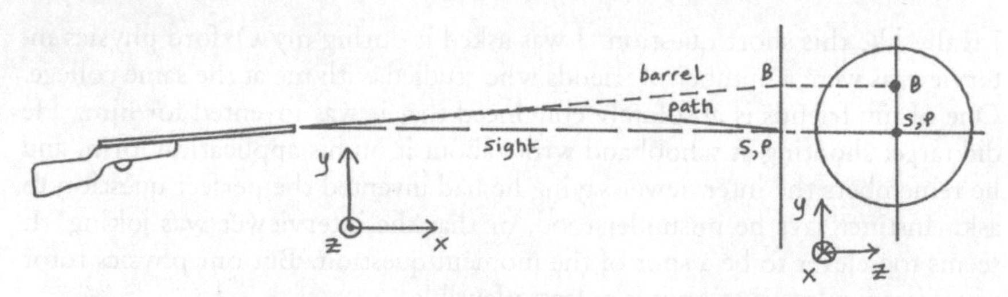

We now rotate the rifle 90° anticlockwise about the line between the gun and the target. Once again we aim the sights at the target. Point S is still in the centre of the target. The relationship between B and S is fixed by

[4]This analysis does not require that the centre of the rifle and the centre of the target are in the same horizontal plane. But it does require that the shot is taken *roughly* horizontally.

geometry, for a given alignment of the rifle sights. The geometry is related to the reference frame of the rifle, not the target, so B rotates about S by 90° anticlockwise, following the rotation of the rifle. The relationship between P and B is unchanged. That is, the bullet falls away by the same amount as it did in the first example. So P is vertically below B.

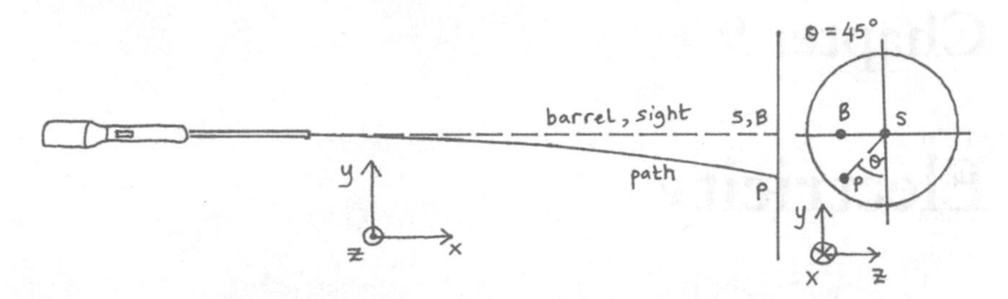

The distances between B and S and between B and P are the same, so P is in the bottom left quadrant at exactly 45° from the vertical (for a very accurate rifle). If we want to hit the target we should aim in the upper right quadrant on a line at 45° from the vertical.

Where it is impossible to calibrate a rifle by firing test shots, marksmen use calibration charts. With these they can correct for the effect of the bullet falling away, the wind speed, and a target which isn't vertically aligned with the rifle.

Chapter 9

Electricity

In this chapter we look at problems involving *electricity*. Some of the problems test the application of very simple concepts, such as Ohm's Law, or the definition of electrical power. You may feel that some of the questions are too simple even for high-school students. They are not, however, and it is surprising how many students struggle to think their way clearly to a cogent answer. Problems like this can be used to determine whether students have a sound knowledge of GCSE-level material on these topics. We must not dismiss them.

I have also included quite a few *resistor puzzles*, which involve groups of resistors either in two-dimensional or three-dimensional arrays. The goal of these is to calculate the overall resistance of the network. Resistor puzzles have been very popular over the years. The classic question is that of the resistor cube, where each edge of the cube has a given resistance (R, say) and the aim is to calculate the overall resistance between two vertices, most commonly the longest diagonal (this is also the simplest variant). The rules for solving these problems are simple, so they reduce almost to logic problems.

I have not included any *unsteady problems*—those which involve the charging and discharging of capacitors, or the time constant of CR circuits, etc.—even though these are also quite popular pre-university puzzles. They are easy to find in most textbooks, and are rather formulaic, so (to me anyway) are therefore less interesting. Nonetheless, this is an important area that students should be familiar with if they've studied physics to pre-university level.

Let's review a few simple definitions that will help us with these problems.

- CURRENT. The current, I, is related to the charge, Q, according to the definition $I = dQ/dt$, where t is time. For steady currents this reduces to $I = Q/t$.

- OHM'S LAW. The current, I, through a conductor is proportional to the potential difference, V, across it, and inversely proportional to the resistance[1] R. We write $I = V/R$.

- **Electrical Power.** The electrical power, P, dissipated in a load of resistance R, is proportional to both the current through the load, I, and the potential difference across it, V. We write $P = IV$. Combining this with Ohm's Law we can write instead $P = I^2R$ or $P = V^2/R$.

- **Resistors in Series.** Resistors in series add. In other words, the total resistance is given by $R_T = R_1 + R_2$.

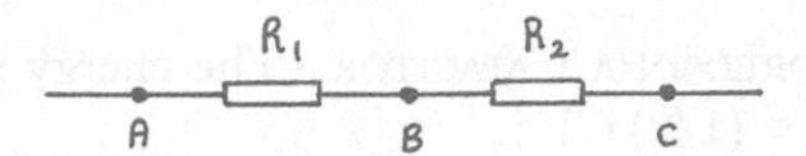

We should prove this relation, because it is used in many of the problems that follow. It is a consequence of Ohm's Law: $V = IR$. The *voltage drops* across R_1 and R_2 happen in series, so the total voltage drop between A and C is given by $V_T = V_1 + V_2$, where V_1 and V_2 are the individual voltage drops. The current flowing from A to B is the same as the current flowing from B to C,[2] so we can rewrite this equation as $IR_T = IR_1 + IR_2$. Dividing the equation by I, we get $R_T = R_1 + R_2$.

- **Resistors in Parallel.** Resistors in parallel obey the following rule.

$$\frac{1}{R_T} = \frac{1}{R_1} + \frac{1}{R_2}$$

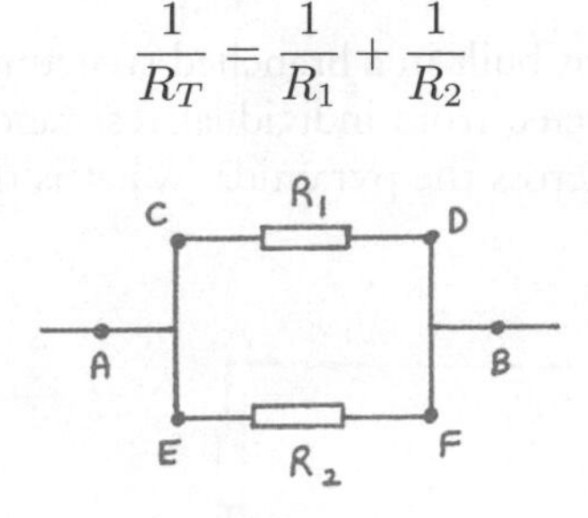

To prove this we note that for a given voltage drop V_{AB} the voltage drop between C and D must be equal[3] to the voltage drop between E and F. That is, $V_{CD} = V_{EF} = V_{AB}$, or $V_1 = V_2 = V$. The current, on the

---

[1]This is only true when the resistance is *not* a function of the current—that is, $R \neq f(I)$. For many real loads in practical situations this assumption may not be valid.

[2]This is because charge is not leaving the circuit, and because there is no place for charge to accumulate. We can think of this as a one-dimensional continuity equation for a conserved quantity (charge).

[3]Taking the wire to have zero resistance, and noting that voltages add in series, we have $V_{CD} = V_{CE} + V_{EF} + V_{FD}$. The resistances between C and E and between F and D are both zero, giving $V_{CE} = 0$ and $V_{FD} = 0$. So $V_{CD} = V_{EF}$.

other hand, obeys an additive law: $I_T = I_1 + I_2$. Combining this with Ohm's Law we get

$$\frac{V}{R_T} = \frac{V}{R_1} + \frac{V}{R_2} \text{ or } \frac{1}{R_T} = \frac{1}{R_1} + \frac{1}{R_2}$$

- Capacitors in Series. Capacitors in series obey $1/C_T = 1/C_1 + 1/C_2$.
- Capacitors in Parallel. Capacitors in parallel obey $C_T = C_1 + C_2$.
- Energy Stored in a Capacitor. The energy stored in a capacitor is given by $E = (1/2)\,CV^2$.

We are now ready to try a few puzzles. As I remarked, the resistor cube is a well-established puzzle that has been circulating for several decades—I've also seen it in at least one physics textbook. The other resistor questions are my own, the result of a productive hour spent at Detroit airport waiting for a connecting flight across the US. Every time the waitress topped up my coffee she asked if I had solved my problem yet. "Yes," I said, "but I keep finding new problems." To which she replied, "Oh dear, honey, you just keep at it then." Good advice for anyone attempting these puzzles.

9.1 Resistor pyramid ★★

A resistor pyramid is built in a branched structure as shown below. If there are N levels, constructed from individual resistances of value R, what is the total resistance R_N across the pyramid? What is the resistance of an infinite pyramid of this type?

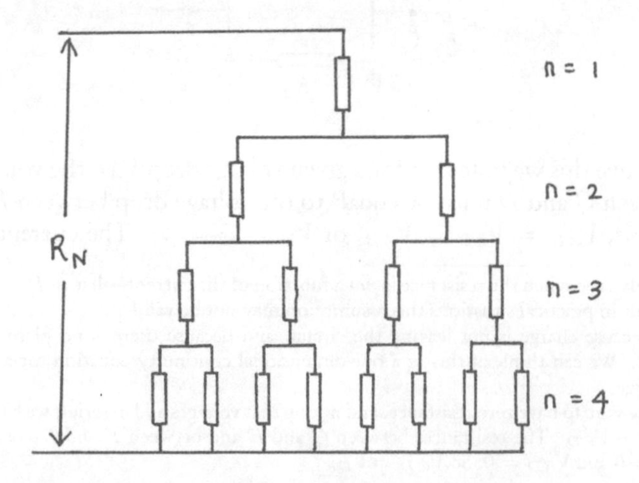

Answer

We can think of the structure of the pyramid as a combination of both series and parallel connections of resistors. From the symmetry of the arrangement we see that all nodes at a given level n in the pyramid are *equipotential points*. We can add *dummy wires* (shown as dotted lines in the diagram) between these equipotential points without changing the current flows in the circuit. So we see that the equivalent circuit has n levels in series, with 2^{n-1} resistors in parallel at the n-th level.

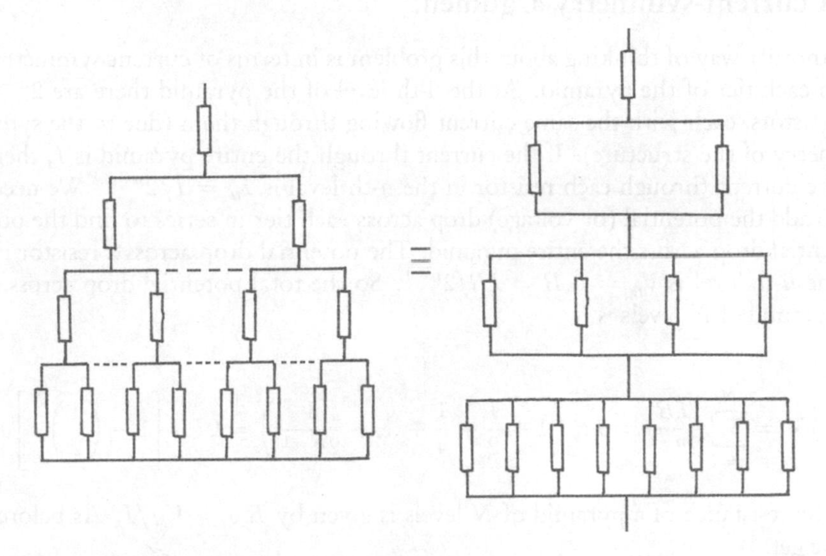

The resistance of the n-th level is therefore

$$R_n = \frac{R}{2^{n-1}}$$

The resistance of a pyramid of size N is the sum of the resistances R_n from $n = 1$ to N. That is,

$$R_N = \sum_{n=1}^{N} \frac{R}{2^{n-1}} = R\left(1 + \frac{1}{2} + \frac{1}{4} + \ldots + \frac{1}{2^{N-1}}\right)$$

The right-hand side of this expression is a geometric series[4] with the sum[5]

$$R_N = 2R\left[1 - \left(\frac{1}{2}\right)^N\right]$$

[4] A geometric series of n terms has the general form $S_n = a(1 + r + r^2 + \ldots + r^{n-1})$, where a is the value of the first term and the common ratio is r. The sum can be expressed simply as follows. We multiply by r to give $rS_n = a(r + r^2 + r^3 \ldots + r^n)$, then subtract this from the previous expression to give $rS_n - S_n = ar^n - a$. Rearranging, $S_n = a(r^n - 1)/(r - 1)$.

[5] Taking $a = R$ and $r = 1/2$.

To calculate the resistance of an infinite pyramid we need the sum to infinity of the series. The common ratio is $r = 1/2 < 1$, so the sum to infinity converges to give

$$R_\infty = 2R$$

It might have been easier to put two resistors in series.

A current-symmetry argument

Another way of thinking about this problem is in terms of current symmetry in each tier of the pyramid. At the n-th level of the pyramid there are 2^{n-1} resistors, each with the same current flowing through them (due to the symmetry of the structure). If the current through the entire pyramid is I, then the current through each resistor in the n-th level is $I_n = I/2^{n-1}$. We need to add the potential (or voltage) drop across each tier in series to find the potential drop across the entire pyramid. The potential drop across a resistor in the n-th level is $V_n = I_n R = IR/2^{n-1}$. So the total potential drop across a pyramid of N levels is

$$V_N = \sum_{n=1}^{N} \frac{IR}{2^{n-1}} = IR\left(1 + \frac{1}{2} + \frac{1}{4} + ... + \frac{1}{2^{N-1}}\right) = 2IR\left[1 - \left(\frac{1}{2}\right)^N\right]$$

The resistance of a pyramid of N levels is given by $R_N = V_N/I$. As before, we get

$$R_N = 2R\left[1 - \left(\frac{1}{2}\right)^N\right] \text{ and } R_\infty = 2R$$

An elegant solution

We now consider an elegant solution that will appeal to the more mathematically minded. A friend came up with the idea, and my brother and I extended it.

In the diagram we represent an infinite pyramid of resistors as a double triangle inscribed with R_∞. If we extend the pyramid by one level, it's the same as having two such pyramids in parallel, in series with a single resistor. If the series converges, the resistance of the extended pyramid must be the same as that of the original pyramid. We write

$$R_\infty = R + \frac{R_\infty}{2}, \quad \text{so} \quad R_\infty = 2R$$

A very elegant solution.

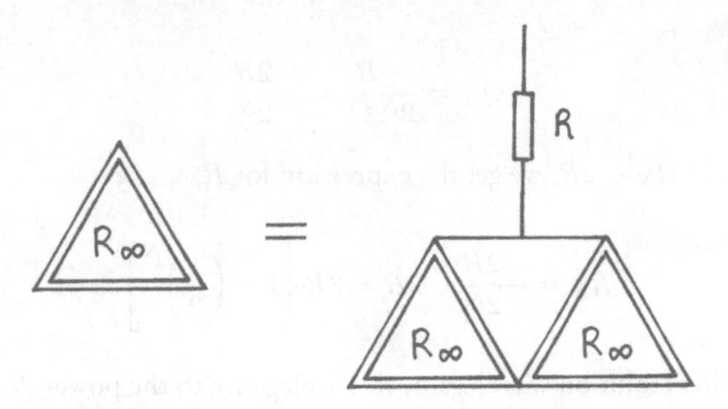

Let's consider how we might extend this to find the resistance of a pyramid with N levels. In the diagram we represent a pyramid with N levels as a double triangle inscribed with R_N. We see that extending the pyramid by one resistor gives

$$R_{N+1} = R + \frac{R_N}{2}$$

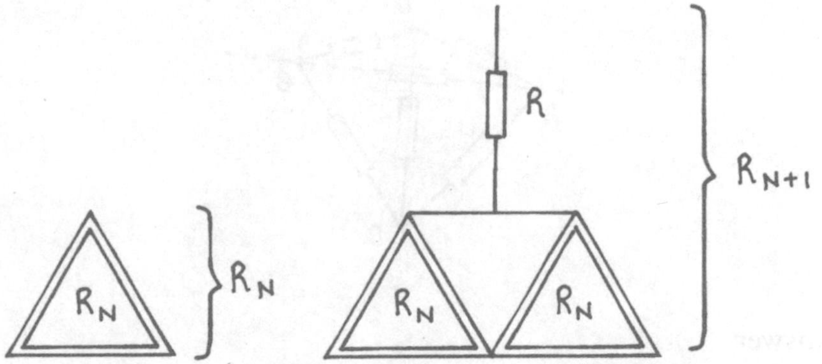

Here we remind ourselves that R_N and R_{N+1} are terms in a series that we seek the expression for. Noting that the sum to infinity of the series is given by $R_\infty = 2R$, we define a new series $Q_N \equiv R_N - 2R$. The series Q_N converges on zero—that is, $Q_\infty = 0$. We substitute the definition of Q_N into the expression for R_{N+1}, giving

$$Q_{N+1} + 2R = R + \frac{(Q_N + 2R)}{2}$$

So

$$Q_{N+1} = \frac{Q_N}{2}$$

Noting that $R_1 = R$, and that therefore $Q_1 = -R$, we see that

$$Q_N = \frac{-R}{2^{N-1}} = -\frac{2R}{2^N}$$

Using $Q_N \equiv R_N - 2R$, we get the expression for R_N.

$$R_N = -\frac{2R}{2^N} + 2R = 2R\left[1 - \left(\frac{1}{2}\right)^N\right]$$

If the previous solution was elegant, this is elegance to the power N.

9.2 Resistor tetrahedron ★

A RESISTOR TETRAHEDRON has individual resistance values R in each leg. What is the total resistance R_T across any pair of nodes in the pyramid?

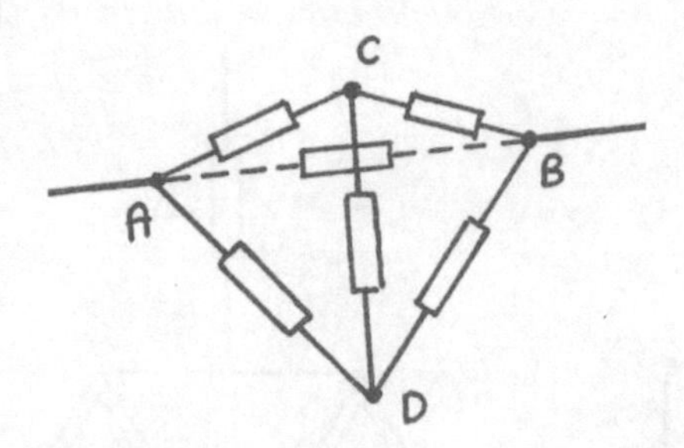

Answer

It is easiest to think about this by flattening the structure onto a plane. Doing this makes it easier to see the essential structure we're dealing with. Let's consider the total resistance between A and B. When it's flattened—as shown in the diagram—we can see that the network reduces to a single resistor (at the top, $A \to B$) in parallel with a group of five resistors (at the bottom, $A' \to B'$). If all the individual resistances take the same value R, the resistance of the bottom group itself becomes R. We see this by noting from the symmetry of the arrangement that the potentials at C and D are the same, so no current flows between those two points. Thus the resistor between C and D can be removed without affecting the currents in the circuit. The circuit is then equivalent to having two resistors of value $2R$ in parallel, giving a resistance of R for this block. The resistance of the top resistor (between A and B) is trivially R. The total resistance between A and B is therefore $R_T = R/2$. By symmetry, this is the resistance between any two nodes.

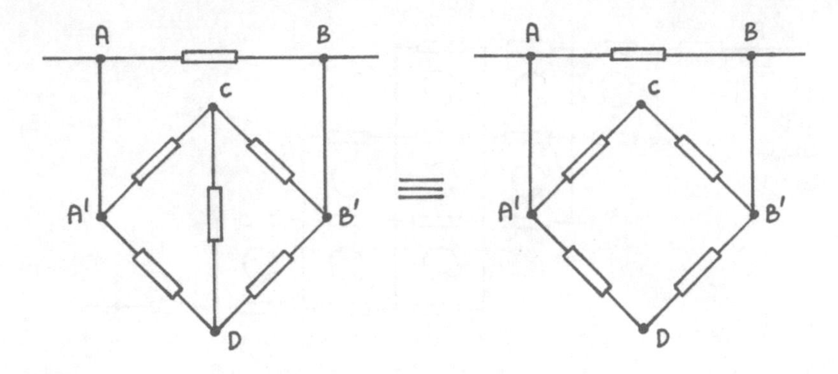

9.3 Resistor square ★★★★

A RESISTOR SQUARE is built with $N \times N$ resistors (the example shows a 4×4 resistor square) with individual resistance values R in each leg. What is the total resistance R_N across the long diagonal of the square (between A and B)?

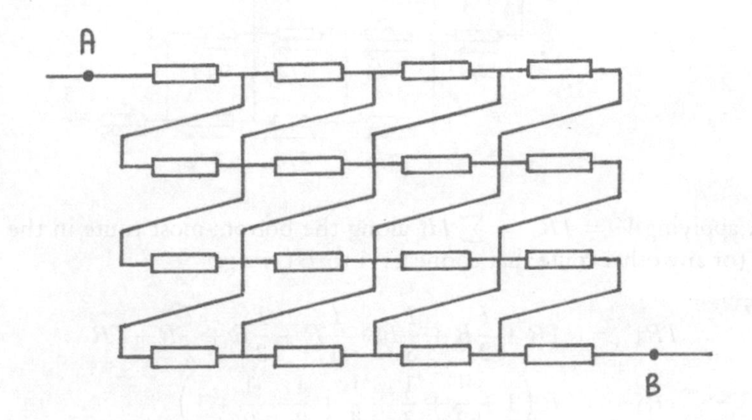

Answer

First we recognise that the $N \times N$ resistor square has a network of the type shown in the example of a 4×4 resistor square.

Consider the closed loops for each cell. All the resistors have the same value (and we assume the wires to have zero resistance), and each loop contains two resistors. So the condition that $\sum IR = 0$ round a closed loop implies that the same amount of current must flow through resistors that belong to the same loop.

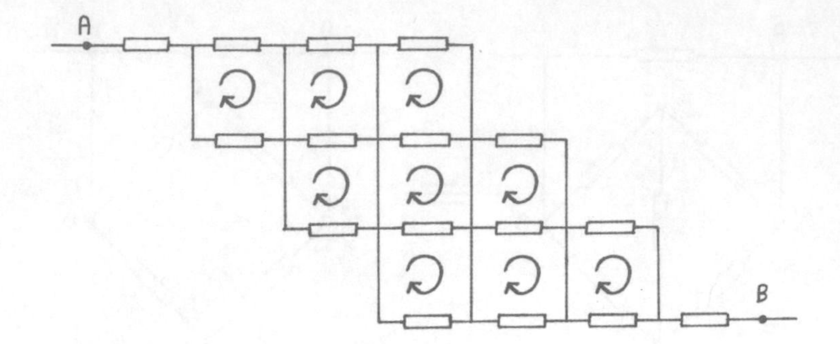

Applying this to each column of cells, we see that, for the network we have drawn, the current must be the same in all resistors in the same column. We get the currents shown in the diagram.

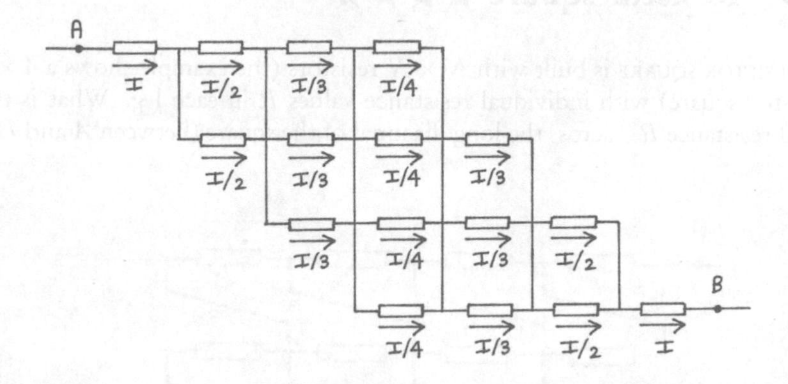

Now, applying $V = IR_4 = \sum IR$ along the bottom-most route in the diagram (or any other route that connects A to B), we get

$$
\begin{aligned}
IR_4 &= IR + \frac{I}{2}R + \frac{I}{3}R + \frac{I}{4}R + \frac{I}{3}R + \frac{I}{2}R + IR \\
R_4 &= R\left(1 + \frac{1}{2} + \frac{1}{3} + \frac{1}{4} + \frac{1}{3} + \frac{1}{2} + 1\right)
\end{aligned}
$$

where R_4 is the overall resistance of a 4×4 resistor square. Generalising this result to a square of size $N \times N$ gives

$$R_N = R\left[\left(2\sum_{r=1}^{N}\frac{1}{r}\right) - \frac{1}{N}\right]$$

This series[6] does not converge, so $R_\infty = \infty$.

[6]The series $\sum_{r=1}^{n}(1/r)$ is known as the *harmonic series*, and has applications in music, where it represents the wavelengths of the overtones of the fundamental wavelength of a string. The

9.4 Resistor cube ★★★

I have tried not to include too many of the most well-known physics puzzles in this book, because they can easily be found elsewhere. This one was too elegant to resist, however.

A RESISTOR CUBE has individual resistance values R in each leg. What is the total resistance R_T across the longest diagonal of the cube?

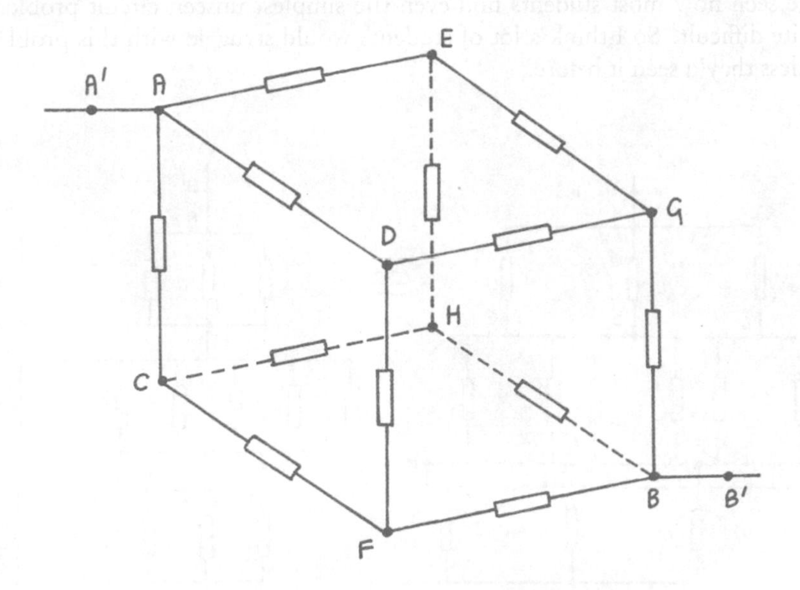

Answer

As with the resistor tetrahedron, I find it very much easier to think about this problem in flattened form, so we can reveal the essential structure. It should be easy to convince yourself that the cube flattens to the equivalent circuit shown. The nodes have been labelled in both figures to allow you

divergence of the series was first proved by Nicole Oresme (c. 1323–1382), a French philosopher of the Middle Ages, but the result was lost for hundreds of years, and was only proved again in the mid-seventeenth century. One way to prove divergence is to compare terms to another diverging series. Separating the first two terms and then grouping the remaining terms in blocks of 2, 4, 8, 16 and so on, we see that each group has a sum greater than 1/2.

$$\begin{aligned}\sum_{r=1}^{n}(1/r) &= 1+\frac{1}{2}+\left(\frac{1}{3}+\frac{1}{4}\right)+\left(\frac{1}{5}+\frac{1}{6}+\frac{1}{7}+\frac{1}{8}\right)+\ldots \\ &> 1+\frac{1}{2}+\left(\frac{1}{2}\right)+\left(\frac{1}{2}\right)+\ldots\end{aligned}$$

The sum of an infinite number of terms equal to 1/2 diverges, so the harmonic series must also diverge.

to cross-reference them. In this form the problem begins to look relatively straightforward. From its symmetry, we see that C, D and E are equipotential points, as are H, F and G. By inserting dummy wires (shown by dotted lines in the diagram) we can reduce the problem to three groups of parallel resistors in series with each other. The three individual groups can easily be shown to have resistances $R/3$, $R/6$ and $R/3$. The total resistance is therefore simply the sum: $R_T = 5R/6$.

Looked at this way, I don't think this is a particularly hard problem. But I've seen how most students find even the simplest unseen circuit problems quite difficult. So I think a lot of students would struggle with this problem unless they'd seen it before.

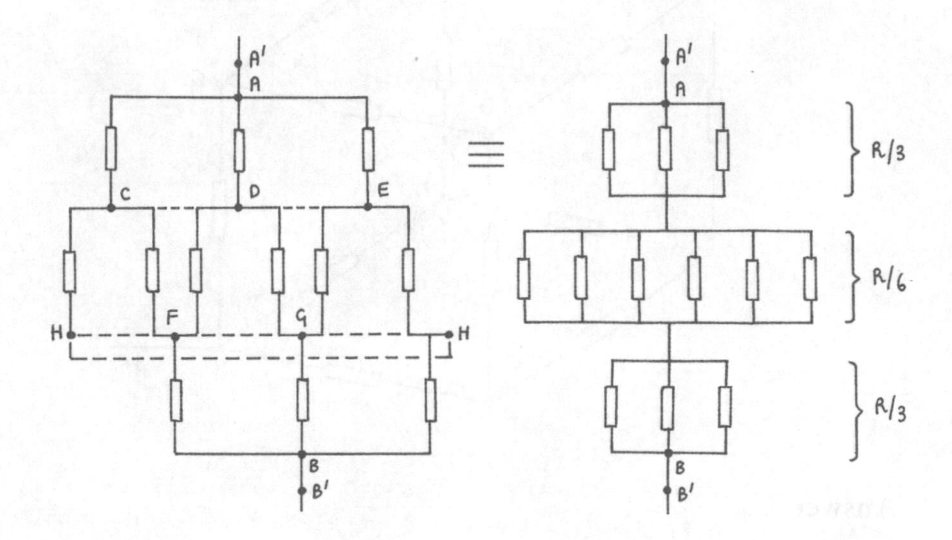

A current-symmetry argument

Now we consider a current-symmetry argument. About a diagonal axis running through points A and B of the cube there is rotational symmetry of order three. The system is totally symmetrical about this axis, so at the first branch the current I through the circuit must split into $I \to I/3 + I/3 + I/3$. Similarly, at the second branch, at each node it must split into $I/3 \to I/6 + I/6$. At the third (recombinative) branch, at each node there is $I/6 + I/6 \to I/3$. At the fourth (recombinative) branch there is $I/3 + I/3 + I/3 \to I$. So, whatever path we take from A to B, the potential difference between A and B is given by

$$V = \sum IR = (I/3)\,R + (I/6)\,R + (I/3)\,R = (5/6)\,IR$$

Thus we have $R_T = (5/6)\,R$, as before.

9.5 Power transmission ★

I was asked this question in my Oxford physics interview. I found it rather confusing at first.

CONSIDER THE FOLLOWING argument. It is well-known that power transmission lines operate at high voltage to minimise the current. Doing this minimises the power loss in the transmission wires. Power loss, P, is given by $P = I^2 R$, where R is the resistance of the cables and I the current we are trying to minimise. On the other hand, we know that $V = IR$, so $P = V^2/R$. Thus we could argue that we should operate at low voltage to minimise power loss. Discuss the apparent paradox.

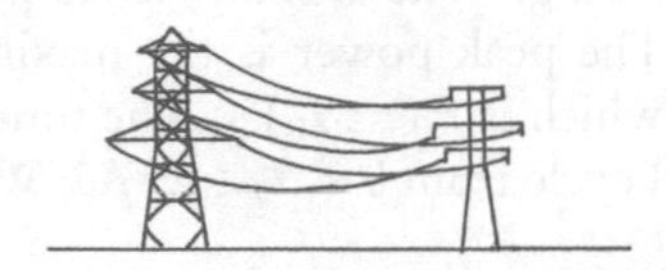

Answer

This is a bit of a trick question. There is no paradox, and we are simply tying ourselves up in knots by using V to represent two entirely different things. The first statement:

> It is well-known that power transmission lines operate at high voltage to minimise the current. Doing this minimises the power loss in the transmission wires. Power loss, P, is given by $P = I^2 R$, where R is the resistance of the cables and I the current we are trying to minimise.

is correct—transmission lines *are* operated at very high voltage indeed.[7] The high voltage V at which we operate the transmission cables is measured with respect to a *system ground*.

Separately, the equation $V' = IR$ (Ohm's Law) relates the current I passing through the wire to the resistance R of the wire and the *potential difference across the ends of the wire*, V'. So the equation $P = (V')^2/R$ is correct (though not particularly helpful) only if V' is taken to be the *voltage drop* in the wire, not the potential with respect to ground. That is, if the input voltage to a wire of resistance R is V, the voltage at the end of the wire is $V - V' \approx V$, where for most efficient transmission systems $V' \ll V$.

[7]Ultra-high voltage (UHV), over 800 kV, and extra-high voltage (EHV), over 230 kV, are used to transmit power over long distances. Medium voltage (MV), between 1 kV and 33 kV, is used on the main transmission routes though urban areas, and low voltage (LV), either 400 V or 230 V, is used in residential areas.

9.6 RMS power ★

IN AN AC CIRCUIT with a *purely resistive load*, voltage and current vary as $V = V_0 \sin(\omega t)$ and $I = I_0 \sin(\omega t)$. Examples of such circuits include an electric kettle and a resistive-element (old-fashioned, incandescent) light bulb. Find the relationship between the peak power P_0 and the average (*time-mean*) power $\bar{P}$.

Answer

This is a rather easy question, and rather dull, but it is a popular textbook proof that is worth knowing. The instantaneous power is given by $P = IV = I_0 V_0 \sin^2(\omega t)$. The peak power is the maximum value the instantaneous power can take, which is $P_0 = I_0 V_0$. The time-mean power, $\bar{P}$, is the average power over a full cycle from $0 < t < 2\pi/\omega$. We write

$$\bar{P} = \frac{\int_0^{2\pi/\omega} P}{(2\pi/\omega)} dt = \frac{I_0 V_0}{(2\pi/\omega)} \int_0^{2\pi/\omega} \frac{1}{2}\left[1 - \cos(2\omega t)\right] dt$$

$$\bar{P} = \frac{I_0 V_0}{(2\pi/\omega)} \frac{1}{2}\left[t - \frac{\sin(2\omega t)}{(2\omega)}\right]_0^{2\pi/\omega} = \frac{I_0 V_0}{(2\pi/\omega)} \frac{1}{2}\left[(2\pi/\omega)\right]$$

$$\bar{P} = \frac{I_0 V_0}{2} = \frac{P_0}{2}$$

The peak power is twice the average power. To extend the question a little, you could plot the power on the same axes as the current and voltage. This is so easy that I leave it as an exercise for you to try.

9.7 Boiling time ★

BOILING A KETTLE with enough water to make four cups of tea takes three minutes using 240 V RMS AC[8] mains power.[9] A sailor decides to take the same kettle on a sailing trip, and modifies the plug for his boat, which has a 24 V DC[10] circuit. Estimate how long the kettle will take to boil on board the boat.

[8] RMS AC means root mean squared, alternating current.

[9] The nominal domestic supply voltage in the UK used to be 240V ±6%, but (following the European Union Low Voltage Directive) is now 230 V +10%/-6%. This directive was designed to harmonise supply voltage and allow equipment throughout Europe to be optimised for smaller voltage range.

[10] Direct current.

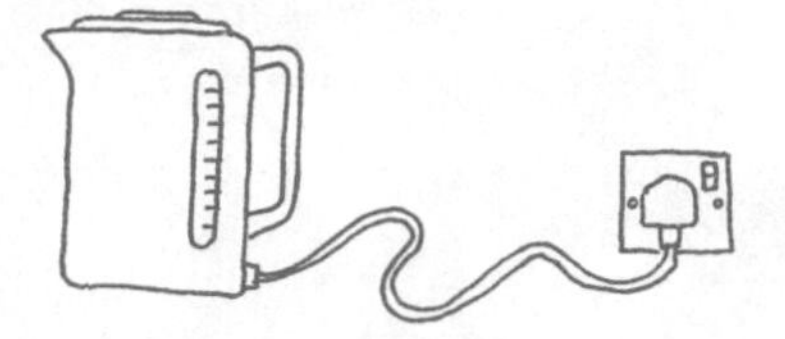

Answer

A kettle is really just a resistance element in good thermal contact with the water in it. The power dissipated is given by $P = I^2 R$, where I is the current and R is the resistance of the element. If we denote I_{240} and I_{24} as the currents with 240 V[11] and 24 V supplies, respectively, using Ohm's Law ($V = IR$) we can show that

$$\frac{I_{24}}{I_{240}} = \frac{24}{240} = \frac{1}{10}$$

We see that the power ratio for the kettle is given by

$$\frac{P_{24}}{P_{240}} = \left(\frac{I_{24}}{I_{240}}\right)^2 = \frac{1}{100}$$

So our initial estimate is that it will take $100 \times 3 = 300$ minutes to boil. In other words, five hours. Of course, a kettle that runs at such low power will never come to the boil, because when it gets relatively warm, heat will be lost (through radiation, for example, or conduction) faster than it can be gained from the heating element. It would clearly be hopeless to attempt to run a normal domestic kettle on a boat.[12]

This is an easy question, but—in my view at least—a rather neat one. I was asked it by a friend who is a sailor. I'm not sure if it is the main reason why electric kettles are not used on boats, but it certainly shows that to boil water at sea you'd need a kettle designed for the purpose.

[11]The voltage of mains power is normally quoted as an RMS value. Thus, the DC equations for power, $P = I^2 R$ and $P = V^2/R$, apply. This is one reason for quoting RMS voltage values.

[12]To make the problem worse, many boats run at 12 V rather than 24 V.

Chapter 10

Gravity

In this chapter we look at problems involving gravity, that strange force of mutual attraction which holds over very great distances. Gravity, by a very long way, is the weakest of the *four fundamental forces,*[1] but it is what glues the universe together over huge length scales. Without it, we could not rotate about our life-sustaining Sun. As with many of the fundamental laws of classical physics, gravity (for most practical purposes at least) obediently follows a very simple equation. The attracting force between two bodies of masses M and m, separated by a distance r, is given by

$$F = \frac{GMm}{r^2}$$

where G is the gravitational constant. Gravity obeys an *inverse square law* and is proportional to the masses of the interacting bodies.

We need to know a number of constants to solve problems that involve gravity. Some of these constants are given below.

---

[1]The four *fundamental forces* are:

1. THE STRONG FORCE. This holds nuclei together, acting against the incredibly strong repulsive forces of protons. The strong force is, as the name suggests, the strongest of the four fundamental forces, but it acts over an extremely short range. The mechanism of this force is still being debated. But in the *Standard Model* it is mediated by the exchange of *gluons* between the quarks that make up protons and neutrons.

2. THE ELECTROMAGNETIC FORCE. This is a combination of the force between charges given by Coulomb's Law, and the magnetic force. Like the gravitational force, it obeys an inverse square law. The electromagnetic force holds atoms and molecules together. The force you feel when you smack your hand against a table is the electrostatic repulsion between the molecules in your hand and those in the table.

3. THE WEAK FORCE. This force acts over a minute range (about 0.1% of the diameter of a proton, or 10^{-18} m) and is responsible for the *transmutation* of quarks from one *flavour* to another. It is essential in explaining nuclear reactions.

4. THE GRAVITATIONAL FORCE. This is considered in the current chapter.

$$\text{Gravitational constant } G \approx 6.7 \times 10^{-11}\,\text{N kg}^{-2}\text{m}^2$$

$$\text{Radii}\begin{cases}\text{Earth } R_E \approx 6.4 \times 10^6 \text{ m}\\ \text{Sun } R_S \approx 7.0 \times 10^8 \text{ m}\\ \text{Moon } R_M \approx 1.7 \times 10^6 \text{ m}\end{cases}$$

$$\text{Distances}\begin{cases}\text{Earth-Sun } R_{ES} \approx 1.5 \times 10^{11} \text{ m}\\ \text{Earth-Moon } R_{EM} \approx 3.8 \times 10^8 \text{ m}\end{cases}$$

$$\text{Masses}\begin{cases}\text{Earth } M_E \approx 6.0 \times 10^{24} \text{ kg}\\ \text{Sun } M_S \approx 2.0 \times 10^{30} \text{ kg}\\ \text{Moon } M_M \approx 7.4 \times 10^{22} \text{ kg}\end{cases}$$

10.1 The hollow moon ★

The force due to gravity acting on a mass m on the surface of a solid sphere of uniform density, mass M, and radius R is given by Newton's Law of Universal Gravitation.

$$F = \frac{GMm}{R^2}$$

Newton showed that the theorem, which was initially developed for point masses, also applied to objects with spherically symmetric mass distributions. This class of objects includes solid spheres of uniform density and spherical shells of uniform density.

USING THE RESULT for a solid sphere, show without using integration that the force due to gravity on the surface of a spherical shell must be the same as that on the surface of a solid sphere, provided they have the same outer radius R and mass M, and are of uniform density. Show also that this result is independent of the inner radius of the spherical shell, r.

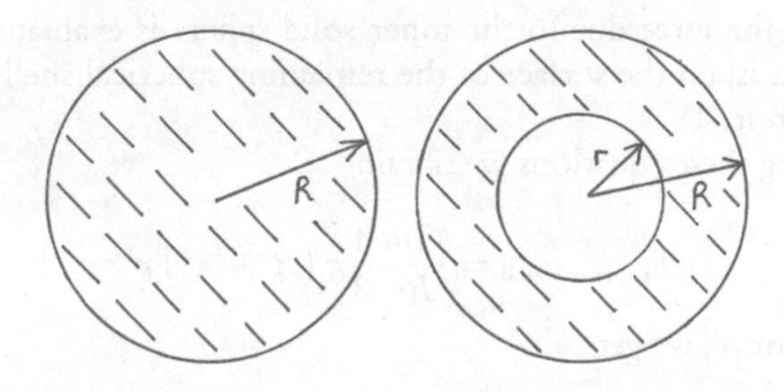

Answer

First consider the solid sphere of uniform density and radius R. We know that the force due to gravity on the surface of a solid sphere of uniform density is given by

$$F_{\text{solid sphere}} = \frac{GMm}{R^2}$$

The mass of the solid sphere is given by

$$M_{\text{solid sphere}} = \frac{4}{3}\pi R^3 \rho$$

where ρ is the density.

Now consider a uniform spherical shell of outer radius R, inner radius r, and density ρ'. The mass of the spherical shell is given by

$$M_{\text{spherical shell}} = \frac{4}{3}\pi \left(R^3 - r^3\right) \rho'$$

Equating the masses of the solid sphere and the spherical shell, $M_{\text{spherical shell}} = M_{\text{solid sphere}} = M$, we obtain an expression for the density ratio.

$$\frac{\rho'}{\rho} = \frac{R^3}{R^3 - r^3}$$

Now that we have an expression for the density of the spherical shell, we devise a way to calculate the force due to gravity on the surface of that shell, using only Newton's Law of Universal Gravitation.

We can think of the spherical shell as a solid sphere of radius R and density ρ' that has had a solid sphere of radius $r < R$ and density ρ' removed from its centre. The force due to gravity on the surface of the spherical shell is therefore the difference between the two gravitational forces. We write

$$F_{\text{spherical shell}} = \frac{GM_R m}{R^2} - \frac{GM_r m}{R^2}$$

where

$$M_R = \frac{4}{3}\pi R^3 \rho' \text{ and } M_r = \frac{4}{3}\pi r^3 \rho'$$

We note that the force due to the inner solid sphere is evaluated at a radius R, not r—that is, on the surface of the remaining spherical shell which is the object of interest.

Combining these equations we obtain

$$F_{\text{spherical shell}} = \frac{Gm}{R^2}\frac{4}{3}\pi \left(R^3 - r^3\right) \rho'$$

Substituting for ρ', we get

$$F_{\text{spherical shell}} = \frac{Gm}{R^2}\frac{4}{3}\pi R^3 \rho = \frac{GMm}{R^2}$$

That is, $F_{\text{spherical shell}} = F_{\text{solid sphere}}$. Provided they have the same mass and radius and are spherically symmetric, a solid sphere (a "solid moon") and a spherical shell (a "hollow moon") have the same gravity.

I rather like this result. It is simple and neat, and shows that once you are outside a spherically symmetric mass distribution, there must be consistency for spherical shells and solid spheres, and by extension any combination of spherical shells and solid spheres including a point mass (try it if you are interested). The simplest way to state the result is that spherically symmetric bodies act on external objects as though their entire mass was concentrated at a central point. Newton worked this out in 1687. It is now called the *shell theorem*. Another interesting consequence of the shell theorem is that there is no net gravitational force inside a shell. It is possible to come to even this latter result by using similar logic to that used above. Try it!

10.2 Lowest-energy circular orbit ★★

TAKE THE EARTH to be a uniform, spherical attracting body of radius R_E. For a satellite in a circular orbit, what orbital radius R_2 has the lowest total energy with respect to a point on the surface of the Earth?

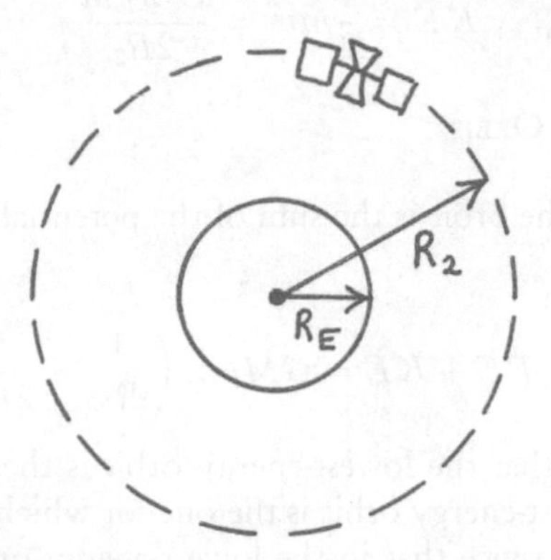

Answer

The total energy E of a particular circular orbit with respect to the surface of the Earth is the sum of the potential energy PE and the kinetic energy KE of the orbit. Let's consider each of these energies in turn.

1) POTENTIAL ENERGY OF ORBIT

We can calculate the potential energy of an orbit of radius R_2 by performing an integral of the force required to overcome gravity between the starting radius R_E and the orbital radius R_2. We write

$$PE = \int_{R_E}^{R_2} F dr = \int_{R_E}^{R_2} \frac{GM_E m}{r^2} dr$$

where G is the gravitational constant, M_E the mass of the Earth, m the mass of the satellite, and r the radius at a point of interest. Evaluating the integral, we have

$$PE = GM_E m\left(\frac{1}{R_E} - \frac{1}{R_2}\right)$$

2) Kinetic Energy of Orbit

Now we consider the kinetic energy of a particular orbit at radius R_2. For a stable circular orbit, the force due to gravity must be equal to the force required to achieve an acceleration v^2/R_2 towards the centre of the orbit. We write

$$\frac{GM_E m}{R_2^2} = \frac{mv^2}{R_2}$$

Using this, we can write the kinetic energy as

$$KE = \frac{1}{2}mv^2 = \frac{GM_E m}{2R_2}$$

2) Total Energy of Orbit

The total energy of the orbit is the sum of the potential and kinetic energies. We write

$$E = PE + KE = GM_E m\left(\frac{1}{R_E} - \frac{1}{2R_2}\right)$$

We immediately see that the lowest-energy orbit is the one with the lowest radius R_2. The highest-energy orbit is the one for which $R_2 \to \infty$, for which the energy is exactly twice that in the lowest-energy orbit, for which $R_2 = R_E$.[2]

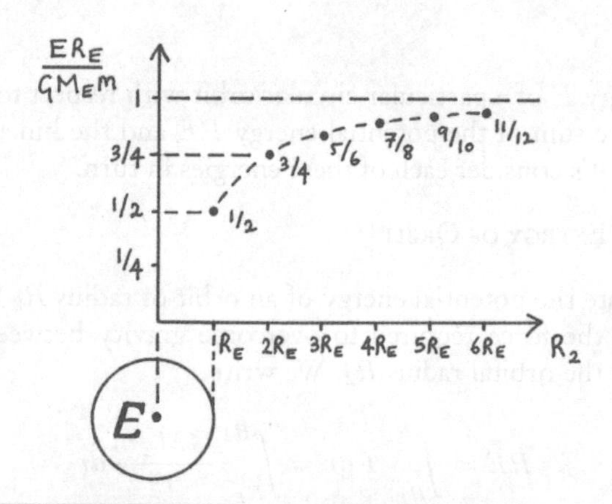

[2]Here we ignore the Earth's atmosphere. In fact, it's a perfectly reasonable first-order approximation for real satellite orbits anyway. The Low Earth Orbit band starts at an altitude of approximately 160 km. This is beyond the outer edge of the atmosphere, but only 2.5% of an Earth radius above the surface of our planet—almost scraping the surface.

10.3 Weightless in space ★★

COSMONAUT BLASTOV (from Russia) and Captain Medallion (from the US) are having a day off at the International Space Station, 400 km above the surface of the Earth. One is a medical doctor and the other is a physicist, and both are highly renowned in their field. Medallion turns to Blastov and says, "look at how my medals float in the air—isn't it amazing to be in such low gravity?" Blastov looks confused for a moment, then turns to Medallion and says, "vot has gravity to do viz ze floating? Ve are having here almost ze same veight as in Moscow. No?"

Show, with calculations, which of the two is the physicist.

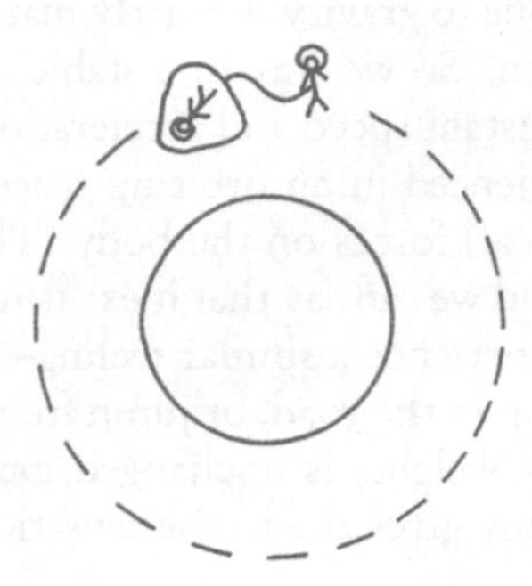

Answer

The apparent "weightlessness" in space is a commonly misunderstood topic. In fact, the force due to gravity on a person—which we define to be the *weight* of a person—is very similar in a Low Earth Orbit to their weight on the surface of the Earth. Before we discuss the problem further, let's calculate the force due to gravity 400 km above the surface of the Earth, F', and compare it to the value on the surface of the Earth, F.

The force due to gravity at an arbitrary radius r is given by

$$F = \frac{GMm}{r^2}$$

So the ratio of the force in a 400 km-high Low Earth Orbit to that on the surface of the Earth is given by

$$\frac{F'}{F} = \left(\frac{R_E}{R_E + 4 \times 10^5}\right)^2 \approx 0.89$$

(where we have taken $R_E \approx 6.4 \times 10^6$). At the height of the International Space Station, a person's true weight is reduced by only 11%. This is a small effect, and not—on its own—enough to cause Captain Medallion's medals to float impressively from his lapel. So Blastov is right: the gravity and therefore the true weight (which we recall is defined as the force due to gravity) are very similar in a Low Earth Orbit to their values in Moscow—or indeed anywhere on the surface of the Earth.

Despite his misguided reasoning, however, Medallion continues to float around the capsule admiring his military honours, and—*apparently* at least—is weightless. Cosmonaut Blastov decides to give him some lessons in basic physiology.

"Medallion, your American medical training is disappointing," says Blastov. "Ze human body is exceptional at sensing non-isotropic stress,[3] which is vot ve feel ven ve have external forces acting on us. Ze external force is vhy ve are feeling so different."

Blastov, as usual, is right. A Low Earth Orbit is a good example of a situation in which the force due to gravity is exactly matched to the force required for centripetal acceleration. So we stay in a stable orbit at a constant height above the Earth, with constant speed and acceleration. The unusual sensation of "weightlessness" experienced in an orbiting spaceship arises because there are no external (mechanical) forces on the body. Thus the body experiences no non-isotropic stress, and we can say that it exists in something close to zero-stress conditions.[4] We experience a similar feeling—for a moment at least—if we go quickly over a hump in the road, or jump from an aeroplane. The gravitational force on us (our weight) is unchanged, but we remove the ground reaction force that normally gives rise to the sensation of weight.

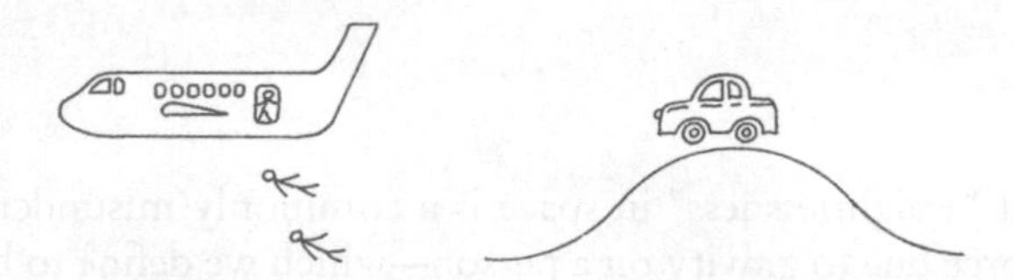

Equally, we can artifically enhance the ground reaction force. This is what we feel when we stand in a lift that is accelerating upwards. The force at our feet not only needs to overcome the gravitational force on our body, it also needs to provide an upward acceleration for the mass of our body. In this example it is only the very top of the head that exists in a zero-stress state, because it neither needs to support any weight above it, nor accelerate any mass above it.

So we can have the sensation of weightlessness whether or not there is an absence of gravitational force on us. This absence can only occur either

[3]It is commonly stated that the sensation of weightlessness we feel in free fall is the result of the body being in a *stress-free state*. I think what we must really mean here is the absence of *non-isotropic stresses*, because the human body seems very insensitive to changes in isotropic stress. Consider, for example, diving at the beach. At a depth of 10 m we have double the normal (atmospheric) pressure acting on us, and double the isotropic stress. But we feel the same as we do on the surface. The current freediving world record (set on 6 June 2012 by Herbert Nitsch, "the deepest man on Earth") was a massive 253.2 meters (830.8 feet). This represents a change in pressure of 25.3 atmospheres, generating huge isotropic stresses within the body, which we don't feel.

[4]Here we ignore any initial stresses within the body that are not due to external forces.

when we are infinitely far from other masses, or when we are between masses such that their gravitational fields cancel. In fact, humans have never been substantially beyond the gravitational field of the Earth. Almost all human space flight has been in Low Earth Orbit, where the true gravitational force is similar to that on Earth. The only exceptions are the Apollo moon missions, in which astronauts have been to quite low-gravity conditions with respect to the Earth, but have not escaped the Earth's gravitational field entirely.

We use the term *g-force* to describe the sensation of weight—g-force is the total mechanical force we experience due to both the acceleration and the resistance of the gravitational force. A man standing on the surface of the Earth experiences a force equal to 1 g. This is due to the mechanical resistance of the force of gravity. The Earth pushes up with a force equal to the gravitational force acting on the man. Equally, we can imagine a man in a spaceship in deep space, far from all sources of gravitational attraction. In this condition he exists in a state of 0 g. We describe the fact that there are no mechanical forces causing a sensation of weight, not the fact, necessarily, that there are no gravitational forces and therefore that there is no actual weight. To illustrate this, now consider that the spaceship accelerates at a rate of 9.8 m s^{-2}. We say that the man experiences 1 g, the same sensation of force as the non-accelerating man on the surface of the Earth. If we were to blindfold the man in the spaceship and the man on Earth, they would not be able to distinguish whether the 1 g sensation, the g-force, was due to the action of gravity or due to acceleration. The sensation of 1 g is the same in both cases.

Let's return now to the case of Low Earth Orbit. This is a situation in which the gravitational force is similar to that on the surface of the Earth. However, the gravitational force causes acceleration towards the centre of the Earth, so the requirement for mechanical resisting forces is removed. A person experiences 0 g, or so-called "weightlessness". The term is quite unfortunate.

Humans seem to have a high tolerance to extremes of g-force for short periods. The average person can tolerate about 5 g for a few seconds without losing consciousness. We used to think that pilots could not stand more than about 6 g for periods of more than a few seconds. But modern pilots can now withstand up to 9 g in sustained turns. Their tolerance is increased by a combination of training, in which they learn to contract certain muscles, and g-suits that inflate rapidly during high-g turns to exert pressure on the legs and abdomen. The aim of both the training and the g-suit is to prevent blood draining from the brain, which can lead to hypoxia. The symptoms of hypoxia are greyout (in which the vision dims), followed by blackout (in which the vision disappears entirely), followed by g-induced loss of consciousness (or g-LOC). A bad thing if you are at the controls of a high speed aircraft! The other extreme, the absence of g-force experienced in orbiting spacecraft (technically micro-gravity), or 0 g, seems to have a completely benign effect on the human body. The main deleterious effects seem to be related to lack of exercise.

The human being who has voluntarily sustained the highest g-force is almost certainly John Paul Stapp (1910–1999), a US Air Force surgeon, who researched the effect of extreme acceleration on the human body. Between 1947 and 1951 he was part of a human deceleration project, and repeatedly climbed aboard a rocket sled that was fired along a railway track at extreme speeds before being decelerated with a water-braking system. Sustaining numerous injuries as the result of his rides on the Gee Whiz (such as broken bones, detached retinas, and blood vessel damage), Stapp survived decelerations of up to 46 g, more than twice the commonly accepted fatal limit of around 18 g. The research was used to develop harnesses for forward-facing seats, but it showed that backward-facing seats afforded much better protection in crashes.[5] Human tolerance of short-duration deceleration was far higher than anyone had thought.

As if that wasn't enough for one man, Stapp also did research into the physiological effects of extreme wind blasts, to see if humans could survive high-speed flight in jet aircraft with damaged canopies. He flew at 570 mph with no canopy and survived. Stapp also had a fondness for inventing amusing aphorisms. By far the most famous one he originated was Murphy's Law: anything that can go wrong will go wrong. Rather nicer is the Stapp Paradox, which states that "the universal aptitude for ineptitude makes any human accomplishment an incredible miracle."

10.4 Jump into space ★★★

ON PLANET X, Mr Bendylegs can jump to a height h_x, where $h_x \ll R_x$, the radius of the planet. "My problem is gravity," says Bendylegs. "Were it not for gravity, I could jump into space." What is the largest planet Y, composed of the same material as planet X, that would allow Bendylegs to escape the gravitational field entirely?

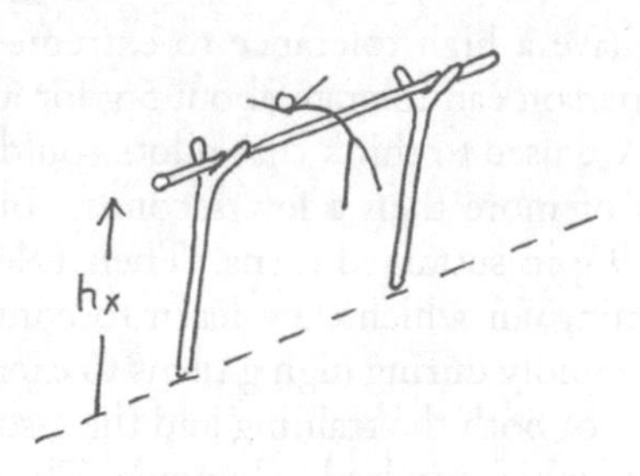

Answer

On planet X, Bendylegs can jump to a height h_x, where $h_x \ll R_x$. When he performs his jump, we can think of Bendylegs working against a uniform gravitational field. His jump is so small compared to the radius of planet

[5] Backward-facing seats are used in some military planes, but have not been adopted for civilian flight, presumably because passengers prefer to face forward.

X that there is almost no change in the force on him due to gravity over the height of his jump. We can calculate the energy of his jump, E, by considering the work done by a hypothetical force that opposes the gravitational force. The work done is this force multiplied by the distance to the highest point of his jump, h_x. We write

$$E = Fh_x = \frac{GM_X m}{R_X^2} h_x$$

where m is the mass of Bendylegs himself.

Now consider the total work, W, that is required to take a mass, m, from the surface of a planet with radius R and move it with a force that acts against the diminishing gravitational pull until the mass is infinitely far from the planet (in other words, until it has escaped the planet's gravitational pull). We can call this work the *escape energy*.

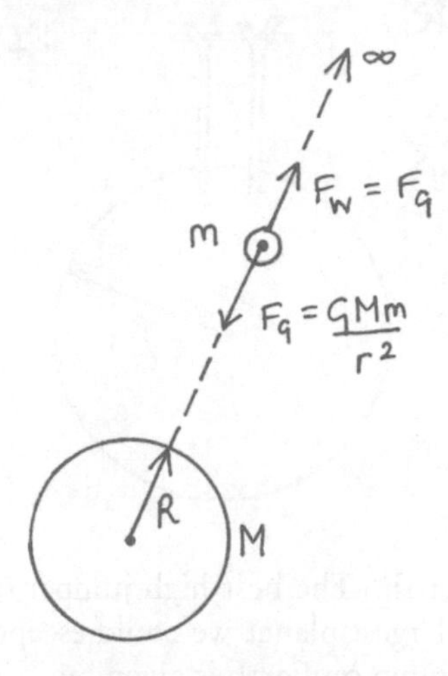

To find the escape energy we need to integrate the hypothetical force required to perform work on the body, F_W, against a gravitational force, F_G. We integrate from the surface of the planet to infinity. The force F_W is equal in magnitude and opposite in direction to the force due to gravity, F_G, at the same radial location, r. We write

$$W = \int_R^\infty F_W dr = GMm \left[\frac{r^{-1}}{-1} \right]_R^\infty = \frac{GMm}{R}$$

To escape the gravitational field of planet Y, the energy Bendylegs puts into his jump must equal the escape energy for planet Y. We already have a measure of the energy Bendylegs can put into his jump on planet X. So we write

$$W = E$$

$$\frac{GM_Y m}{R_Y} = \frac{GM_X m}{R_X^2} h_x$$

If the planets are composed of the same material—that is, they have the same density—then we can substitute for M_X and M_Y using the equation for the mass of a solid sphere, $M = \frac{4}{3}\pi R^3 \rho$. We can then write

$$R_Y = \sqrt{R_X h_x}$$

which is the answer Bendylegs needs.

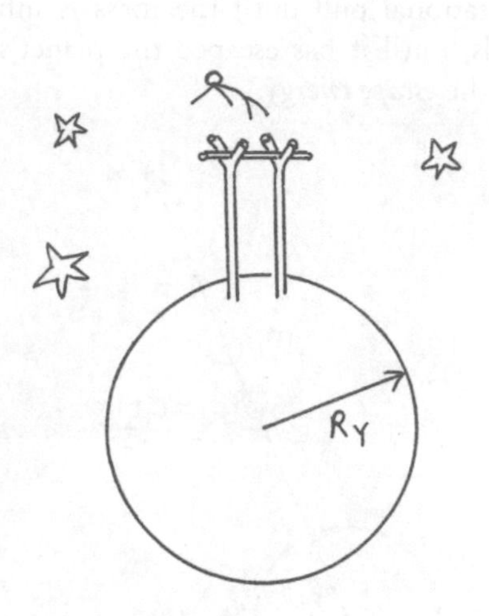

Let's return to planet Earth. The best high jumper can jump approximately 2 m. The radius of the largest planet we could escape from using the energy required for a 2 m high jump on Earth is given by

$$R_Y = \sqrt{R_E 2}$$

where R_E is the radius of the Earth ($R_E \approx 6.4 \times 10^6$ m). This gives $R_Y \approx 3,521$ m, or 3.5 km. The answer, as we might expect, is independent of the mass of the person performing the jump—if it's not obvious why this should be so, think about it for a bit.

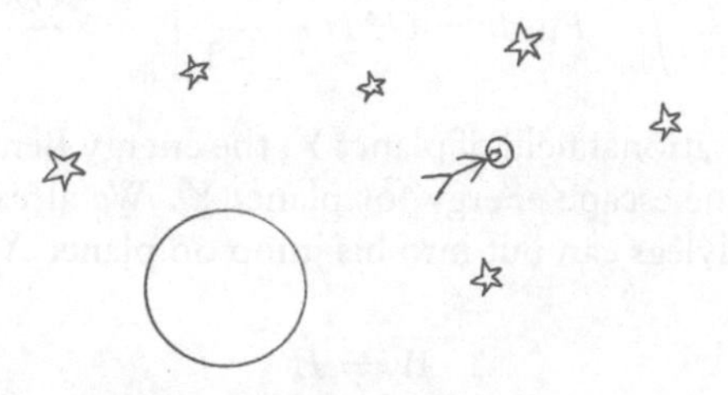

Further discussion

When we worked out the energy of the high jumper's jump, E, we glossed over a subtlety that I'd like to return to for a moment.

Consider what happens when you bend your legs at the knee, lowering your centre of gravity, and then straighten them again. As you straighten them—and we could call this motion the *power stroke*—you work against gravity and potentially also act to accelerate your body. If you do it slowly enough you have no kinetic energy at the top of the stroke and you simply return to a standing position with zero upward velocity. The amount of work you have done is $\int F_W dx = mg\Delta x$, where Δx is the difference in height of your centre of gravity between the beginning and the end of the power stroke. This work done does not help us escape the gravitational field of the planet, it simply restores the potential energy we had before we bent our legs. The motion we have described is one in which the force performing work is approximately equal to the force on our body due to gravity, so $F_W \approx F_G$. Our acceleration during this process is very small.[6]

We see that the force that gives rise to acceleration is the difference between F_W and F_G. We write

$$F_W - F_G = m\ddot{r}$$

where $\ddot{r}$ is the acceleration in the radial direction, and $F_G = mg$. For $F_W \gg F_G$, most of the work we perform in the power stroke goes into kinetic energy at the point we leave the ground, which allows us to gain height. For F_W only slightly greater than F_G, almost no energy is available as kinetic energy at the end of the power stroke.

This analysis describes one aspect of how humans do work when jumping. Our physiology limits the maximum force that we can impose on the ground during the power stroke, and the total displacement during the power stroke. So the total energy put into the system is quite well-defined. But it is well-defined only with respect to the correct *zero-energy datum*. This is the position of the man's centre of mass when his legs are bent at the very start of the power stroke, not his centre of mass when he's standing.

We specified the height the man could jump with reference to the ground as $h_x = 2$ m. We now see that we used the wrong *datum height*. The rational datum height in both systems (X and Y)—now we are thinking in terms of the system energy—should be the height of the man's centre of mass when his legs are bent, at the beginning of the power stroke. This changes our answer a little. Now we understand the principles, I'll leave the more precise calculation for you to try.

---

[6]To initiate the motion at the start of the stroke we require F_W to be very slightly greater than F_G so that we achieve some finite velocity of our centre of mass upwards. Towards the end of the stroke we require F_W to be very slightly smaller than F_G so that we decelerate to zero velocity again.

10.5 Space graveyard ★★★

The change in velocity of a spacecraft during a *burn*[7] is governed by the Tsiolkovsky rocket equation.[8]

[7]Here we mean a very short period of time in which the rocket engine is fired and the velocity of the rocket is changed quickly. In many practical cases, in terms of the change in trajectory, the *burn* can be regarded as instantaneous.

[8]Here is a short derivation of the Tsiolkovsky rocket equation. Consider a rocket with mass m and velocity V as measured in the lab frame. We can also represent the rocket in its own frame of reference, in which the velocity is zero.

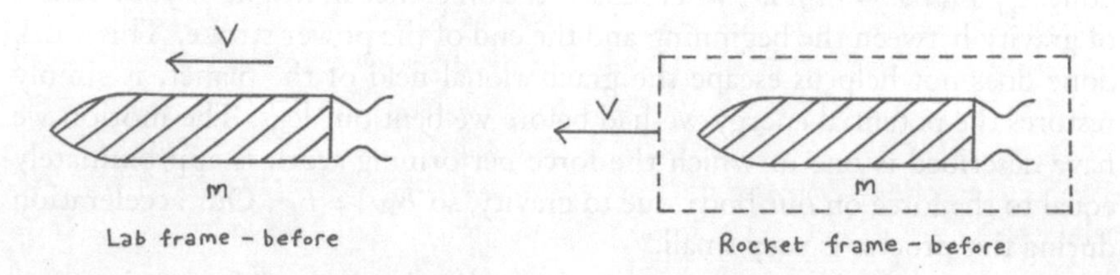

Allow the rocket to eject a small mass element dm at a velocity $-V_e$ with respect to the rocket (that is, as measured in the rocket frame). This increases the velocity of the rocket to $V + dV$ in the lab frame. The velocity of the mass element dm in the lab frame is $V + dV - V_e$.

The sum of the external forces on the rocket and the ejected mass element is zero, so the momentums before and after ejecting the mass element are the same. In the lab frame of reference we write

$$\begin{aligned} \text{Momentum before} &= \text{Momentum after} \\ mV &= (m - dm)(V + dV) + dm\,(V + dV - V_e) \end{aligned}$$

Here we are taking momentum to the left as positive. Expanding and simplifying, we get

$$0 = mdV - dmV_e$$

This looks like a differential equation in m and V, which we can integrate.

$$V_e \frac{dm}{m} = dV$$

Before we integrate, however, we note that positive dm, as defined in the question (that is, a positive mass element *leaving* the rocket), leads to a reduction in m. To relate dm and m in a rational way requires us to introduce a minus sign (that is, to swap dm for $-dm$). Doing this, and integrating between initial and final masses m_i and m_f, and between initial and final velocities V_i and V_f, we have

$$\Delta V = V_e \ln \frac{m_i}{m_f}$$

where V_e is the exhaust velocity of the rocket, and m_i and m_f are the initial and final masses of the rocket. Strictly speaking, it applies only in situations where no other forces act, so doesn't apply when there is *gravity drag*[9] or aerodynamic drag. But we can think of it as a good approximation when the thrust is very much larger than any external force.

The equation is derived by considering conservation of momentum for mass ejection from a rocket with constant relative exhaust velocity. Rocket engineers refer to the quantity ΔV as the "delta-v" required to perform a particular orbital manoeuvre. In mission planning at a conceptual stage, a rocket engineer will draw up a *delta-v budget*, listing the delta-v requirements of every planned manoeuvre in the mission. This defines the overall delta-v requirement of the rocket. If a rocket has to take a fixed payload m_f on a mission with a particular delta-v requirement, taken together this defines the rocket's required exhaust velocity and initial mass.

COSMONAUT BLASTOV AND Captain Medallion are in an experimental spaceship in a circular Earth orbit. The craft has malfunctioned, and to safely dispose of it, and to preserve honour in both Russia and the US, the men have only two options. The first is to slow the craft until they can perform a *dead drop* to the centre of the Earth (path A), and the other is to achieve escape velocity and effectively leave the Earth's gravitational influence (path B). The rocket's engine is very powerful, so the impulse that the craft will be given during the burn will occur very quickly. Fuel is limited, however, so they are unsure how much delta-v they can achieve. Medallion turns to Blastov and says, "we are in Earth orbit, we should take advantage of the gravitational field and go with option A."

"On ze contrary," says Blastov. "Viz such lee-tle thrust left ve are more likely to manage path B."

Which path, A or B, requires the smaller amount of fuel?

$$V_e \int_{m_i}^{m_f} \frac{-dm}{m} = \int_{V_i}^{V_f} dV \text{ or } V_e \ln \frac{m_i}{m_f} = V_f - V_i = \Delta V$$

which is the Tsiolkovsky rocket equation.

Rather more rigorous derivations are plentiful on the internet, and you should consult them if you are interested.

[9]Gravity drag describes the thrust used to support the weight of a rocket when thrusting in a gravitational field. This thrust does not contribute to acceleration. Gravity drag is minimised in trajectory design by reducing the time period over which the impulse is applied, and thrusting as perpendicular as possible to the local gravitational field.

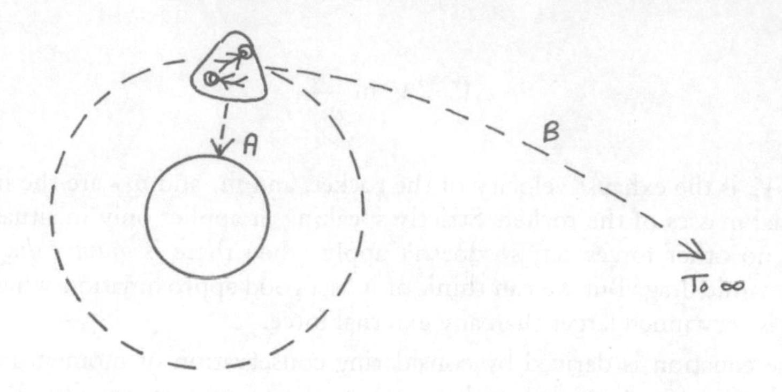

Answer

For a circular Earth orbit at a given radius R, the gravitational force must be equal to the force required for centripetal acceleration.

$$F = \frac{GM_E m}{R^2} = \frac{mV_i^2}{R}$$

This defines the initial velocity of the spaceship that Medallion and Blastov are travelling in, V_i, in terms of the radius of the orbit and the mass of the Earth, M_E.

$$V_i^2 = \frac{GM_E}{R}$$

Let's consider the delta-v required for paths A and B in turn. Path A is a dead drop towards the centre of the Earth. To execute this manoeuvre we need a ΔV that leaves us with zero final velocity with respect to the centre of the Earth. We take the ΔV to be always a *postive quantity* (the impulse given to the rocket in the *desired* direction) and write

$$\Delta V_A = |V_f - V_i| = |0 - V_i| = V_i$$

Path B is an acceleration to the escape velocity of the Earth. The escape energy is the integral from R to ∞ of a hypothetical force, F_W, required to overcome the force of gravity, F_G. Here F_W is equal in magnitude and opposite in sign to F_G. We write

$$W = \int_R^\infty F_W dr = GM_E m \left[\frac{r^{-1}}{-1}\right]_R^\infty = \frac{GM_E m}{R}$$

Using the expression for the orbital velocity (the initial velocity) we see that

$$W = mV_i^2$$

When the kinetic energy of the spaceship is equal to the escape energy, the spaceship can just escape the Earth's gravitational field.

$$\frac{1}{2}m\left(V_i+\Delta V_B\right)^2=W=mV_i^2$$

Using the expression $V_i=\Delta V_A$, we get

$$2\Delta V_A\Delta V_B=\Delta V_A^2-\Delta V_B^2$$

So

$$\Delta V_A-\Delta V_B=\frac{2\Delta V_A\Delta V_B}{\Delta V_A+\Delta V_B}$$

In the above analysis, ΔV_A and ΔV_B are both defined as positive. We conclude that $\Delta V_A>\Delta V_B$. Blastov is right: the delta-v requirement is lower to take path B than to take path A. In other words, to send the malfunctioning satellite out of the Earth's gravitational influence, rather than to perform a dead drop to the centre of the Earth. Interestingly, the result is independent of the orbital radius.

Further discussion

For exactly the reasons discussed in this question, it is quite common to push satellites at the end of their operational lives higher into what is known as a *graveyard orbit*, rather than to perform a de-orbit manoeuvre and allow them to break up in the atmosphere. Of course, atmospheric drag as the satellite re-enters the Earth's atmosphere helps contribute to the necessary delta-v for de-orbit. But for satellites in relatively high orbits, such as geosynchronous orbits,[10] the energy required to go higher into a graveyard orbit is significantly less than to de-orbit. According to the recently published Handbook of Space Engineering:[11]

> When a satellite in geosynchronous orbit nears the end of its life and is almost out of propellant, an attempt may be made to place it into a graveyard orbit. These orbits are slightly higher in altitude than a geosynchronous orbit by a few hundred kilometres. When a spacecraft is failing, there is a desire to make sure it will not cause issues for other, active satellites. Ideally, it would be de-orbited and allowed to burn up on re-entry, removing a potential piece of space debris. However, space-craft in geosynchronous orbits are so far from the Earth that it would take a significant amount of velocity change in order to reach the atmosphere: about 1,500 m/s. Instead, the spacecraft is sped up slightly to raise it into a graveyard orbit, requiring only around 10 m/s of velocity change.

[10] A special case of a geosynchronous orbit is the geostationary orbit above the Earth's equator. These orbits have a radius of approximately 42,000 km, or about 6.6 Earth radii.

[11] Darrin, A., O'Leary, B.L., 2012, "Handbook of space engineering, archaeology, and heritage (Advances in Engineering Series)," CRC Press, ISBN-10: 1420084313; ISBN-13: 978-1420084313.

> Once in the orbit, there is very little drag from the Earth's atmosphere, so only the solar wind has any effect at all. This means that the dead spacecraft will stay in the graveyard orbit for a very long time, several hundred years at least.

The graveyard orbit is necessary to reduce the build-up of space debris, and to avoid the possibility of the Kessler syndrome. This is a runaway event in which collisions lead to a sudden increase in space debris (an "avalanche") in a particular orbit. Donald Kessler's original 1978 paper[12] first proposed the possibility of the avalanche effect. In an interesting 2010[13] paper that reviewed their original predictions, Kessler and his co-authors wrote:

> The term "Kessler Syndrome" is an orbital debris term that has become popular outside the professional orbital debris community without ever having a strict definition. The intended definition grew out of a 1978 JG paper predicting that fragments from random collisions between catalogued objects in low Earth orbit would become an important source of small debris beginning in about the year 2000, and that afterwards, "... the debris flux will increase exponentially with time, even though a zero net input may be maintained". The purpose of this paper is to clarify the intended definition of the term, to put the implications into perspective after 30 years of research by the international scientific community, and to discuss what this research may mean to future space operations. The conclusion is reached that while popular use of the term may have exaggerated and distorted the conclusions of the 1978 paper, the result of all research to date confirms that we are now entering a time when the orbital debris environment will increasingly be controlled by random collisions. Without adequate collision avoidance capabilities, control of the future environment requires that we fully implement current mitigation guidelines by not leaving future payloads and rocket bodies in orbit after their useful life. In addition, we will likely be required to return some objects already in orbit.

At 300 km above the geosynchronous altitude, the solar radiation pressure is low enough that hundreds of years elapse before dead satellites pass back through the main satellite band. But objects also need to be removed from Low Earth Orbit. A 2005 article published by the European Space Agency[14]

[12] Kessler, D.J., Cour-Palais, B.G., 1978, "Collision frequency of artificial satellites: the creation of a debris belt," Journal of Geophysical Research, Vol. 83, No. A6, pp. 2,637–2,646.

[13] Kessler, D.J., Johnson, N.L., Liou, J.C., Matney, M., 2010, "The Kessler Syndrome: implications to future space operations," 33rd Annual AAS Guidance and Control Conference, Paper AAS 10-016, Breckenridge, CO, February 6–10, 2010.

[14] European Space Agency, 2005, "Space debris mitigation: the case for a code of conduct," European Space Agency website, 15 April 2005.

proposes a compromise solution to the problem of removing Low Earth Orbit satellites from space.

> In LEO, the solution is even more straightforward. For example, ESA's ERS satellite orbits at about 800 km altitude. Ideally, if it were slowed and lowered at mission end to 200 km altitude it would naturally de-orbit and burn up in about 24 hours; but this would take a lot of fuel. "For a craft the size of ERS, however, it will de-orbit naturally within 25 years if we merely bring it down to 600 km," says Dr Jehn, "so this [altitude] is a fuel-saving compromise."

It's an interesting problem, and one that according to the literature (despite the understated comments in the recent paper by Donald Kessler) is getting significantly worse. I, for one, would feel rather uncomfortable sitting in the International Space Station knowing that there were about 20,000 objects larger than 5 cm in size, and over 300,000 objects smaller than 1 cm, orbiting the Earth at relative velocities of tens of thousands of kilometres per hour.

The significance of the problem was illustrated rather vividly on 10 February 2009, when two communications satellites (the US Iridium 33 and the Russian Kosmos-2251) met by chance on almost exactly perpendicular trajectories with a relative velocity of $42,000 \text{ km h}^{-1}$ above Siberia. In approximately 0.1 ms, the satellites, with a combined mass of one and a half tonnes, were turned into thousands of pieces of space debris. About 2,000 of those pieces are actively being tracked by NASA.[15] On 24 March 2012, one of those tracked objects passed close enough to the International Space Station that the crew were ordered into a smaller docked rendezvous craft while the missile flew past. Gulp.

10.6 Newton's cannonball ★★

Sir Isaac Newton was undoubtedly one of the most influential scientists of all time, making wide contributions to pure mathematics, optics, mechanics and gravitation. He lived to the remarkable age of 84, and died in London in 1727, a bachelor and, reputedly, a virgin. On that latter subject, Voltaire later commented that Newton "had neither passion nor weakness". It is certainly true that he seemed primarily devoted to the sciences. Famously modest about his achievements, Newton said of himself, "I seem to have been only like a boy playing on the sea-shore, and diverting myself in now and then finding a smoother pebble or a prettier shell than ordinary, whilst the great ocean of truth lay all undiscovered before me." Newton published many books, the most famous of which was Philosophiæ Naturalis Principia Mathematica,

[15] According to reports, about 25% of the debris has burnt up in the outer edge of the atmosphere. This is because the orbital height of the collision was quite low, and because some of the resulting debris left the collision with lower velocity than the satellites.

referred to now simply as The Principia, in which he described the rules of gravitation, and what are now known as Kepler's Laws of planetary motion. A year after his death, in 1728, A Treatise of the System of the World was published in London. In this posthumous monograph Newton describes the famous thought experiment now known as *Newton's cannonball*, in which he imagines projecting a body from a high mountain at greater and greater speed.

> Let A F B represent the surface of the Earth, C its centre, VD, VE, VF, the curve lines which a body would describe if projected in an horizontal direction from the top of an high mountain, successively with more and more velocity. And, because the celestial motions are scarcely retarded by the little or no resistance of the spaces in which they are performed; to keep up the parity of cases let us suppose either that there is no air about the Earth, or at least that it is endowed with little or no power of resisting. And for the same reason that the body projected with a less velocity, describes the lesser arc VD, and with a greater velocity, the greater arc VE, and augmenting the velocity, it goes farther and farther to F and G; if the velocity was still more and more augmented, it would reach at last quite beyond the circumference of the Earth, and return to the mountain from which it was projected.

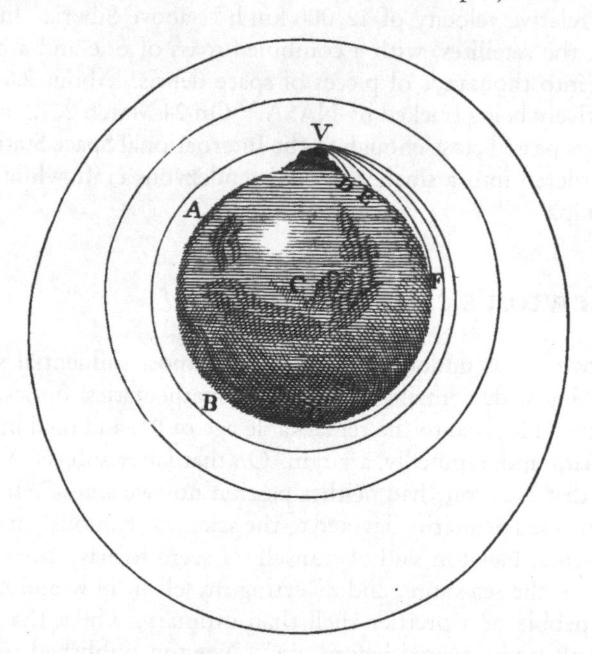

Newton was as loquacious as he was brilliant, and describes how, in the absence of air resistance, higher and higher projectile speeds lead first to a circular orbit, and finally to an *eccentric* or elliptical orbit. Newton was way ahead of his time with the concept of the *space gun*. It wasn't until over 100 years later that the idea was picked up again. The most famous example may be that of the *Columbiad space gun*, which appears in Jules Verne's 1865 novel De la Terre à la Lune (From the Earth to the Moon)—the story of a group of people trying to fire themselves to the Moon with an enormous gun. Jules Verne's science-fiction writing foreshadowed numerous important scientific developments, and he was credited with inspiring generations of influential submariners, aviators and rocket engineers. It is said that he did calculations to ensure he specified the requirements of the space gun correctly.

If you think this is fanciful, consider that another 100 years later the concept was picked up by government research agencies. The High Altitude Research Program (HARP) was started by the US and Canadian governments in 1961. A team led by Canadian ballistics engineer Gerald Bull built and tested a gun designed to fire payloads into the upper atmosphere to conduct re-entry studies associated with the intercontinental ballistic missile programme. The 40 m gun projected out over the Atlantic Ocean from the Caribbean island of Barbados. In November 1966 the gun fired a 180 kg projectile at muzzle velocity of 3.6 km s^{-1}. The projectile went beyond the edge of the upper atmosphere and performed a sub-orbital space flight, setting an altitude record of 180 km. The record still stands today.

Gerald Bull's real aspiration, however, was to fire objects into orbital flights. When the HARP programme was cancelled, to continue his work in the field he started designing guns for South Africa and Iraq. In 1988 Bull convinced Saddam Hussein (then President of Iraq) to start work on Project Babylon, ostensibly a satellite launcher gun. A 46 m Baby Babylon gun weighing 102 tonnes was completed and tested with lead projectiles. A much larger gun was planned with a barrel length of 156 m and a weight of 2,100 tonnes, which would have been capable of firing projectiles into orbit around the Earth. Big Babylon was never completed, however, because in March 1990 Bull was assassinated, allegedly by Israeli or Iranian intelligence agencies, who were worried about his involvement in other Iraqi military programmes. Research in the area continued with the Super High Altitude Research Project (Super HARP), another US government-funded programme that ran between 1985 and 1995. It had plans for a gun with a 3.5 km barrel that could produce orbital projectiles. The estimated cost of the gun was $1 billion, a sum that was never raised. We have, however, come a remarkably long way towards realising Jules Verne's dream. Time will tell if the space gun ever becomes a practicable launch vehicle.[16]

One problem with ballistic flights has not been addressed in any of these projects. Without in-flight trajectory correction, an object launched at less than the escape velocity from a point on the surface of the Earth will return

---

[16]The suggested application is to use space guns to launch satellites and supplies, not humans. The acceleration required to achieve a realistic launch speed of, say, 7.5 km s^{-1} in a 3.5 km barrel is $8,035 \text{ m s}^{-2}$. This is approximately 820 times the force per unit mass due to the Earth's gravity (820 g)—an instant death!

to the same point,[17] having performed an elliptical flight, possibly at great altitude. Generally speaking the idea is that our satellites stay up there, rather than falling back to Earth 90 minutes after launch (the time for a low-altitude circular orbit). There is still something for the Jules Vernes of the future to work on.

Whatever the practicalities of ballistic launches, the basic problem is interesting. We can calculate, as Newton did, the velocities required—at least theoretically—to put a cannonball into a circular orbit around the Earth.

CONSIDER TWO CANNONS A and B. Cannon A is at the North Pole. Cannon B is at the Equator pointing east. Assuming the Earth to be a perfect sphere, if we ignore atmospheric drag, calculate the velocities v_A and v_B (measured relative to the surface of the Earth) required to put cannonballs into circular orbits very close to the Earth's surface.

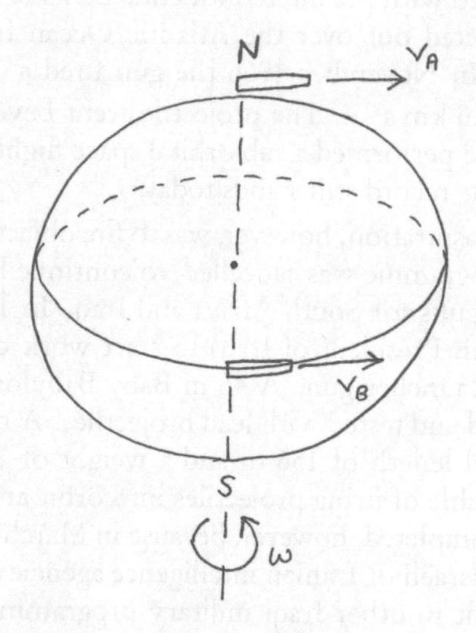

Answer

For a circular Earth orbit at a given radius R, which we assume to be approximately equal to the radius of the Earth ($R \approx R_E$), the gravitational force must be equal to the force required for centripetal acceleration.

$$F = \frac{GM_E m}{R_E^2} = \frac{mv^2}{R_E}$$

So, with respect to the centre of the Earth, the required velocity for a circular orbit at the same distance as the radius of the Earth is given by

$$v = \sqrt{\frac{GM_E}{R_E}}$$

[17] Here we speak rather loosely, neglecting the rotation of the Earth about its own axis.

Taking $G = 6.7 \times 10^{-11}$ N kg^{-2}m^2, $M_E = 6.0 \times 10^{24}$ kg, and $R_E = 6.4 \times 10^6$ m, we get $v \approx 7.93$ km s^{-1}. If we launch from the North Pole there is no beneficial effect from the Earth's rotation, so we require a velocity $v_A \approx 7.93$ km s^{-1}. Consider the case of launching from the Equator, however. We can calculate the equatorial velocity v_{surf} (the speed of the surface of the Earth about its centre) from the angular frequency of rotation of the Earth about its own axis.

$$\omega_E = \frac{2\pi}{24 \times 60 \times 60} \approx 7.27 \times 10^{-5} \text{ rad s}^{-1}$$

We get $v_{surf} = \omega_E R_E \approx 0.47$ km s^{-1}. If we choose to launch east from the Equator using cannon B, the required velocity is only $v_B = v - v_{surf} = 7.46$ km s^{-1}. The energy requirement, which is proportional to v^2, is reduced by almost 12%.

10.7 De la Terre à la Lune ★★

In Jules Verne's 1865 novel De la Terre à la Lune (From the Earth to the Moon), a group of gun enthusiasts build the Columbiad space gun, a powerful ballistic cannon, with which they fire themselves to the Moon in an unpowered capsule. During the journey they notice the following:

> From the moment of leaving the earth, their own weight, that of the projectile, and the objects it enclosed, had been subject to an increasing diminution...As it distanced the earth, the terrestrial attraction diminished: but the lunar attraction rose in proportion. There must come a point where these two attractions would neutralize each other: the projectile would possess weight no longer.

The amateur astronauts experience diminishing weight until a point is reached somewhere between the Earth and the Moon when the gravitational pulls "of the two orbs" effectively cancel each other out. At this point the travellers feel weightless.[18] As they journey further towards the Moon they feel that *up* and *down* have switched over, and gravity slowly increases again.

PLOT A QUALITATIVE graph of the forces due to gravity on a unit mass along a line drawn between the Earth and the Moon. Find the distance of the point (if such a point exists) at which the net force (due to the combined gravitational fields of the two orbs) is zero. It may be convenient to express the distance as *non-dimensionalised* by the Earth-Moon separation. If we were fired by a cannon along this line (neglecting the effects of the atmosphere and the fact that the initial acceleration would kill us), is Jules Verne's description of the weight we feel during the journey (and its absence) accurate?

[18] "...being attracted equally by both orbs, and not being drawn more toward one than toward the other."

Answer

The attracting force between two bodies of masses M and m, separated by a distance r, is given by $F = GMm/r^2$, where G is the gravitational constant. The gravitational force on unit mass ($m = 1$ kg), or the *acceleration* of the mass m, is simply $F_U = a = GM/r^2$. If we arbitrarily take the gravitational force on a unit mass due to the Earth (F_E) as positive (that is, positive directed towards the centre of the Earth), and due to the Moon (F_M) as negative, we can express the total force on a unit mass (F_T) as

$$F_T = F_E - F_M = G\left(\frac{M_E}{r_1^2} - \frac{M_M}{r_2^2}\right)$$

where r_1 is the distance from the centre of the Earth and r_2 is the distance from the centre of the Moon.

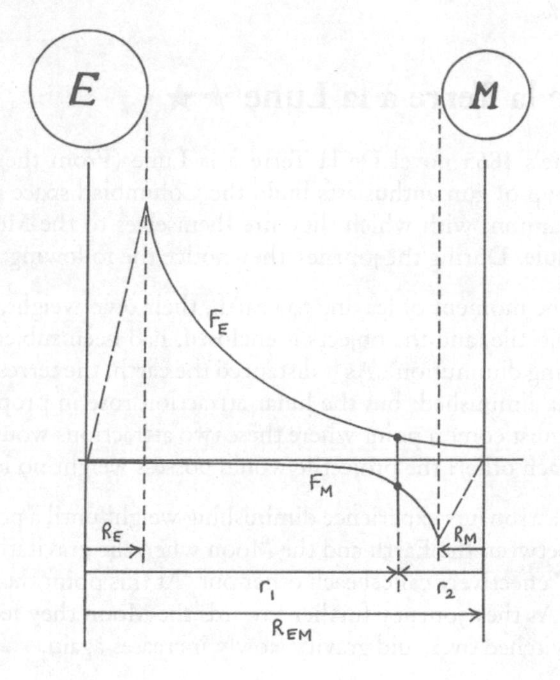

Noting that $r_2 = R_{EM} - r_1$, where R_{EM} is the separation of the Earth and the Moon, the condition for $F_T = 0$, or $|F_E| = |F_M|$, is

$$\frac{M_E}{r_1^2} = \frac{M_M}{(R_{EM} - r_1)^2}$$

This is a quadratic in r_1, which is expressed most compactly in the non-dimensional co-ordinates $x \equiv r_1/R_{EM}$ and $c \equiv M_E/M_M$. In this form we have

$$x^2(1-c) + x(2c) - c = 0$$

Solving for x, we get

$$x = \frac{r_1}{R_{EM}} = \frac{-c \pm \sqrt{c}}{1 - c}$$

For $M_E \approx 6.0 \times 10^{24}$ kg and $M_M \approx 7.4 \times 10^{22}$ kg, we have $c \approx 81.1$, giving $x^+ \approx 0.900$ and $x^- \approx 1.125$. The x^- solution is *unphysical* (both the Moon and the Earth would have gravitational pulls in the same direction), giving a single physically meaningful solution: $r_1/R_E = x^+ \approx 0.900$.

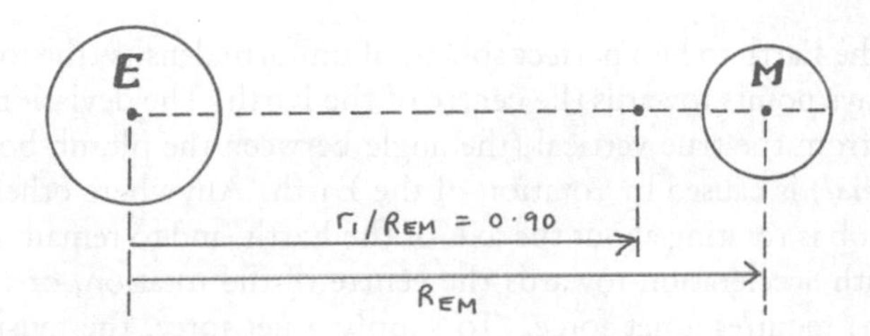

So, almost exactly nine-tenths of the way to the Moon, the net force due to gravity is zero.

Let's return briefly to Jules Verne's description. He describes first a sensation of diminishing weight as we move away from the Earth. This is followed by a feeling of weightlessness at a neutral point. Then we feel that *up* and *down* have swapped over, and then we feel our weight increasing again as we approach the Moon. This is certainly *not* what we would feel on a journey of this type. In an unpowered flight, once we are released from the cannon (ignoring the effect of the atmosphere), the entire journey is in free fall until the moment we crash-land on the suface of the Moon. So we would *feel* weightless during the entire journey. We feel weightless because there are no external mechanical forces on our bodies. We float around inside the capsule.

10.8 Professor Plumb's Astrolabe-Plumb ★★★

PROFESSOR PLUMB HAS invented a device he calls the Astrolabe-Plumb. It allows him to measure very accurately the difference between the true vertical (that is, a radial line passing through the centre of the Earth) and the apparent vertical as measured by his *plumb* (or *bob*), which is a weight on the end of a string. Professor Plumb refers to the difference between these two measures of the vertical as the *apparent gravity deviation angle*, α. Assuming the Earth to be a perfect sphere of uniform density, calculate α as a function of latitude, θ.

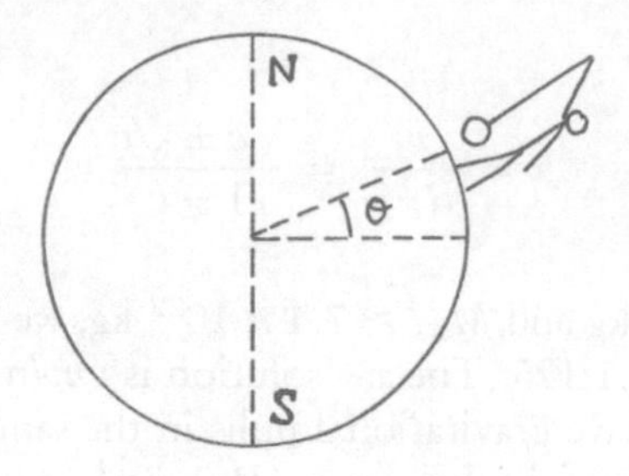

Answer

Assuming the Earth to be a perfect sphere of uniform density, the force due to gravity always points towards the centre of the Earth. The deviation of *apparent gravity* from the true vertical (the angle between the plumb bob and the *surface normal*) is caused by rotation of the Earth. Anywhere other than the Poles, the bob is rotating about the axis of the Earth, and to remain in circular motion (with acceleration towards the centre of the rotation, or centripetal acceleration) requires a net force. To supply a net force, the tension in the string supporting the bob cannot be exactly equal in magnitude and opposite in direction to the gravitational force on the bob. The difference between these two terms is the net force required for centripetal acceleration. Consider three points on the surface of the Earth: A, B and C.

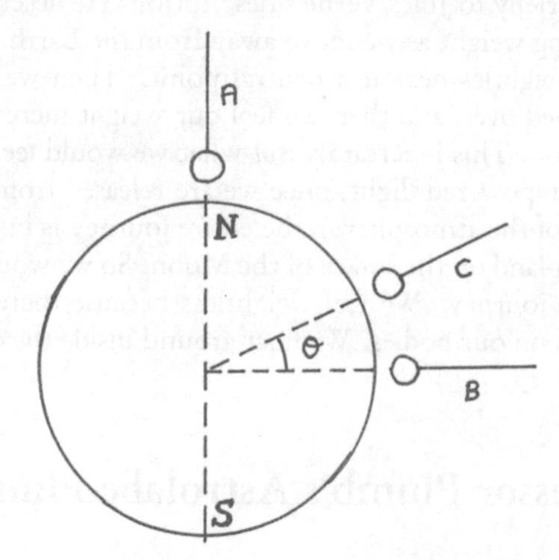

- At point A, when the plumb is at the North Pole, the apparent gravity and the true gravity are aligned. They both point towards the centre of the Earth, and the plumb is aligned with the true vertical. The centripetal acceleration is zero, because the tangential velocity of the Earth at the North Pole is zero.
- At point B, the Equator, the tangential velocity of the Earth (which is rotating about an imaginary line through the Poles) and the centripetal

acceleration, are both at a maximum. The centripetal acceleration is directed towards the axis of the Earth. The net force required to perform this acceleration is the difference between the gravitational force and the string tension. Because the centripetal acceleration and true gravity are aligned, so also is the apparent gravity. The tension in the string is less than the gravitational force on the bob by an amount equal to the force required for the centripetal acceleration—there is an apparent reduction in weight. All three vectors point towards the centre of the Earth and there is no deviation of the plumb.

- Consider now a general point C, at a latitude θ above the Equator. It is helpful at this stage to consider the forces on the plumb, and its acceleration. There is a centripetal acceleration, a, *towards the axis of rotation*. There are two forces on the plumb: the gravity force, mg, which acts towards the centre of the Earth, and a tension in the string, T, which acts (in general) at an angle α to the gravitational force, as shown below. The difference between these two forces, which we denote F, is the force required to perform the centripetal acceleration.

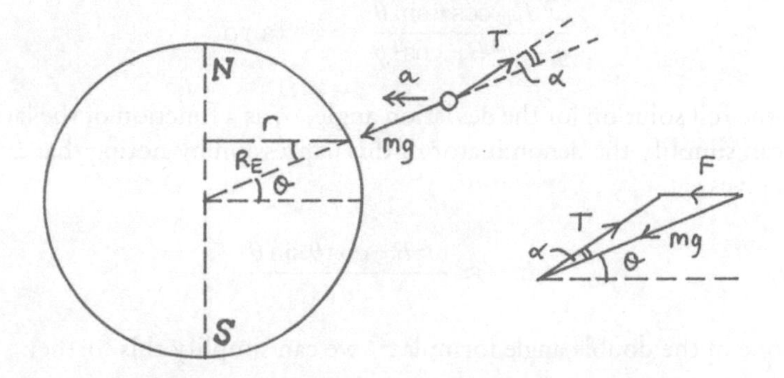

Let us now examine the forces at a general point C in more detail. The centripetal acceleration is of magnitude $a = \omega^2 r$, where r is the radius from the axis of rotation, and ω is the angular speed of the Earth (in radians per second). Looking at the geometry of the Earth we see that $r = R_E \cos\theta$, where R_E is the radius of the Earth. Combining these equations gives

$$a = \omega^2 R_E \cos\theta$$

Using Newton's Second Law we see that the net force on the plumb required to achieve this acceleration must be

$$F = ma = m\omega^2 R_E \cos\theta$$

We now consider how the forces due to gravity, mg, and string tension, T, combine to give a net force on the plumb, F. Allowing the string tension to act at an angle α to the force due to gravity, we see from the vector diagram (using the sine rule) that

$$\frac{\sin(\pi - (\alpha + \theta))}{mg} = \frac{\sin\alpha}{F}$$

Rearranging, substituting for F, and noting that

$$\sin(\pi - (\alpha + \theta)) = \sin(\alpha + \theta)$$

we have

$$\sin(\alpha + \theta)\,\omega^2 R_E \cos\theta = g\sin\alpha$$

Expanding using one of the trigonometric addition formulae,[19] we have

$$\begin{aligned} [\sin\alpha\cos\theta + \sin\theta\cos\alpha]\,\omega^2 R_E\cos\theta &= g\sin\alpha \\ \omega^2 R_E \cos\theta\sin\theta\cos\alpha &= \left(g - \omega^2 R_E\cos^2\theta\right)\sin\alpha \\ \frac{\omega^2 R_E\cos\theta\sin\theta}{g - \omega^2 R_E\cos^2\theta} &= \tan\alpha \end{aligned}$$

This is the full solution for the deviation angle, α, as a function of the latitude, θ. We can simplify the denominator of this expression by noting that $\omega^2 R \ll g$. This gives

$$\tan\alpha \approx \frac{\omega^2 R_E\cos\theta\sin\theta}{g}$$

Using one of the double-angle formulae[20] we can simplify this further.

$$\tan\alpha \approx \frac{\omega^2 R_E\sin(2\theta)}{2g}$$

The maximum deviation is at $\theta = 45°$, for which we get

$$\alpha_{max} \approx \tan^{-1}\left(\frac{\omega^2 R_E}{2g}\right) \approx \frac{\omega^2 R_E}{2g}$$

The angular speed of the Earth is 2π in 24 hours, or $\omega \approx 7.3 \times 10^{-5}$ rad s^{-1}. Taking $g = 9.8$ m s^{-1} and $R_E = 6.4 \times 10^6$ m, we get $\alpha_{max} \approx 0.10°$. If Professor Plumb's Astrolabe-Plumb is 1 m long (assuming it to be a hand-held device), the deviation from the vertical at a latitude of $\theta = 45°$ would be 1.74 mm, significant enough to be very easily measurable.

[19] $\sin(A + B) = \sin A\cos B + \sin B\cos A$

[20] $\sin(2A) = 2\sin A\cos A$

10.9 Jet aircraft diet ★★

TWO JET AIRCRAFT fly with a ground speed of $1,674$ km h^{-1} along the line of the Equator just above sea level. One pilot heads due east and the other heads due west. Before getting in their planes both pilots weigh in at exactly 100 kg (including the weight of the equipment they are wearing). Do the pilots have the same *sensation of weight* as they pass the same point on the Equator in opposite directions?

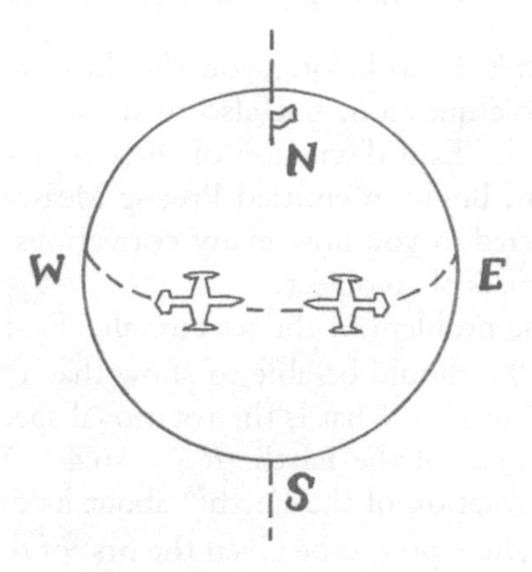

Answer

This is my own modification of a well-known puzzle about whether we weigh more at the Equator or the North Pole. The answer to that question, of course, depends on how we account for the Earth's rotation. I never really liked the North Pole question. This was partly because, although it is often reasonable to assume that the Earth is a perfect solid sphere[21] of constant density (and therefore has a constant value of gravity on its surface), in this particular question the effect of variations in the value of gravity due do non-uniformities[22] are larger than the difference we are encouraged to distinguish. The principle of the question is quite nice—it's just a shame it doesn't work out that neatly in the real world.

I also think that the term *weight* is used a little loosely in the North Pole question as I have heard it asked. The term weight is generally reserved to refer to the force due to gravity, not the reaction force you read off common scales. For that latter quantity, the scientist (or at least the pedantic scientist who is asking a question about gravity) should use the term *sensation of weight*. The sensation of weight is what we *feel*, and what is registered by our bathroom scales. If we are on a rotating sphere, the sensation of weight is the force

[21] There is, after all, the famous and almost certainly apocryphal physicist's story of a spherical cow. If you're unfamiliar with it you should look it up. The Earth, you might argue, is every bit as spherical as even the most well-fed cow.

[22] Non-uniformity of both sphericity and density.

due to gravity minus the hypothetical force[23] required to cause centripetal acceleration towards the centre of the Earth. We write

$$F_{total} = mg - R = mv^2/r$$

where m is our mass, R the reaction force (or our sensation of weight), v the tangential velocity around the axis of the Earth, and r our radius from the axis of rotation. Our sensation of weight is therefore given by

$$R = mg - mv^2/r$$

Assuming you couldn't do so before, you should now be able not only to answer the North Pole question, but also to do so without terminological inexactitude.[24] For a brilliant discussion of these topics it is worth referring to a paper by Richard Boynton entitled Precise Measurement of Mass.[25] It probably never occurred to you how many corrections you have to make to establish the precise mass of an object.

Let's return to the problem of the jet aircraft. First we will convert the speed into SI units. We should be able to show that $1,674$ km h^{-1} is equal to approximately 465 m s^{-1}. What is the rotational speed of the Earth at the Equator? It is the radius of the Earth, $R_E = 6.4 \times 10^6$ m, multiplied by the angular speed of rotation of the Earth[26] about its own axis, $\omega = 7.27 \times 10^{-5}$ rad s^{-1}. We might expect to be given the first of these numbers, and we can easily calculate the second. So the surface speed of the Earth (relative to the centre of the Earth) is given by

$$|v_{surf}| = R_E\omega \approx 465 \text{ m s}^{-1}$$

Thus the speed of the east-going aircraft, $|v_1|$, the speed of the west-going aircraft, $|v_2|$, and the speed of the surface of the Earth about its own centre, $|v_{surf}|$, are all equal. We write $|v_1| = |v_2| = |v_{surf}|$.

Now consider the aircraft in the frame of reference of an observer on the surface of the Earth, at a point A over which both aircraft pass simultaneously. The observer at A sees the planes travelling in opposite directions at speeds equal to $|v_{surf}|$. So far so good.

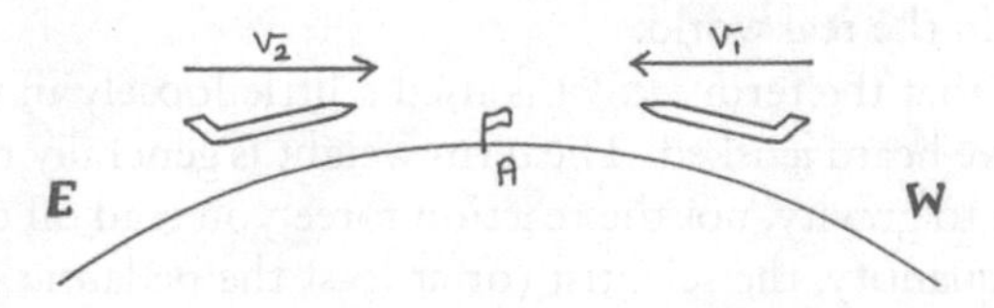

[23]This force comes from the force due to gravity.

[24]A fine phrase first coined by Winston Churchill during the 1906 election. It is favoured by parliamentarians as a polite way of suggesting that someone might by lying. A modern translation is *economical with the truth*. All very English indeed.

[25]Boynton, R., 2001, "Precise measurement of mass," 60th Annual Conference of the Society of Allied Weight Engineers, Arlington, Texas, 21–23 May, 2001, Paper No. 3,147.

[26]The time period of the Earth's rotation is 24 hours, or $T = 24 \times 60 \times 60 = 86,400$ s. The angular speed, ω, is given by $\omega = 2\pi/T = 7.27 \times 10^{-5}$ rad s^{-1}.

Now consider the aircraft in the frame of reference of a *non-rotating observer* looking down on the North Pole from outer space. This observer sees the point A travelling around the centre of the Earth at a speed equal to $|v_{surf}|$. Instantaneously, point A is travelling eastwards. In this frame of reference the east-going aircraft is travelling at a speed $|v_1'| = 2\,|v_{surf}|$. This is the velocity of aircraft A about the centre of the Earth. There are two forces acting on the pilot. The first is a downward force due to gravity, which is equal to mg. The second is an upward force from the seat of the aircraft, R_1, that is equal to the gravitational force less the component of that force that causes an acceleration towards the centre of the Earth. We write

$$R_1 = m\left(g - \frac{v_1'^2}{R_E}\right) = mg\left[1 - \frac{(2v_{surf})^2}{gR_E}\right]$$

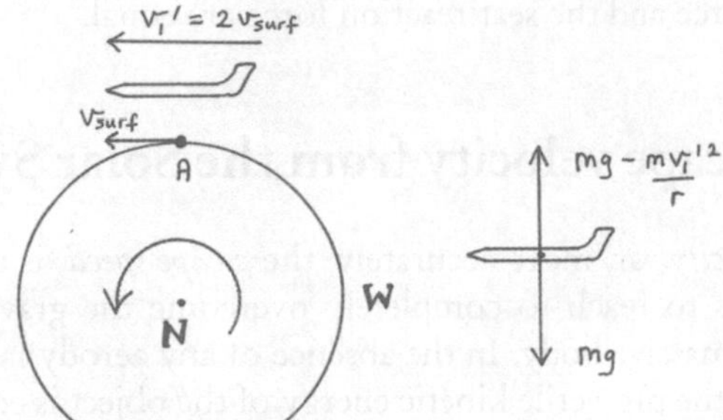

Now consider the west-going aircraft. In our frame of reference, from where we look down on the North Pole from outer space, the aircraft is travelling at a speed $|v_2'| = |v_2| - |v_{surf}| = 0$. Viewed from a non-rotating frame of reference the aircraft is stationary! The plane has zero velocity about the centre of the Earth. Now the downward force due to gravity, mg, is exactly equal to the upward force from the seat of the aircraft, R_1. In our non-rotating frame of reference the aircraft has zero acceleration. We write

$$R_2 = m\left(g - \frac{v_2'^2}{R_E}\right) = mg$$

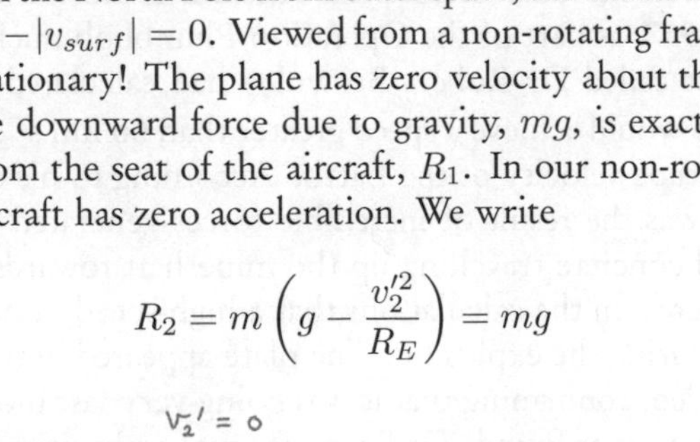

Let's compare the reaction forces. These forces, we recall, give the pilots their sensation of weight.

$$\frac{R_1}{R_2} = 1 - \frac{(2v_{surf})^2}{gR_E} \approx 0.9862$$

The east-going pilot *feels* approximately 1.4% lighter than the west-going pilot. The effect is due to the fact that the east-going pilot is undergoing circular motion (about the centre of the Earth), whereas the west-going pilot is stationary about the centre of the Earth. In effect, the east-going pilot is in partial free fall, in which part of the gravitational force is responsible for acceleration towards the centre of the Earth, and the reaction force provided by the seat is reduced. The west-going pilot, however, has no such acceleration (his acceleration is zero in the frame of reference of the observer in outer space), so the gravitational force and the seat reaction force are equal.

10.10 Escape velocity from the Solar System ★★★

The *escape velocity*, or, more accurately, the *escape speed*, is the ballistic speed an object needs to reach to completely overcome the gravitational force of attraction of a massive body. In the absence of any aerodynamic drag, it is the speed at which the projectile kinetic energy of the object is equal to the change in potential energy needed to overcome gravitational attraction.

Some claim that the first object to escape the gravitational influence of the Earth was an armour steel plate blown off the mineshaft of an underground nuclear test site in Nevada. Before the Pascal-B explosion (which occurred on 31 August 1957, as part of the Operation Plumbbob nuclear tests[27]) a US nuclear scientist called Dr Robert Brownlee had calculated that the 900 kg steel cover plate would achieve a speed greater than 66 km s^{-1}, approximately six times the escape velocity of the Earth. According to his calculations, this extreme speed was the result of incredible forces generated by a hypersonic jet of vaporised concrete travelling up the mineshaft towards the cap. There was enough interest in the calculations that a high-speed camera was set up to view the plate during the explosion. The plate appeared in only one frame of the resulting video, confirming that it was going very fast indeed. Needless to say, the plate was never found. Dr Brownlee was a physicist interested in the forces involved in underground nuclear containment and the shock reflection effects on capping. The original point of his calculation—which has become a legend of sorts—was simply to show the huge forces involved. In an article he wrote in 2002[28] he made it clear that he and his team did not expect the

[27] Carothers, J., 1995, "Caging the dragon: the containment of underground nuclear explosions," Technical Report, Technical Information Centre, Oak Ridge Tennessee, June 1995.

[28] Brownlee, R.R., "Learning to contain underground nuclear explosions," June 2002 (published online).

plate to be projected into space. It is almost certain that the remains of the vaporised plate are deposited over a wide expanse of the Nevada desert.

The first man-made object to escape the gravitational influence of the Earth was Luna I, which was launched by Russia from the desert in Kazakhstan at 4:41 pm on 2 January 1959. It was intended to collide with the Moon, but due to a programming error it missed its target, passing within about 3,700 miles of the Moon's surface 34 hours after takeoff. For the past 50 years it has been orbiting the Sun, at an orbital radius between that of the Earth and Mars. It has been dubbed the First Cosmic Rocket, because it was the first man-made object to achieve escape velocity from the Earth.

The first spacecraft to achieve escape velocity from the Solar System (which for us humans effectively means the Earth and the Sun[29]) was Pioneer 10, which the US launched on 3 March 1972 from Cape Canaveral in Florida. The probe passed Jupiter in November 1973, taking photographs as it passed within about 130,000 km of the surface. It stayed in contact with the Earth for almost 30 years, sending its final signal on 23 January 2003, at a distance of 12 billion kilometres from the Sun. Pioneer is still heading into deep space. In September 2012 it was estimated to be almost 16 billion kilometres from the Sun, travelling away at approximately 12.0 km s^{-1} (relative to the Sun). It is so far away that sunlight now takes almost 15 hours to reach it.

FOR A PROBE on the surface of the Earth to escape the Solar System, the significant gravitational bodies are the Earth ($\boldsymbol{E}$) and the sun ($\boldsymbol{S}$). Taking account of the rotation of the Earth about the Sun, and the rotation of the Earth about its own axis, calculate the minimum and maximum probe escape velocities from the Solar System, v_{P1} and v_{P2}, measured in the reference frame of the surface of the Earth. Ignore the effect of the atmosphere.

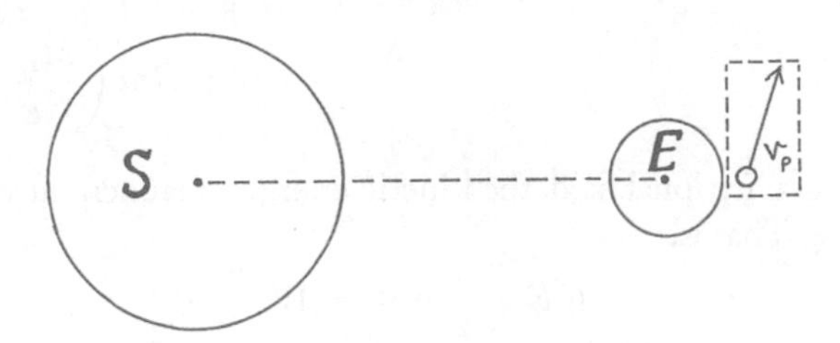

Answer

So that we understand the basic physics involved, let's first consider a simple system in which we take the Earth to be stationary with respect to the Sun.

[29]When we calculate the energy required to leave the solar system from the surface of the Earth, the gravitational influence of the other planets—even massive ones like Jupiter and Saturn—is very weak, and we can effectively ignore it. This is because, from the Earth's surface, we are at the bottom of the *potential well* of the Earth's gravitational field, but not so for other planets. However, the Sun is so massive that its gravitational pull is very significant in the calculation, even though we're a long way from it.

We need to find the work, W, required to remove the probe from the surface of the Earth to $r = \infty$. To remove the probe from the solar system, we need to supply a hypothetical force, F_W, which is equal in magnitude and opposite in sign to the force due to the gravitational attraction of *both* the Earth and the Sun, F_G. The work done is the integral of this force over a displaced distance between the surface of the Earth and $r = \infty$. In defining the integral we need to consider the starting distance of the probe from the centres of the Earth and the Sun. The initial distance from the centre of the Earth is simply the radius of the Earth, R_E. For the Sun the initial distance is the Earth-Sun radius plus the radius of the Earth: $R_{ES} + R_E$. Because $R_E \ll R_{ES}$, we approximate $R_{ES} + R_E \approx R_{ES}$.

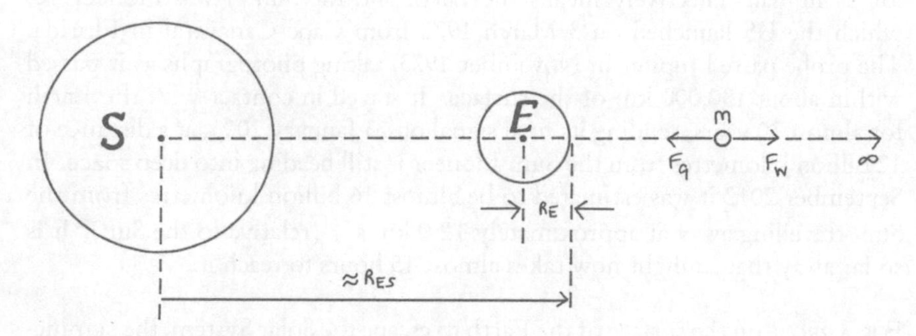

The integral becomes

$$W = \int_{r_1}^{r_2} F dr = GM_E m \left[\frac{r^{-1}}{-1}\right]_{R_E}^{\infty} + GM_S m \left[\frac{r^{-1}}{-1}\right]_{R_{ES}}^{\infty}$$
$$= Gm\left(\frac{M_E}{R_E} + \frac{M_S}{R_{ES}}\right)$$

To escape the gravitational field, the kinetic energy of launch must be equal to the work done. That is,

$$KE = \frac{1}{2}mv_P^2 = W$$

where v_P is the projectile velocity we need to escape the gravitational field of the Solar System.

Combining our equations we get

$$v_P = \sqrt{2G\left(\frac{M_E}{R_E} + \frac{M_S}{R_{ES}}\right)}$$

Taking $G = 6.7\times10^{-11}$ N kg^{-2}m^2, $M_E = 6.0\times10^{24}$ kg, $M_S = 2.0\times10^{30}$ kg, $R_E = 6.4\times10^6$ m, and $R_{ES} = 1.5\times10^{11}$ m, we get $v_P \approx 43.7$ km s^{-1}. For our hypothetical system in which the Earth is stationary with respect to the

Sun, and not rotating about its own axis, this is the escape velocity we need to achieve in the frame of reference of the surface of the Earth.

Now consider the velocity of the Earth around the Sun, v_E (taking a circular orbit approximation). Also consider the equatorial velocity of the Earth's surface about the centre of the Earth due to rotation about its own axis, v_{surf}.[30] We calculate the velocity of the Earth around the Sun by considering the angular frequency of the orbit.

$$\omega_{ES} = \frac{2\pi}{365 \times 24 \times 60 \times 60} \approx 1.99 \times 10^{-7} \text{ rad s}^{-1}$$

Using this we obtain $v_E = \omega_{ES} R_{ES} \approx 29.9 \text{ km s}^{-1}$. We perform a similar calculation to determine the equatorial velocity of the surface of the Earth about the centre of the Earth. The angular frequency of rotation of the Earth about its own axis is given by

$$\omega_E = \frac{2\pi}{24 \times 60 \times 60} \approx 7.27 \times 10^{-5} \text{ rad s}^{-1}$$

We obtain $v_{surf} = \omega_E R_E \approx 0.47 \text{ km s}^{-1}$. We need to know if v_E and v_{surf} act in the same or opposite directions. There is no real excuse for not being able to work out the direction of rotation of the Earth about its own axis. Since time immemorial the Sun has risen in the east and set in the west. This means that when we look down on the Earth from a point above the North Pole, as we do in the diagram, the Earth must be rotating counter-clockwise about its own axis. We mark this on the diagram as a vector pointing out of the page—that is, a dot with a circle around it at the centre of the Earth.[31]

We can be excused for not knowing the direction of rotation of the Earth about the Sun. Either we were in that particular astronomy class or we weren't. It happens that it is in the same direction as the rotation of the Earth about its own axis. In other words, when viewed from a point above the north pole of the Sun, the Earth appears to be rotating counterclockwise. When viewed from a point above the north pole of the Sun, the velocities look like this.

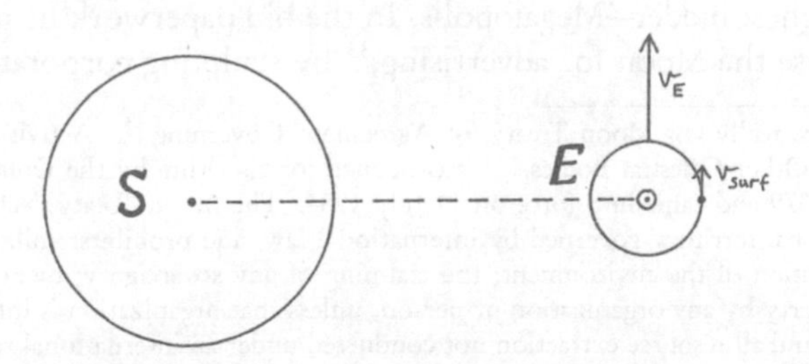

[30]For this, ignore the tilt of the axis—that is, assume that the axis of rotation is perpendicular to the plane in which the Earth orbits the Sun.

[31]This comes from the *right-hand rule*. Clockwise rotation around the North Pole in the same view would be a vector pointing into the page, represented by a circle with a cross in it.

So the combined velocity of a point on the surface of the Earth with respect to the Sun is given by $v_T = v_E + v_{surf} \approx 30.4 \text{ km s}^{-1}$.

Let's return to the problem of calculating the minimum and maximum probe escape velocities from the Solar System, $v_{P1}$and v_{P2}. We're interested in the velocities *relative to the surface of the Earth*, which is where we intend to launch from. If we launch in a direction such that the rotational velocities favour us, the Earth-relative launch velocity is given by

$$v_{P1} = v_P - v_T \approx 13.3 \text{ km s}^{-1}$$

If we launch in a direction such that the combined rotations disadvantage us, our Earth-relative launch velocity is

$$v_{P2} = v_P + v_T \approx 74.1 \text{ km s}^{-1}$$

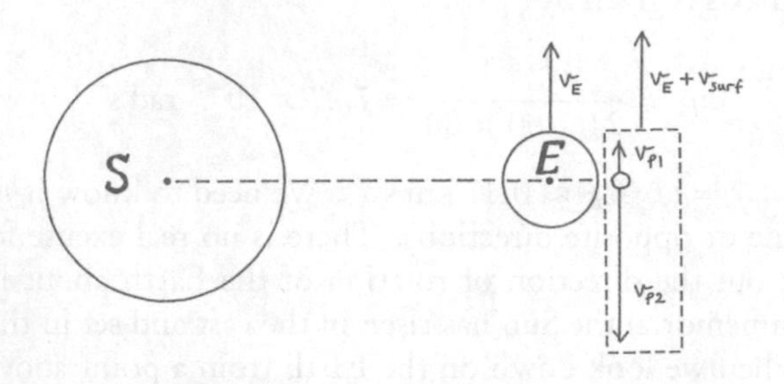

If we need to make a quick getaway (over the next thirty years or so), we would do well to attempt our launch in the correct direction.

10.11 Mr Megalopolis' expanding Moon ★★★

It is the year 2050, and Mr Megalopolis has bought the Moon. The one that rotates around the Earth. The one that used to subtend an angle of 0.52°in the night sky. Not for much longer. After the collapse of the Moon Treaty[32] and failed attempts to extract minerals, a bankrupt United Nations sold the Moon off to the highest bidder—Megalopolis. In the bid paperwork he maintains he is going to use the Moon for advertising,[33] by sculpting corporate logos into

[32]In fact there really is a Moon Treaty, or Agreement Governing the Activities of States on the Moon and Other Celestial Bodies. It was opened for signature by the United Nations on 18 December 1979 and came into force on 11 July 1984. The Moon Treaty declares the Moon to be international territory governed by international law, and prohibits: military activity of any kind; alteration of the environment; the claiming of any sovereignty; ownership of extraterrestrial property by any organization or person, unless that organization is international and governmental; and all resource extraction not conducted under an international regime. Twenty years later not a single space power has ratified the treaty, making it rather meaningless.

[33]There are several barmy schemes to monetise space advertising. Schemes involving the Moon include projecting images onto the lunar surface, and landing solar powered vehicles on the Moon to create "shadow farms" visible from the Earth. On 20 Aug 2013, the US patent office awarded a patent to the Moon Publicity Corporation, with the title Shadow Shaping to Image Planetary or

the surface using massive machinery operating on the lunar dust. Everyone thinks the Moon is too small for this, but Megalopolis plans to *expand* the Moon. How? Well by extracting the core, of course, and by distributing it across the surface. When Megalopolis Moon Enterprises is floated on the global stock exchange, he tells investors that he will extract literally the entire mass of the Moon and spread it evenly over the surface, creating a hollow Moon. "I will do it as many times as we need", he says, "until the Moon is a huge billboard in the night sky." The audience gasps. "At Megalopolis Moon Enterprises we think big!" says Megalopolis, showing computer generated images of what the New Moon™ will look like.

Megalopolis has a team of seismologists and geophysicists to consult for the project. The geophysicists have their doubts. To begin with, they don't believe that a hollow Moon would be structurally stable. But even if it was, they say, it would have a dangerously weak gravitational field if its shell became too thin, making it unsafe to work on. "But I'm not removing the mass, I'm just moving it around", says Megalopolis. "If you want to work for me you need to think big!"

TAKE THE MOON to be a solid sphere of uniform density. If we hollowed it out from the centre to create a thick spherical shell with a hollow middle the size of the original Moon, by what proportion would gravity have changed, if at all? If the mass of the spherical shell were moved to the surface a second, third or n-th time, by what proportion would gravity have changed, if at all?

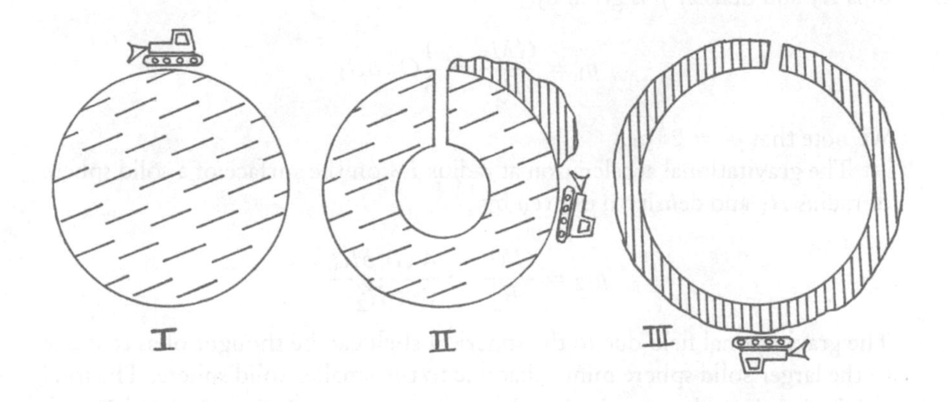

Lunar Surfaces (patent number US 8515595 B2). The abstract reads:

> A method is disclosed for forming a shadow pattern on a planetary or lunar surface, including providing a rough terrain vehicle having a plurality of wheels capable of imparting to the planetary or lunar surface shadow shaping components to produce a shadow pattern capable of being seen from a distance.

Answer

Let the original radius be R_1. The new radius, R_2, is calculated by considering the equation for the volume of a sphere. Letting the new sphere have twice the volume of the original sphere, we get

$$\frac{4}{3}\pi R_2^3 = 2\left(\frac{4}{3}\pi R_1^3\right)$$

So

$$R_2 = 2^{\frac{1}{3}} R_1$$

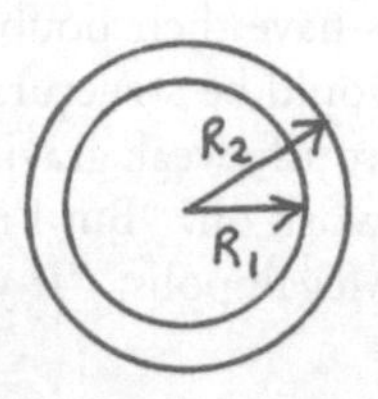

The mass of a sphere of density ρ and diameter R is given by $M = (4/3)\pi R^3 \rho$. The strength of the gravitational field on the surface of a solid sphere of radius R_2 and density ρ is given by

$$g_2 = \frac{GM_2}{R_2^2} = \frac{4}{3}G\pi\rho R_2$$

Similarly, the gravitational acceleration on the surface of a solid sphere of radius R_1 and density ρ is given by

$$g_1 = \frac{GM_1}{R_1^2} = \frac{4}{3}G\pi\rho R_1$$

We note that $g_2 = 2^{\frac{1}{3}} g_1$.

The gravitational acceleration at radius R_2 on the surface of a solid sphere of radius R_1 and density ρ is given by

$$g_{12} = \frac{GM_1}{R_2^2} = \frac{4}{3}\frac{G\pi\rho R_1^3}{R_2^2}$$

The gravitational field due to the spherical shell can be thought of as that due to the larger solid sphere minus that due to the smaller solid sphere. The total gravitational acceleration is given by $g_T = g_2 - g_{12}$. Using $R_2 = 2^{\frac{1}{3}} R_1$, we get

$$g_T = \frac{g_2}{2} = \frac{g_1}{2^{2/3}}$$

So the gravitational field strength has reduced by about 37%. Mr Megalopolis should indeed be careful about expanding his empire too far.

Let's now consider what happens on the n-th iteration of moving all the material of the Moon to the surface. The *n-th Moon*, as we will call it, has a volume V_n, which is related to the volume of the original Moon, V_1, by

$$V_n = nV_1$$

It's easy to see this, because we have moved the entire volume of the original Moon n times. So we can write

$$R_n = n^{1/3} R_1$$

If we follow the logic above,[34] we see that where the mass of a spherically symmetric object is unchanged, the gravitational field strength on the surface of that object is inversely proportional to the square of the radius of the object. The gravitational field strength on the surface of the n-th Moon, g_n, is related to the gravitational field strength on the surface of the original Moon, g_1, by

$$g_n = \left(\frac{R_1}{R_n}\right)^2 g_1 = \frac{g_1}{n^{2/3}}$$

10.12 Asteroid games ★★★★

Cosmonaut Blastov and Captain Medallion are playing a game of catch. This is not some idle pastime, however—they have a wager on how a ball is going to behave when dropped through a solid asteroid and a hollow asteroid. The game of catch is to resolve an argument they had the previous day concerning Newton's Shell Theorem.

"It's perfectly simple," said Medallion. "If we can find two asteroids with the same radius and mass, then according to the shell theorem they will both behave as though the mass is concentrated at the centre. If we drop a ball into holes drilled through their centres, in both cases the ball should take the same time to get to the other side."

"You are for-get-ting one zing, my friend", said Blastov. "Ze shell zeorem says zere is zero force inside the shell. You vill find zat it vill take very much longer to vall zrough ze hollow azteroid."

"I too have read the theorem, Blastov, and I'm happy to bet one of my medals on the times being the same," said Medallion.

"And I vill bet an entire bottle of wudka on the fact ze times are different," said Blastov.

And so the wager was set.

TWO ASTEROIDS A AND B have the same mass and the same outer radius R_1. Asteroid A is solid and of uniform density. Asteroid B is a thin shell with internal radius R_2, also of uniform density. By thin, we mean $0.9 \leq R_2/R_1 < 1$.

[34] There is a more formal proof in the problem The Hollow Moon.

Balls are dropped from the surfaces, at $r = R_1$, into holes bored through the centres of the asteroids. Write expressions for the *time periods*[35] of the balls, T_A and T_B, and sketch T_B/T_A as a function of R_2/R_1. Use approximations where appropriate. Use Newton's Shell Theorem, which can be stated as follows: spherically symmetric bodies act on external objects as though their entire mass were concentrated at a central point; and there is no net gravitational force inside a shell.

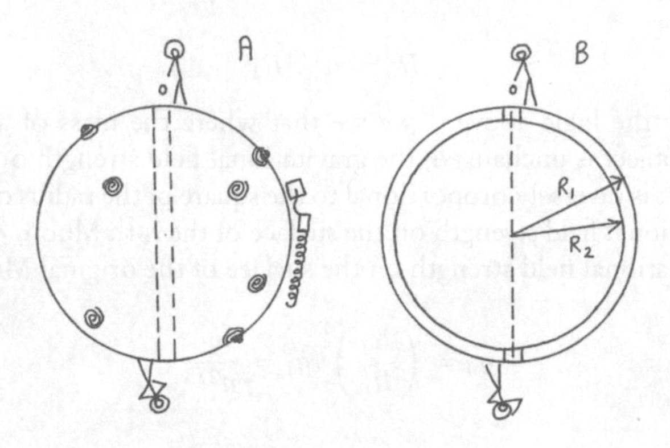

Answer

We can start with Newton's Law of Universal Gravitation. The force on a ball of mass m at radius r from the centre of a spherically symmetric body with mass M is

$$F = \frac{GMm}{r^2}$$

We now consider the asteroids in turn.

1) ASTEROID A: SOLID UNIFORM SPHERE

At a general point r within a solid sphere of uniform density, radius R_1, and mass M, we can think of the system as being composed of an inner attracting sphere ($0 \le r' \le r$) and an outer shell that does not attract ($r \le r' \le R_1$). The mass in the inner attracting sphere, M_{rA}, is given by

$$M_{rA} = M\left(\frac{r}{R_1}\right)^3$$

So, the force due to gravity at a radius $r \le R_1$ is given by

$$F_A = \frac{GM_{rA}m}{r^2} = \frac{GMm}{R_1^3}r = k_A rm$$

[35]The *time period* is the time taken to complete a full oscillation from 0 to 2π.

where $k_A = GM/R_1^3$. In vector form, noting that the force due to gravity acts in the opposite direction to displacement from the centre (that is, it is a restoring force), we can write

$$\boldsymbol{F}_A = -k_A \boldsymbol{r} m$$

We recognise this as simple harmonic motion. Using Newton's Second Law ($\boldsymbol{F} = m\ddot{\boldsymbol{r}}$) we get

$$\ddot{\boldsymbol{r}} = -k_A \boldsymbol{r}$$

This is the form of the governing equation for simple harmonic motion, and has solution $r = R_1 \cos(\omega_A t)$, where the angular frequency $\omega_A = \sqrt{k_A} = \sqrt{GM/R_1^3}$. The period, T_A, is the time taken to perform one complete cycle of 2π, and is defined by $T_A \omega_A = 2\pi$. So

$$T_A = \frac{2\pi}{\sqrt{k_A}} = 2\pi\sqrt{\frac{R_1^3}{GM}}$$

The time period is the time required for a round trip, from one side to the other and back again.

2) Asteroid B: Thin Shell Between $R_2 \leq r \leq R_1$

Now consider asteroid B, a thin shell of uniform density with outer and inner radii R_1 and R_2 respectively. At a general point $R_2 \leq r \leq R_1$, we can think of the system as being composed of an inner attracting shell ($R_2 \leq r' \leq r$) and an outer shell that does not attract ($r \leq r' \leq R_2$). In the region $r < R_2$ there is no force on the ball within the shell.

The mass of asteroid B can be expressed as

$$M = \frac{4}{3}\pi\left(R_1^3 - R_2^3\right)\rho = \frac{4}{3}\pi\left((R_2 + H)^3 - R_2^3\right)\rho$$

which we expand to give

$$M = \frac{4}{3}\pi\left(R_2^3 + 3R_2^2 H + 3R_2 H^2 + H^3 - R_2^3\right)\rho$$

where ρ is the density of the shell, and $H = R_1 - R_2$ is the thickness of the shell. For $H \ll R_2$ we can make the approximation $M \approx 4\pi R_2^2 H\rho$. We recognise this as the surface area of a shell ($4\pi R^2$) multiplied by the shell thickness and density.

The mass of the inner attracting shell ($R_2 \leq r' \leq r$) is

$$M_{rB} = (4/3)\,\pi\left(r^3 - R_2^3\right)\rho$$

which by the same argument can be simplified to the approximation

$$M_{rB} \approx 4\pi R_2^2 h\rho$$

where $h = r - R_2$; that is, h is the thickness of the inner attracting shell. We can now write

$$M_{rB} \approx M\left(\frac{h}{H}\right) \quad \text{for} \quad 0 \leq h \leq H$$

So, the force due to gravity in the interval $R_2 \leq r \leq R_1$ is given by

$$F_B = \frac{GM_{rB}m}{r^2} \approx \frac{GMm}{r^2}\left(\frac{h}{H}\right) \approx \frac{GMm}{R_1^2}\left(\frac{h}{H}\right) \approx k_B hm$$

where $k_B = GM/\left(R_1^2 H\right)$. In this approximation we set $r^2 \approx R_1^2$ in the interval $R_2 \leq r \leq R_1$, which is true for $R_2 \approx R_1$ or $H \ll R_1$. Once again, in vector notation, noting that the force opposes the displacement from the centre, we can write

$$\boldsymbol{F}_B \approx -k_B \boldsymbol{h} m \quad \text{for} \quad 0 \leq h \leq H$$

$$\boldsymbol{F}_B = \boldsymbol{0} \quad \text{for} \quad -R_2 \leq h \leq 0$$

This is a combination of (approximately) SHM in the region $R_2 \leq r \leq R_1$ (or $0 \leq h \leq H$) interrupted by a large region of constant velocity motion in the region $0 \leq r \leq R_2$ (or $-R_2 \leq h \leq 0$) where there is no force on the ball. The SHM has solution $h = H\cos(\omega_B t)$, with angular frequency $\omega_B = \sqrt{k_B} = \sqrt{GM/\left(R_1^2 H\right)}$ and amplitude H (interrupted every quarter cycle by a transit of the hollow centre of the asteroid at constant velocity). This gives a period for this part of the motion, T_{B1}, which is defined by $T_{B1}\omega_B = 2\pi$. So

$$T_{B1} = \frac{2\pi}{\sqrt{k_B}} \approx 2\pi\sqrt{\frac{R_1^2 H}{GM}}$$

Now consider the transit of the hollow centre of the asteroid at uniform velocity. We can determine the velocity acquired by the ball when it reaches the inner radius, R_2, by integrating the force acting on the ball between $R_2 \leq r \leq R_1$, or, in our alternative co-ordinate system, between $0 \leq h \leq H$. If the ball has zero kinetic energy when it is dropped into the hole at $r = R_1$ ($h = H$), then at $r = R_2$ ($h = 0$) the integral with respect to displacement of the force acting in the range $R_2 \leq r \leq R_1$ is equal to the kinetic energy at $r = R_2$. We write

$$\int_0^H F_B dh = k_B m \int_0^H h dh = k_B m \left[\frac{h^2}{2}\right]_0^H = \frac{1}{2}mv^2$$

We get $v = H\sqrt{k_B}$. The distance the ball needs to travel in one direction is $2R_2$. To complete a transit and return to the starting point it needs to do this twice. The time taken to complete the double transit at uniform velocity is

$$T_{B2} = \frac{4R_2}{v} = \frac{4R_2}{H\sqrt{k_B}}$$

The total period of the motion is given by $T_B = T_{B1} + T_{B2}$. We have

$$T_B = T_{B1} + T_{B2} = \frac{1}{\sqrt{k_B}}\left(2\pi + \frac{4R_2}{H}\right)$$

This is the time period for a ball dropped through asteroid B.

3) Comparison of Time Periods T_A and T_B

We now compare the time periods T_A and T_B. The ratio T_B/T_A is

$$\frac{T_B}{T_A} = \sqrt{\frac{k_A}{k_B}}\left(1 + \frac{2R_2}{\pi H}\right)$$

Recalling that $k_A = GM/R_1^3$ and $k_B = GM/\left(R_1^2 H\right)$, this reduces to

$$\frac{T_B}{T_A} = \sqrt{\frac{H}{R_1}} + \frac{2R_2}{\pi\sqrt{HR_1}}$$

For the arbitrary particular ratio $R_2/R_1 = 0.9$, we have $R_2 = (9/10)\,R_1$ and $H = R_1 - R_2 = (1/10)R_1$, giving

$$\frac{T_B}{T_A} = \sqrt{(1/10)} + \frac{2\,(9/10)}{\pi\sqrt{(1/10)}} \approx 2.1$$

It takes just over twice as long to traverse a hollow asteroid with $R_2/R_1 = 0.9$ as it does to traverse the solid asteroid. The force profiles through the thin-walled shell and the solid asteroid can both be taken as approximately linear over their respective ranges. But the force acts over only a small distance for the thin-walled asteroid. Once again Blastov is right, and has the better understanding of Newton's Shell Theorem.

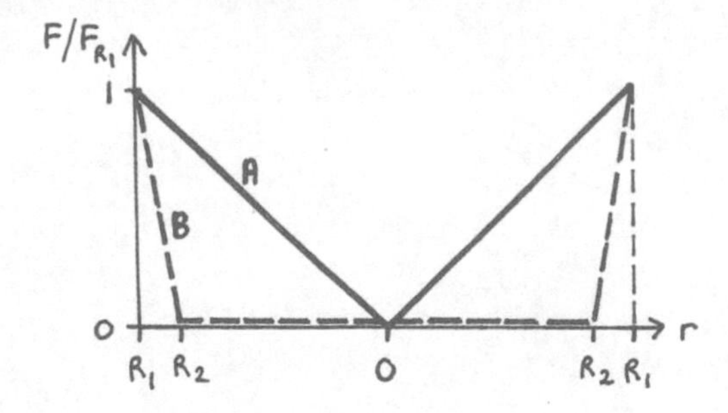

The time ratio T_B/T_A increases as we make the shell thinner and thinner. Our approximations become more accurate as we do this. For time ratios $R_2/R_1 = 0.9,\ 0.99$ and 0.999, we have $T_B/T_A = 2.1,\ 6.4$ and 20.1. In the limit of a very thin shell, the work done on a ball as it is dropped through the surface becomes very small. For thicker shells—that is, with values in the range $R_2/R_1 < 0.9$, our approximation becomes very poor, and we must resort to full integration rather than a first-order approximation. In the limit $R_2/R_1 \to 0$, we find that $T_B/T_A \to 1$. That is, the solutions for the shell and the solid sphere agree, as expected. The general behaviour is shown in the graph below.

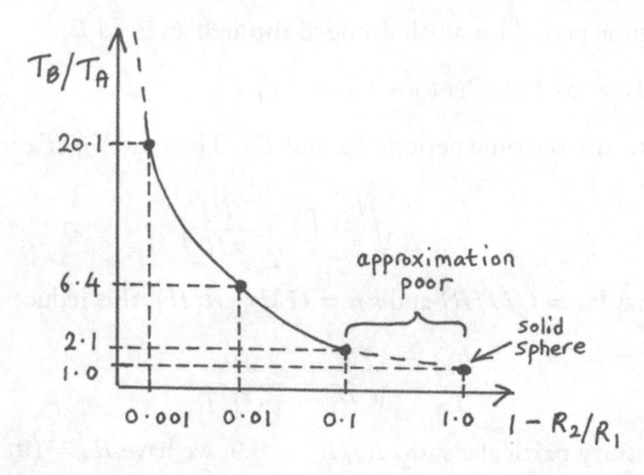

Chapter 11

Optics

In this section on *optics* we look at problems built on simple principles of reflection and refraction. The underlying material is no more than GCSE level in difficulty, but the questions are designed to be challenging and unusual. Some are well-known questions, and others are my own invention.

- LAW OF REFLECTION. The angle between the incident ray and the normal is the same as the angle between the reflected ray and the normal.

- SNELL'S LAW OF REFRACTION. A ray of light passing through an interface between two different media of refractive indices n_1 and n_2 obeys Snell's Law, $n_1 \sin\theta_1 = n_2 \sin\theta_2$, where θ_1 and θ_2 are the angles to the normal of the incident and refracted rays.

11.1 Mote in a sphere ★

I have only ever heard the word *mote* used in two places. The first was the Bible,[1] in which one appears in a person's eye. The second was in my Oxford physics interview, during which I had to answer the following optics puzzle. Fortunately, at my school a passing knowledge of the Bible was inescapable, so I knew that a mote was simply a small speck of something—a dust particle, for example. However, I'd not seen any unusual optics questions before then—but I suppose that was the point. The question is as follows.

A MOTE IS at the very centre of a perfect sphere of glass. Where, if anywhere, do you see the mote?

[1]It appears in the Sermon on the Mount, as a moral lesson against hypocrisy and self-righteousness. The word *mote* comes from Greek, and means a very small dry body. Today we use the word *speck* more commonly.

Answer

We are, of course, expected to draw a ray diagram to demonstrate the solution. Snell's Law tells us that refraction occurs according to the rule $n_1 \sin\theta_1 = n_2 \sin\theta_2$. Because the mote is at the very centre of the sphere, all emitted rays have an angle of incidence of $\theta_1 = 90°$ with the glass-air interface. So the rays are not diffracted by passing through the interface, and the mote appears to be exactly where it is. In the ray diagram we represent the eye with a converging lens, showing it focussing the mote down to a point on the retina.

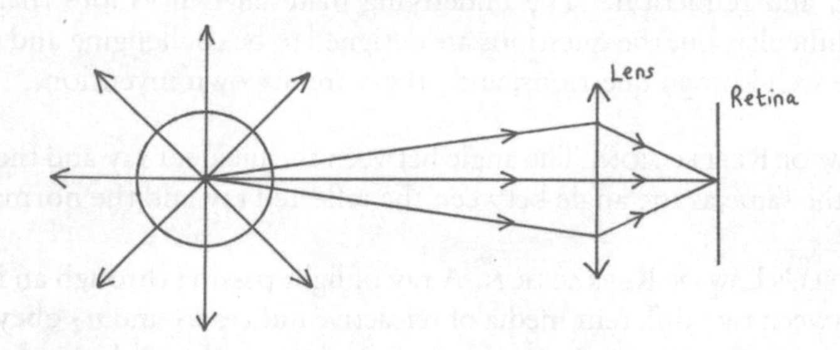

11.2 Diminishing rings of light ★★★

I first heard this question about 15 years ago, though I think it is much older than that. My PhD supervisor was also a great fan of physics puzzles, and this was one of his favourites. I used to think the question was too hard for high-school students, but tried it on a few recently and found that with enough hints students made surprisingly good progress, and most were able to complete the question with a bit of help. I have also tried it on a number of PhD students, who did no better. I attributed this to them forgetting most of their high-school physics: they were all engineers. It is a very enjoyable question to ask, and best done with an experiment. First I would ask students to describe what they saw before deriving equations from first principles to explain it. Students really enjoy this, and find it very satisfying when they manage to derive equations for the interesting optical display. When you attempt the question, remember that it works best with an experiment and someone to give you hints.

YOU ARE GIVEN a long, thin polished metal tube (a 1 m length of the copper pipe used in domestic plumbing, for example) and asked to put your eye at one end and to point the other end at a diffuse light source (a well-illuminated wall, for example). What you see is a pattern of rings. At the centre there is a bright circle, which is the diffuse light source seen directly through the hole in the end of the tube. Surrounding this is a ring of light that is less bright than the central circle. Around that there's another ring that's less bright still. And so on. The rings become successively less bright until the pipe appears dark. The rings appear to have approximately the same width as they get larger. You are asked to derive an equation for the angle subtended at the eye by the n-th ring.

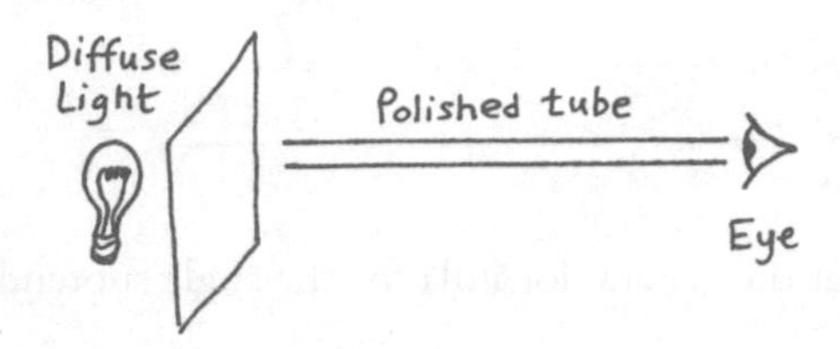

Answer

When approaching this question, it may help to first sketch a diagram of the tube, with the centre marked as a dotted line, in preparation for drawing ray diagrams. Analyse first the light that is directly incident at the eye from the diffuse light source without reflection, then consider the first-order reflection, second-order reflection, and so on.

We start by drawing the ray that subtends the highest angle at the eye (which is at the centre of the tube). We see that where the diameter of the tube is D, and where the length of the tube is L, the highest angle that can be subtended at the eye without reflection is given by $\tan\theta_1 = (D/2)/L$. For a long tube, taking the *small-angle approximation*, we have $\theta \approx D/2L$.

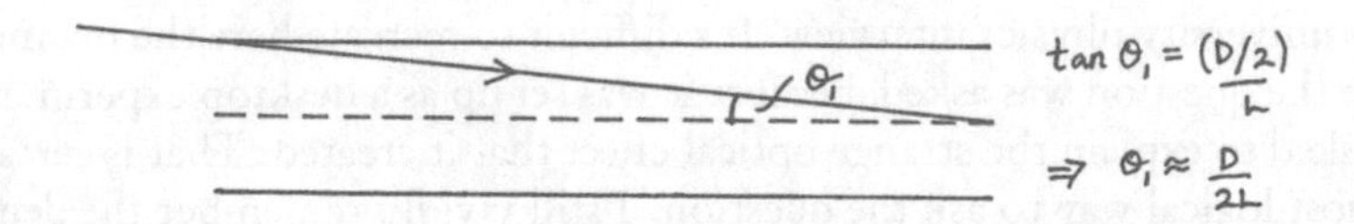

We then consider the first reflected ray. To do this, the only additional thing we need to know is the law of reflection, which is simply that the incident and reflected rays make the same angle with the wall. We see that the ray that subtends the highest angle at the eye, and which is reflected once, divides the tube into three. We have $\tan\theta_2 = (D/2)/(L/3)$. That is, $\theta_2 \approx 3D/2L$.

We then consider the second reflected ray. The ray that subtends the highest angle at thc cye, and which is reflected twice, divides the tube into five. We have $\tan\theta_3 = (D/2)\,/(L/5)$. That is, $\theta_3 \approx 5D/2L$.

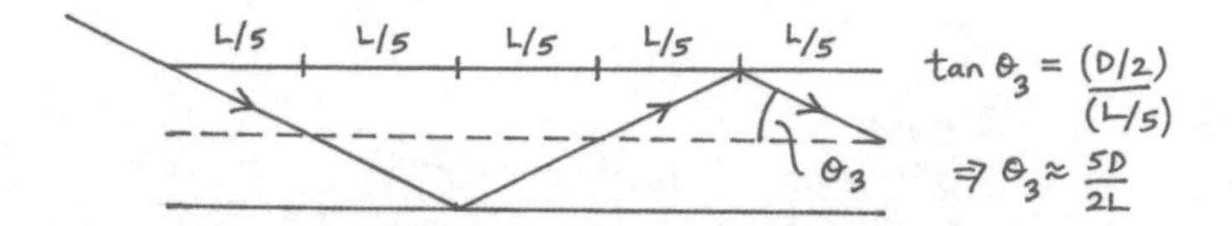

We quickly see that the general formula for the angle subtended at the eye by the n-th ring is

$$\tan\theta_n = \frac{(D/2)}{(L/\,(2n-1))}$$

Taking the small-angle approximation, we see that

$$\theta_n \approx \frac{(2n-1)\,D}{2L}$$

This is a rather satisfying question, and one that works very well with the accompanying demonstration. If you haven't already done so, track down a length of copper pipe a meter or so long (from your local plumber's shop if necessary) and try it. It is very neat indeed.

11.3 Floating pigs ★★★

If you've read the introduction to this book, you'll know that I was asked this in my university physics interview. It's difficult to recreate here the manner in which the question was asked, because it was set up as a desktop experiment. I was asked to explain the strange optical effect that it created. That is certainly the most logical way to ask the question. I still vividly remember the demonstration. The interviewer took two bowls from a drawer, dusted them off with a handkerchief, and placed them proudly—and with something of a flourish—on the table in front of me. They were about the size of cereal bowls, and their insides were perfect mirrors. The top bowl had a hole about two inches in diameter in its base, and was set upside-down on top of the lower bowl, creating a mirrored cavity which could be viewed through the hole. Before that, though, the interviewer placed a small, bright pink plastic pig—about the size of the last joint of my little finger—at the bottom of the cavity (in the centre of the lower bowl). When the mirrored lid was added, a pig appeared, floating above the hole. It looked very real, and very three-dimensional. It felt like you could reach out and touch it.

THE OPTICAL ARRANGEMENT in the drawing below creates a floating image of any small object placed in the base of the lower mirror, when viewed obliquely from above, as shown. Explain how the device works.

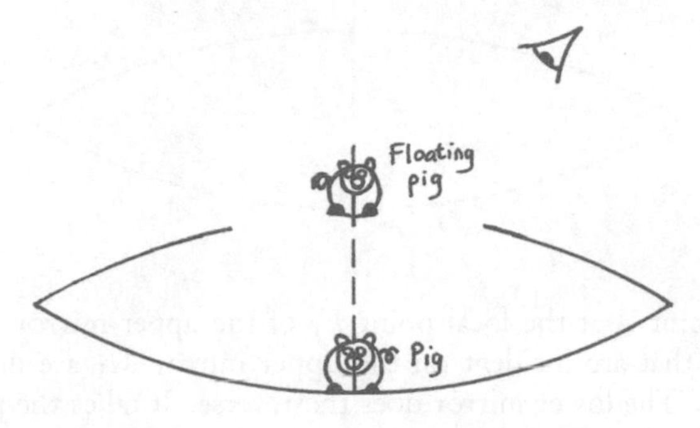

Answer

To answer this, we need to be familiar with the function of parabolic mirrors. They are designed to take parallel rays of light and focus them to a common point, known as a *focal point*, or *focus*. Parabolic mirrors take the form of a *parabola* (a trough bent out of a single sheet of flexible metal, for example), or a *paraboloid* (a bowl- or dish-like shape). The mirror can be any segment of such a paraboloid, but it normally contains the *vertex* (the lowest point of the bowl), so it is symmetrically weighted about the axis of symmetry. Among other things, paraboloid mirrors are used for satellite receivers and solar concentrators. Their function is sometimes reversed to create a parallel beam, such as in a car headlight, for example, in which the light source is placed near the point of focus.

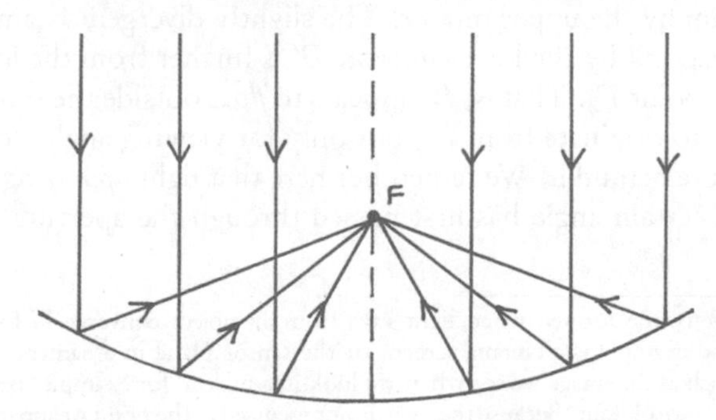

Consider two parabolic mirrors, M_1 and M_2, with focal points F_1 and F_2. We will assume that the correct optical system is one in which the mirrors are arranged such that the maximum separation (the distance between the vertices) is equal to the separation of the focal points. We will justify this with ray diagrams. We draw the focal point of the upper mirror, F_1, at the vertex of the lower mirror, and vice versa. I've left the aperture in the upper mirror out of the drawings that follow so you can see the symmetry of the system more clearly.

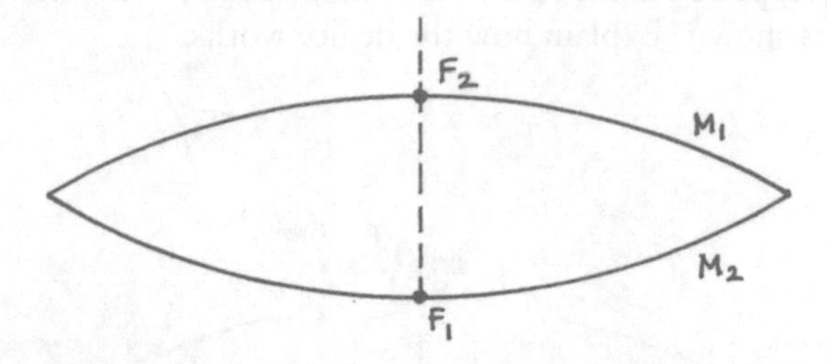

Consider a point A at the focal point F_1 of the upper mirror M_1. Rays of light from A that are incident on the upper mirror M_1 are directed into a parallel beam. The lower mirror does the reverse. It takes the parallel beam and refocusses it as an image A', at the focal point F_2 of the lower mirror M_2. A' is a *real image*[2] of A. That is, A' behaves as though light is originating from it.

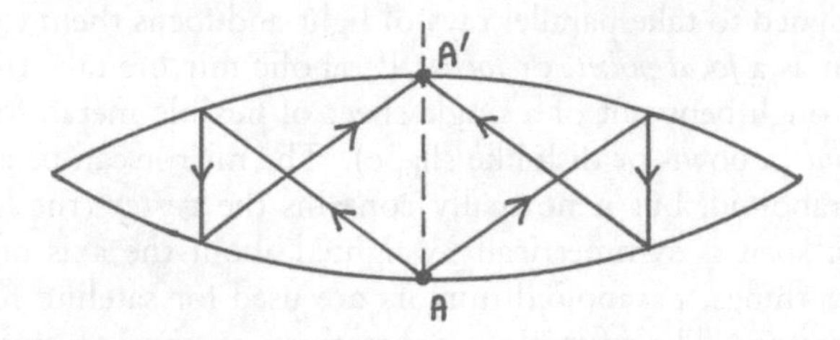

Consider a second point, B, on the axis of the mirrors, but closer to the upper mirror than the focal point F_1. Light from B is directed into a slightly divergent beam by the upper mirror. The slightly divergent beam is focussed into a real image B' by the lower mirror. B' is further from the lower mirror than the focal point F_2. That is, B' appears to *float* outside the mirror system. Light appears to originate from B', but only for viewing angles for which B' has the aperture behind it. We remember here that light *appearing to originate* from B' at a certain angle has first passed through the aperture at the same angle.

[2]A *real image* is one formed when light rays from an object converge in focus at a given location. A good example is a cinema screen, or the sensor plane in a camera. In contrast, a *virtual image* (such as the image we see when we look in a mirror, for example) is seen at a point of *apparent divergence* of light. Because the light is not focussed at the point of apparent divergence (and may not even pass through that point), virtual images cannot be directly imaged on a screen at their apparent location. Virtual images can, however, be seen with the eye or a camera, and have definite size and location.

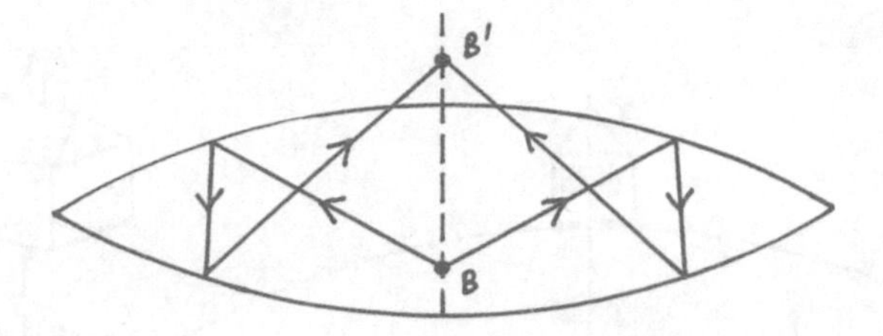

Lastly, consider a point, C, on the surface of the lower mirror but displaced from the axis of the mirror system. C is approximately the focal distance from the upper mirror, but is not at the focal point. We see that rays of light originating from C and incident on the upper mirror lead to an almost parallel beam which is tilted away from the axis. When this beam is incident on the lower mirror it is refocussed at C', a point just above the upper mirror (and therefore outside the mirror system) and off-axis in the reverse direction to C. Once again, C' is a real image of C. The system has rotational symmetry about the axis of the mirrors, so we conclude that C' is a rotation of C by an angle π about the axis.

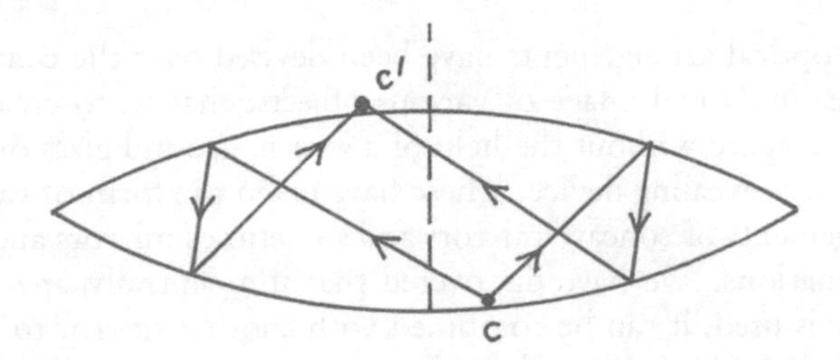

Consider again that A, B and C form the real images A', B' and C'. By extension, we see that if we put a solid object in the centre of the lower mirror (and therefore relatively close to the focal point of the upper mirror), it forms a *three-dimensional real image* outside the mirror system, just above the upper mirror. Points on the object that are further from the focal point of the upper mirror are more significantly distorted in the real image. The image is rotated by π about the axis of the mirrors. It's difficult to represent the rotation of a three-dimensional object in a two-dimensional sketch. For a two-dimensional mirror system and two-dimensional object, the effect is to reflect the image about the vertical axis, with increasing distortion the further you move away from the focal point of the upper mirror. The general effect is shown below. There is an excellent article on the modelling of this mirror system that's well worth reading.[3]

[3]Lingguo, B., 2010, "Modeling the Mirascope using dynamic technology," Loci, November 2010, DOI: 10.4169/loci003595.

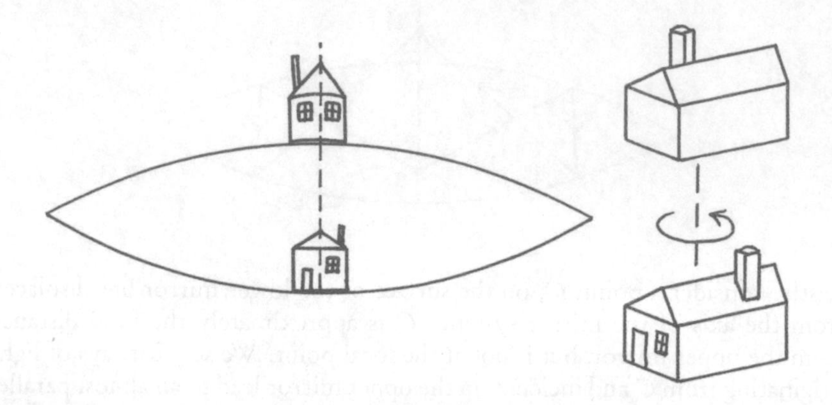

The story goes that the effect was first noticed by someone cleaning searchlight reflectors at the University of California, Santa Barbara, who realised he was attempting to clean a real image of dust rather than dust itself. Apparently he teamed up with someone in the physics department to develop a commercial product. This was as recently as 1970, when the first patent citing this impressive optical phenomenon was filed.[4]

> Many optical arrangements have been devised over the centuries for creating a real image of various objects; that is, to create an image in space without the help of a screen, ground glass or similar image-revealing device. These have taken the form of various arrangements of concave mirrors and sometimes mirrors and lens combinations...We have discovered that if a centrally apertured mirror is used, it can be combined with another mirror to cause a real image of an object placed between the mirrors to be projected through the aperture. The image may appear in space at the aperture, or just below it, or may appear above the aperture.

It's interesting how many scientific discoveries are made by accident. You can now buy the illusion as an executive toy or science demonstration. It's available from museums and toy shops, and from plenty of places online. Several manufacturers make similar products in a range of sizes and prices, and some are very inexpensive and really quite fun. The inventors, however, seem to have had something entirely different in mind.

> The display arrangement is well suited for the display of valuable objects such as jewellery because the object itself can be placed under glass, out of reach of an acquisitive viewer. The image shows the object in its true three-dimensional form, making visible the top, bottom and sides of the object.

[4] Elings, V.B., Landry, C.J., 1970, "Optical display device," US Patent Number US3647284 A.

11.4 The Martian and the caveman ★ or ★★★★

ZAD AND UG are standing on the shore of Lake Hopeful, on planet Earth, trying to catch some fish. Ug is a caveman, so only has a spear to hunt with. Zad is a Martian, and can simply fire his laser eye at his prey. They see a fish. It appears to be a distance $2d$ away from them in the horizontal direction, and a distance $2d$ away in the vertical direction. That is, the fish appears to subtend an angle of 45° from the horizontal at their optical organs, which for both Zad and Ug are at a height d above the surface of the lake. Take the refractive index of air to be $n_1 = 1$, and of water to be $n_2 = 4/3$.

- For ★, should Zad and Ug aim their weapons *above* the fish, *below* the fish, or *at* the fish?
- For ★★★★, ignoring the variation in refractive index with wavelength (of light), calculate the exact location of the fish.

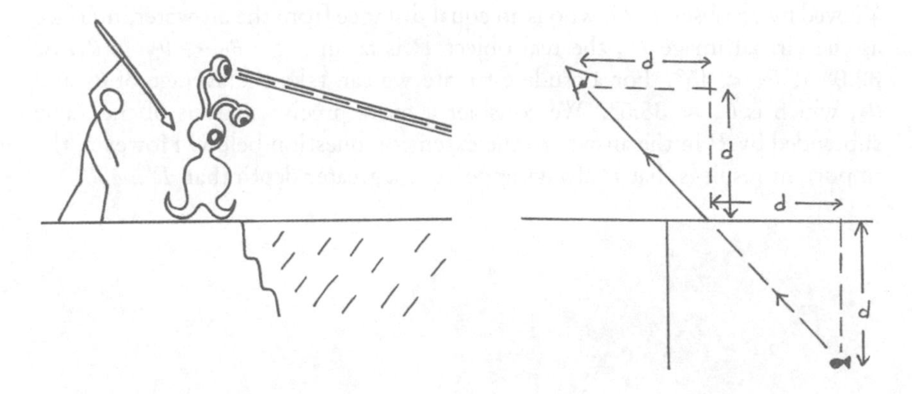

Qualitative answer for ★

If a ray of light emanating from an underwater object P is incident on the water-air interface, then according to Snell's Law the light is refracted away from the normal. That is, the angle of the refracted ray to the surface normal, θ_1, is greater than the angle of the incident ray to the surface normal, θ_2. Snell's Law is given by

$$n_1 \sin\theta_1 = n_2 \sin\theta_2$$

where n_1 and n_2 are the refractive indices of the media. Taking the refractive indices for air and water to be $n_1 = 1$ and $n_2 = 4/3$, respectively, we see that for $\theta_1 = 45°$, $\sin\theta_2 = 3/\left(4\sqrt{2}\right) = 3\sqrt{2}/8$. That is, $\theta_2 \approx 32.0°$. Viewed from a point C, just above the point where the ray from P intersects the surface of the water, a virtual image P' at an angle $\theta_1 = 45°$ from the vertical corresponds to a real object P at an angle $\theta_2 \approx 32.0°$ from the vertical—that is, at a greater depth.

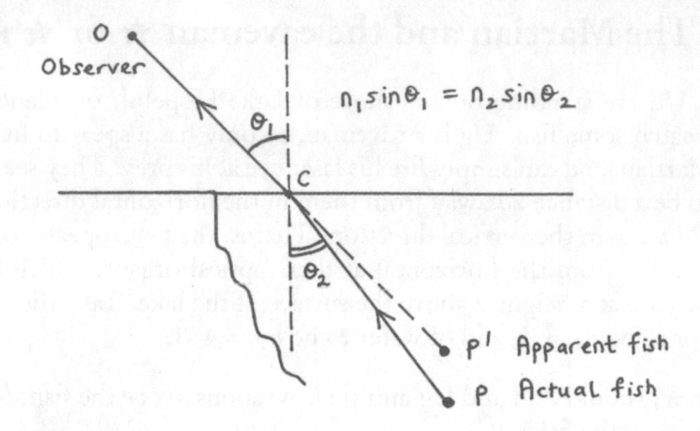

Viewed by an observer O, who is an equal distance from the air-water interface as the virtual image P', the real object P is at an angle $\theta_2 < \theta_3 < \theta_1$, or $32.0° < \theta_3 < 45°$. For a crude estimate we can take the average of θ_1 and θ_2, which is $\theta_3 \approx 38.5°$. We consider a more precise analysis of the angle subtended by P in the answer to the extension question below. However, the important result is that P always appears at a greater depth than P'.

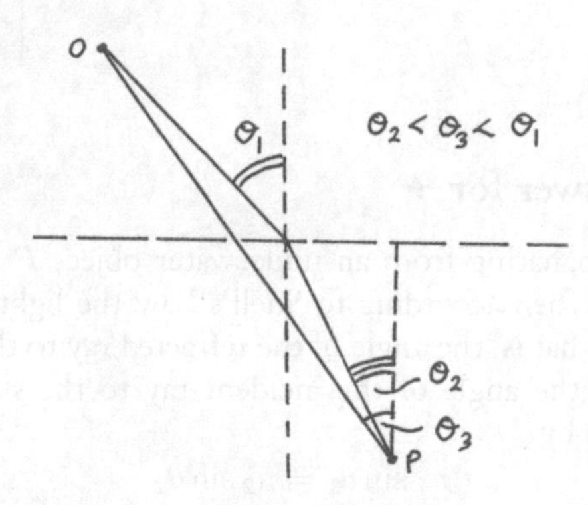

Ug, the caveman, should direct his spear into deeper water than the apparent position of the fish at P'. He should aim *below* the virtual image he sees. We saw that a crude estimate of the angle he should aim at is $\theta_3 \approx 38.5°$ for $\theta_1 = 45°$.

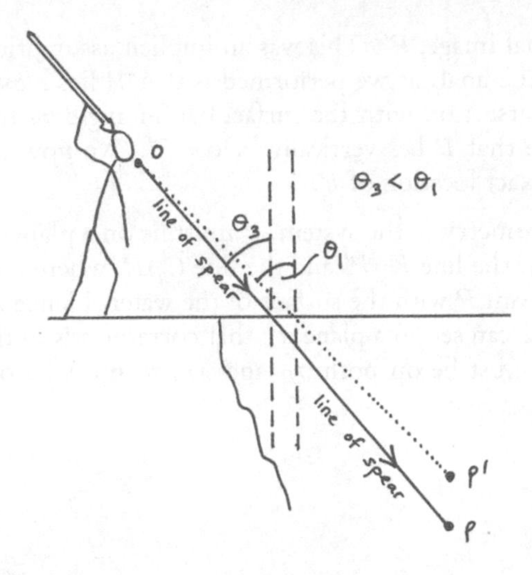

In contrast, Zad, the Martian, should direct his laser eye exactly *at* the virtual image of the fish, P'. When the laser is directed at the virtual image, the beam is refracted at the interface and hits the real fish, P. We do not need to know the exact position of P. The laser simply follows the reverse path of the light emanating from the real fish. Zad needs to aim at exactly[5] $\theta_1 = 45°$.

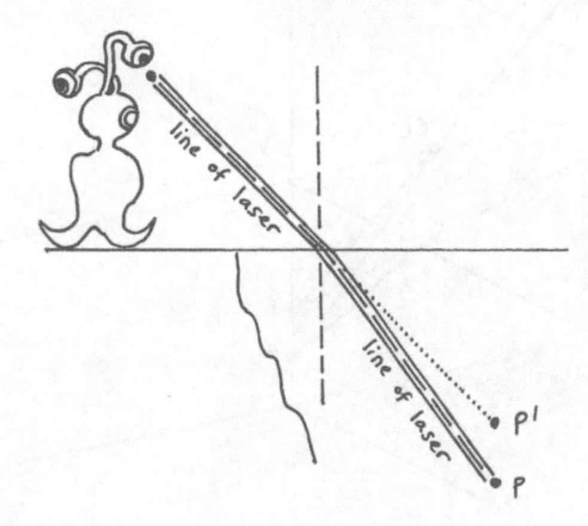

Exact answer for ★★★★

We now want to determine the exact location of the fish, P. In the previous drawings we represented the real object, P, as being vertically below the cor-

[5]The refractive index also depends on the wavelength of light, an effect that leads to *dispersion*, causing different colours (wavelengths) to be refracted more or less, and to diverge slightly when passing obliquely through an interface separating two regions with different refractive indices. It's what you see when white light passes through a prism, or when water droplets cause rainbows. The effect also causes *chromatic aberration* in lenses (when we see ghosts of alien colour at the edges of objects).

responding virtual image, P'. This was an implicit assumption, however. All we know from the analysis we performed is that P lies *along the line* drawn from C (the intersection with the surface) at an angle θ_3 from the vertical. In fact, it is true that P lies vertically below P'. We now prove this before calculating the exact location of P.

From the symmetry of the system, P must lie on a plane π_R that contains the right eye E_R, the line E_RP' and the line C_RP, where C_R is the intersection of the ray from P with the surface of the water. Hence π_R is vertical. In the same way, we can set up a plane π_L that corresponds to the light entering the left eye. P must be on both π_R and π_L, so must be on a vertical line through P'.

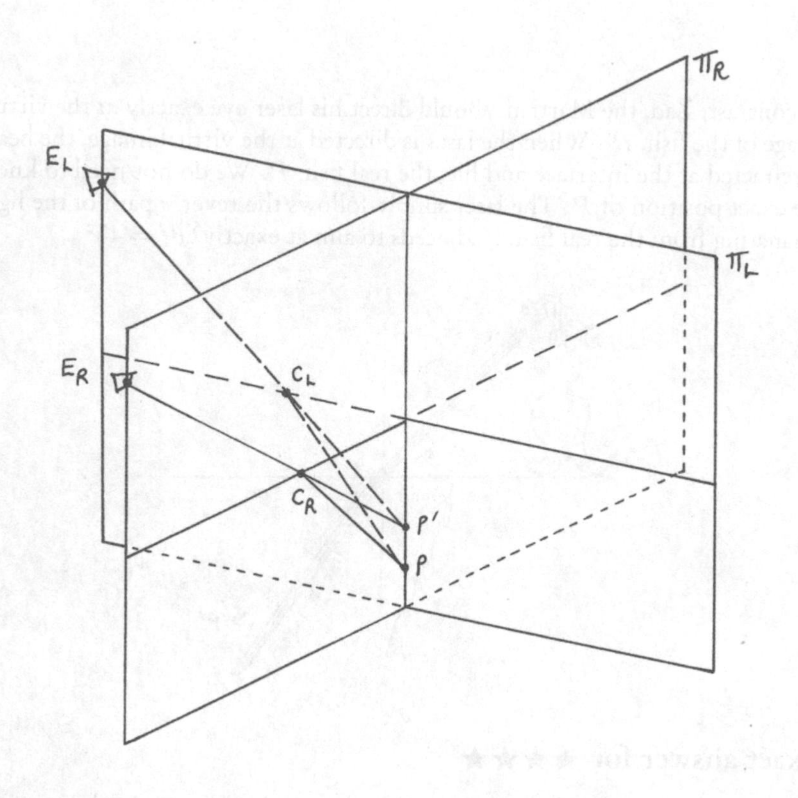

Consider the construction of P and P' on the same vertical line, at depths h and d respectively. We get

$$\tan\theta_3 = \frac{2d}{d+h} = \frac{2d}{d + d\cot\theta_2} = \frac{2}{1+\cot\theta_2}$$

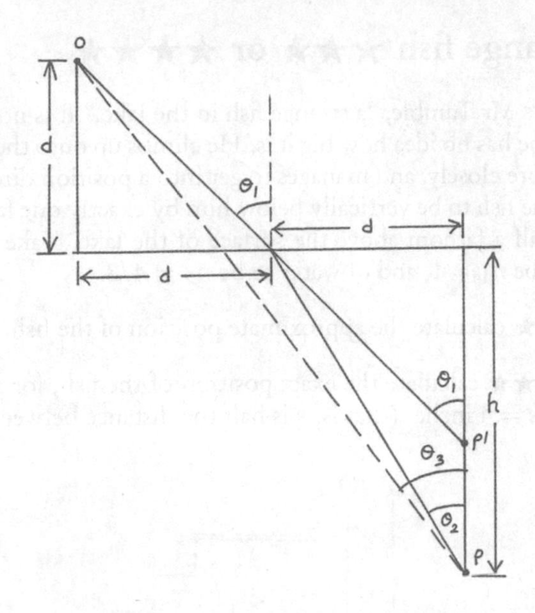

From Snell's Law we have

$$\sin\theta_2 = \frac{n_1}{n_2}\sin\theta_1 = \frac{3\sqrt{2}}{8}$$

And from the geometry of a right-angled triangle containing θ_2 we see that $\cot\theta_2 = \sqrt{23}/3$.

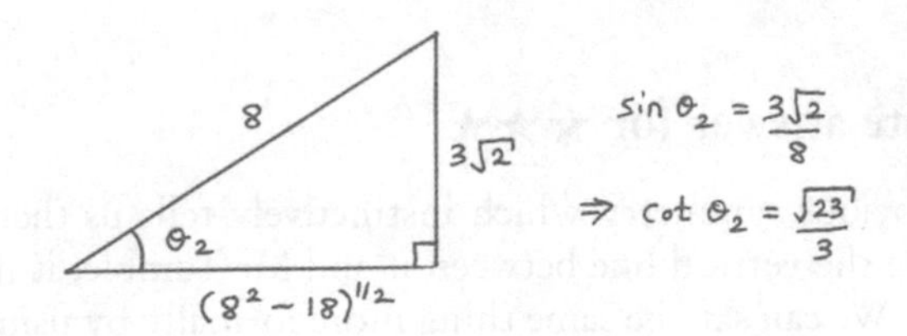

So

$$\tan\theta_3 = \frac{2}{1+\cot\theta_2} = \frac{2}{1+\sqrt{23}/3} = \frac{3\sqrt{23}-9}{7}$$

This gives $\theta_3 \approx 37.6°$, fairly close to our linear approximation of $\theta_3 \approx 38.5°$. From this we can calculate the exact location of the fish, which is exactly $2d$ away in the horizontal direction, and approximately $2.60d$ away in the vertical direction (that is, $1.60d$ below the surface of Lake Hopeful). Although Zad would have no trouble sucessfully hitting the fish with his weapon, we need the more accurate calcuation to give Ug the same chance of a successful catch.

11.5 Strange fish ★★★ or ★★★★

"THERE IS," SAYS Mr Tumble, "a strange fish in the lake." It is not a fish he has seen before so he has no idea how big it is. He climbs up onto the diving board to inspect it more closely, and manages to get into a position directly above it. He perceives the fish to be vertically below him by exactly one fathom,[6] when he is exactly half a fathom above the surface of the lake. Take the refractive index of air to be $n_1 = 1$, and of water to be $n_2 = 4/3$.

- For ★★★ calculate the approximate position of the fish.
- For ★★★★ calculate the exact position of the fish, for an eye separation of $2s = 4$ inches (that is, s is half the distance between the eyes).

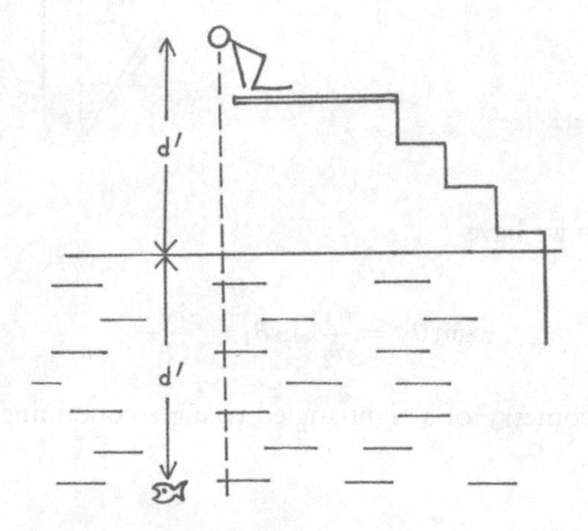

Approximate answer for ★★★

There is an obvious symmetry which instinctively tells us that if the fish is perceived along the vertical line between it and Mr Tumble, it must lie along that same line. We can say the same thing more formally by using Snell's Law ($n_1 \sin\theta_1 = n_2 \sin\theta_2$) for a ray of light travelling across the water-air interface. We see that for a ray of light arriving at the water-air interface with $\theta_2 = 0$, Snell's Law is satisfied for $\theta_1 = 0$. That is, there is no refraction of a ray that's perpendicular to the interface. If we see the fish vertically below us, the fish *is* vertically below us. This does not tell us *where* the fish is along the vertical line, however. We'll examine that now.

Consider a ray of light emanating from a point of interest P, incident on the water-air interface at an angle θ_2, and refracted to a higher angle $\theta_1 > \theta_2$. Multiple rays forming either a fan of light, in two dimensions, or a cone of

[6] A fathom was traditionally based on the distance between a man's outstretched arms, and is a nautical term used primarily for measuring the depth of water. It was eventually formalised at 6 feet or 1.83 m.

light, in three dimensions, spread out on passing through the interface. If we consider a small angular band of rays, they appear to come from a point P', where a *virtual image* of the fish is formed.

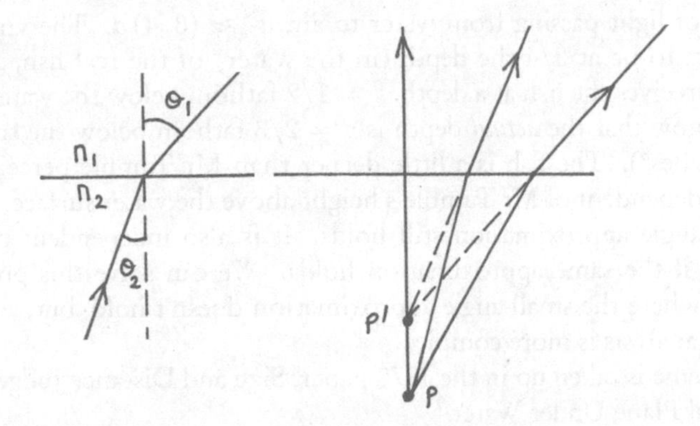

For an observer who is not directly above an underwater object, and who is viewing it through the air-water interface at some large angle θ_1, it's quite difficult to calculate the actual position of the object from the apparent position.[7] For an observer who is directly above an underwater object, however, if the eye separation is small compared to the height of the eyes above the water-air interface, the problem reduces to very small angles θ_1. For very small θ_1 we can take the small-angle approximation, $\sin\theta \approx \theta$, and reduce Snell's Law to

$$\theta_1 \approx \frac{n_2}{n_1}\theta_2$$

Taking $n_1 = 1$ for air and $n_2 = 4/3$ for water, we see that $\theta_1 \approx (4/3)\,\theta_2$. Consider a ray of light emanating from a point P at depth d, but appearing to emanate from a virtual origin P'. For small angles we can show that the apparent depth d' of P' is given by

$$d' \approx d\frac{\theta_2}{\theta_1}$$

θ_1 x d' θ_2 P' d P

$\tan\theta_2 = x/d$

$\tan\theta_1 = x/d'$

small θ: $\theta_1/\theta_2 \approx d/d'$

[7] Try it if you're interested!

Using the reduced form of Snell's Law we have

$$d' \approx d\frac{n_1}{n_2}$$

For a ray of light passing from water to air, $d' \approx (3/4)\,d$. The virtual fish, P', appears to be at 3/4 the depth (in the water) of the real fish, P. If Mr Tumble perceives the fish at a depth $d' = 1/2$ fathom below the water surface, then we know that the *actual* depth is $d = 2/3$ fathom below the surface (or $d = 48$ inches[8]). The fish is a little deeper than Mr Tumble perceives. The result is independent of Mr Tumble's height above the water surface, provided the small angle approximation still holds. It is also independent of his eye separation if the same approximation holds. We can solve this problem in situations where the small-angle approximation doesn't hold, but, as we have noted, the analysis is more complex.

The theme is taken up in the 1970 paper, Size and Distance Judgements in the Vertical Plane Under Water.[9]

> The underwater environment affects the distance perception of divers in various ways. One of the best known effects is the optical distortion due to the difference in refractive index between water and air. The diver looks into water through the air trapped in his face-mask, and sees a virtual image which is optically located at about 3/4 of its physical distance.

As we'd expect, this agrees with the calculations we've just performed. However, the situation is a little more complex in terms of our perception. The authors continue:

> ...he does not necessarily perceive objects to be at their optical distance, since there are many other factors affecting distance perception. One of the most important of these is reduced brightness-contrast, which leads to the overestimation of distance both in water and in a fog on land. Low contrast does not only act as a cue to increased distance—it also obliterates or reduces other useful cues such as stereopsis and linear perspective.

The impression of depth and three-dimensionality is highly complex, as are most things associated with the visual system. These sensations occur through a range of visual cues. These can be classed as *oculomotor cues* (the ability to sense position and focus through muscle tension within the eye), *monocular cues* (the relative size of objects, atmospheric perspective, shadows, occlusion and movement-produced cues), and *binocular cues* (binocular disparity, or the effect by which each eye sees a different image of an object because the eyes are separated).

[8] A fathom is 6 feet. A foot is 12 inches.

[9] Ross, H.E., King, S.R., Snowden, H., 1970, "Size and distance judgements in the vertical plane under water," Psychologische Forschung (Psychological Research), 33, pp. 155–164.

Not that long ago I was invited to a strange party at a remote location in Devon. The sole purpose of the party was to get together for two days to solve mathematical puzzles (of the recreational variety) and to play mathematical games. The host had gathered some quite prominent mathematicians, magicians and puzzle makers, some of whom had flown from distant shores for the occasion. There was a surprise lecture by a man well-known in London for supplying England with the largest single collection of illusions, scientific toys and curiosities. His lecture was on a bizarre chapter in the life and work of Charles Wheatstone. Wheatstone (1802–1875) is probably most famous for explaining the importance of what is now known as the Wheatstone bridge.[10] He performed scientific work in a wide range of fields, however, including research on stereo vision, or *stereopsis*. For this purpose Wheatstone invented the *stereoscope*, an arrangement of mirrors that allowed left-eye and right-eye images of the same scene (which were hand-drawn on card) to be presented only to the correct eye. By doing so, Wheatstone demonstrated the ability of the eyes to *fuse* the images and create the sensation of depth from two flat drawings. It was a landmark in our understanding of how we interpret what we see. Many of Wheatstone's results were included in his brilliant 1838 paper Contributions to the Physiology of Vision.[11] It was a time of great discovery, and this paper is well worth reading. Wheatstone begins:

> When an object is viewed at so great a distance that the optic axes of both eyes are sensibly parallel when directed towards it, the perspective projections of it, seen by each eye separately, are similar, and the appearance to the two eyes is precisely the same as when the object is seen by one eye only. There is, in such case, no difference between the visual appearance of an object in relief and its perspective projection on a plane surface; and hence pictorial representations of distant objects, when those circumstances which would prevent or disturb the illusion are carefully excluded, may be rendered such perfect resemblances of the objects they are intended to represent as to be mistaken for them; the Diorama is an instance of this. But this similarity no longer exists when the object is placed so near the eyes that to view it the optic axes must converge; under these conditions a different perspective projection of it is seen by each eye, and these perspectives are more dissimilar as the convergence of the optic axes becomes greater. This fact may be easily verified by placing any figure of three dimensions, an outline cube for instance, at a moderate distance before the eyes, and while the head is kept perfectly steady, viewing it

[10] An electrical circuit designed to determine the value of an unknown resistance by balancing a bridge circuit with other known resistances. Its invention is credited to Samuel Hunter Christie in 1833 but it was popularised by Charles Wheatstone in 1843.

[11] Wheatstone, C., 1838, "Contributions to the physiology of vision. Part the first. On some remarkable, and hitherto unobserved, phenomena of binocular vision," Philosophical Transactions of the Royal Society, London, January 1, 1838, doi:10.1098/rstl.1838.0019.

with each eye successively while the other is closed.

Wheatstone also invented the *pseudoscope*. This was another arrangement of mirrors (easily constructed from materials available in hardware shops) that reversed the images appearing at each eye. If you view a normal scene through the device, it appears *hollow*. Our depth perception of a face, for example, is inverted, making it look like a mask viewed from the inside. There is an interesting paper on the topic by Jearl Walker.[12]

> Branches at the rear of a tree seem closer than branches at the front. The sight is eerie because I realize that the front branches partially block my view of the rear branches. Depth is also inverted when I look at an object that can easily be reversed mentally. For example, a pot hung bottom out on the kitchen wall suddenly appears to bulge inward rather than outward.

Walker also describes using the *hyperscope*, an instrument that effectively increases the separation of the eyes, and makes everything look more three-dimensional.

> The hyperscope also alters the apparent height and width of nearby objects...Seen through the hyperscope, an object looks smaller because the angle of convergence required to see it through the mirrors is larger than normal. Many other familiar objects take on a strange appearance through the hyperscope. For example, a person's face looks thinner and seems to have a prominent nose.

I have also seen demonstrations of a *monoscope*, a variation of the hyperscope in which the eyes are effectively brought to a single point. Depth perception vanishes and although the scene retains all its correct proportions, it is as though we are looking at a photograph. Everything is flat. It's interesting to think that our perception of the world depends so much on the spacing of our eyes.

Exact answer for ★★★★

Let the separation of the eyes be $2s$. Mr Tumble perceives the fish to be at a point P', which is a distance d' below the surface of the lake. Mr Tumble is a distance d' above the surface of the lake. From the geometry of the problem,

$$\tan\theta_2 = \frac{s/2}{D}$$

where D is the depth of the fish P. Rearranging we have an expression for the depth.

$$D = \frac{s}{2}\cot\theta_2$$

[12] Walker, J., 1986, "The hyperscope and the pseudoscope aid experiments on three-dimensional vision," Scientific American, Vol. 255, No. 3, pp. 134–138.

We now find $\cot\theta_2$ by considering the geometry of the problem.

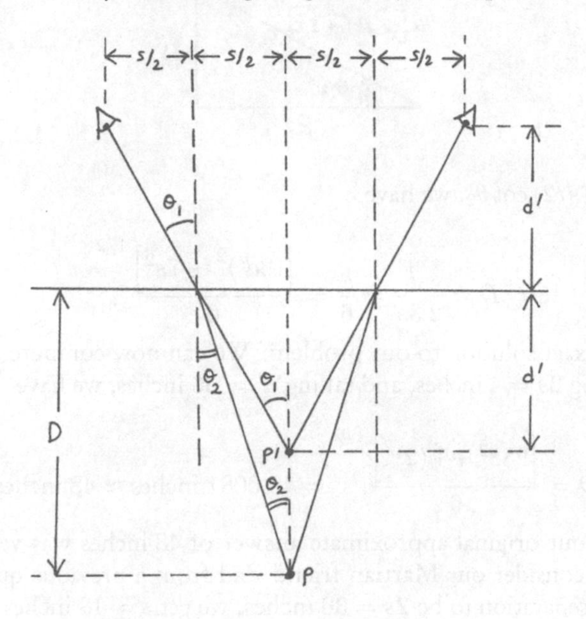

By considering a right-angled triangle with one angle equal to θ_1, we see that

$$\sin\theta_1 = \frac{s}{\left[(2d')^2 + s^2\right]^{1/2}}$$

Using this result, and Snell's Law ($n_1 \sin\theta_1 = n_2 \sin\theta_2$), where we take $n_1 = 1$ and $n_2 = 4/3$, we have

$$\sin\theta_2 = \frac{3}{4}\sin\theta_1 = \frac{3s}{4\left[(2d')^2 + s^2\right]^{1/2}}$$

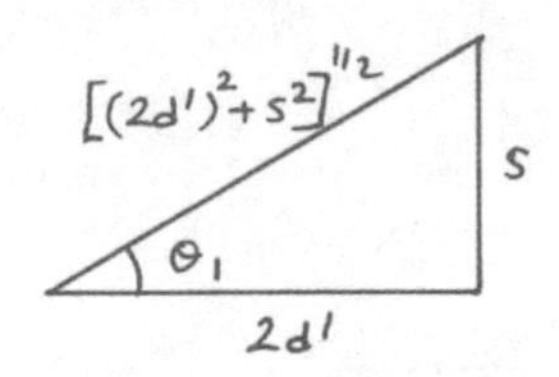

Using this result, and considering a right-angled triangle with one angle equal to θ_2, we see that

$$\tan\theta_2 = \frac{3s}{k}, \text{ where } k = \left[(8d')^2 + 7s^2\right]^{1/2}$$

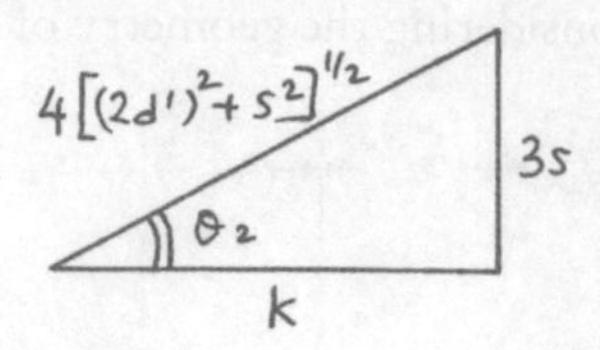

Using $D = (s/2)\cot\theta_2$, we have

$$D = \frac{s}{2}\frac{k}{3s} = \frac{k}{6} = \frac{\left[(8d')^2 + 7s^2\right]^{1/2}}{6}$$

This is the exact solution to our problem. We can now compute a numerical result. Taking $2s = 4$ inches, and taking $d' = 36$ inches, we have

$$D = \frac{\left[288^2 + 7\,(2)^2\right]^{1/2}}{6} = 48.0081 \text{ inches} \approx 48 \text{ inches}$$

We see that our original approximate answer of 48 inches was very good indeed. If we consider our Martian friend Zad from a previous question, and take his eye separation to be $2s = 36$ inches, we get $s = 18$ inches. This gives

$$D = \frac{\left[288^2 + 7\,(18)^2\right]^{1/2}}{6} = 48.65 \text{ inches}$$

Amazingly, despite Zad's huge eye separation, which is equal to his distance above the surface ($2s = d'$), the result is still very close to 48 inches.

Chapter 12

Heat

In this short section on *heat* we look at conceptual problems associated with the thermal expansion of solids, and time-dependent or *transient* heat transfer problems. For these you won't need more than a basic intuitive knowledge of several things: what heat is, how it's stored, and how it affects solids at the macroscopic level. You should also know that it's transferred across temperature gradients from hot to cold, at a rate proportional to the temperature difference and area. Before we look at the problems, however, we should state the most generically useful concepts more formally.

- HEAT CAPACITY. The heat capacity per unit mass, or *specific heat capacity*, c_p, is a measure of the ability of a substance to store heat. So the heat, Q, required to raise a mass, m, by a temperature difference, ΔT, is given by $Q = mc_p\Delta T$.

- FOURIER'S LAW OF HEAT CONDUCTION. The rate of heat flow to an object, dQ/dt, is proportional to the surface area, A, the temperature gradient, $\Delta T/\Delta x$, and the thermal conductivity, k, according to

$$\frac{dQ}{dt} = kA\frac{\Delta T}{\Delta x}$$

- NEWTON'S LAW OF CONVECTIVE COOLING/HEATING. The rate at which a body loses or gains heat is proportional to the surface area of the body, A, the temperature difference between the body and its surroundings, $T_1 - T_2$, and the heat transfer coefficient, h. The heat transfer coefficient is a constant of proportionality for a particular physical situation. We write

$$\frac{dQ}{dt} = hA(T_1 - T_2)$$

  In general, T_1 and T_2 can be functions of time.

- STEFAN-BOLTZMANN LAW. The heat transferred by radiation between a body at a temperature T_1 and its surroundings at temperature T_2 is

$$\frac{dQ}{dt} = \varepsilon\sigma A\left(T_1^4 - T_2^4\right)$$

  where A is the surface area of the body, ε is the *emissivity* of the body, and σ is the Stefan-Boltzmann constant.

- THERMAL EXPANSION COEFFICIENT. The thermal expansion of length is described by the coefficient of linear thermal expansion, α_L, according to

$$\alpha_L = \frac{1}{L}\frac{dL}{dT}$$

  where T is the temperature and L is the *characteristic length*.

12.1 The heated plate ★

This is another question I was asked in my university physics interview. It's a neat question but one that gives relatively limited scope for discussion. It is posed very straightforwardly as follows.

A STEEL PLATE has a hole cut out of it. We heat the plate up. Does the hole get smaller, bigger, or stay the same?

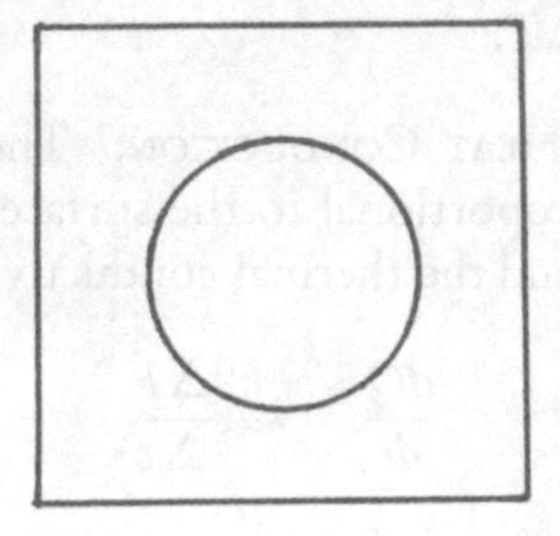

Answer

For an *isotropic* material which expands on heating, every length scale gets bigger by the same amount, so everything, including the hole, gets bigger. The thermally expanded (hot) object looks like a larger version of the cold object.

This is probably all we need to say to answer the qualitative question that was posed. In the discussion, we now quantify the magnitude of change one might expect for a particular temperature difference.

Further discussion

Almost all substances expand on heating. The reason is that the average kinetic energy of the atoms is increased and, as a result, they maintain slightly greater mean separation from each other. The amount by which substances expand on heating is described by the coefficient of linear thermal expansion, α_L, which is defined by

$$\alpha_L = \frac{1}{L}\frac{dL}{dT}$$

where L is the length scale and T is the temperature. Here α_L simply describes the *proportional change* in length with temperature. Values of α_L for common substances typically come from experiments.

We distinguish between essentially isotropic materials (most materials), which have the same α_L in all directions, and anisotropic materials (some crystals and composites, for example), which can have different α_L according to direction within the material.

On heating, a sheet of an isotropic material will expand by the same amount in all directions. If we were to draw a grid on the material like that on a chess board, after heating the grid would look liked a scaled-up version of itself. The scaling factor, or the ratio of length in the heated condition, L', to length in the unheated condition, L, is given simply by

$$\frac{L'}{L} = \frac{L + \Delta L}{L} = 1 + \alpha_L \Delta T$$

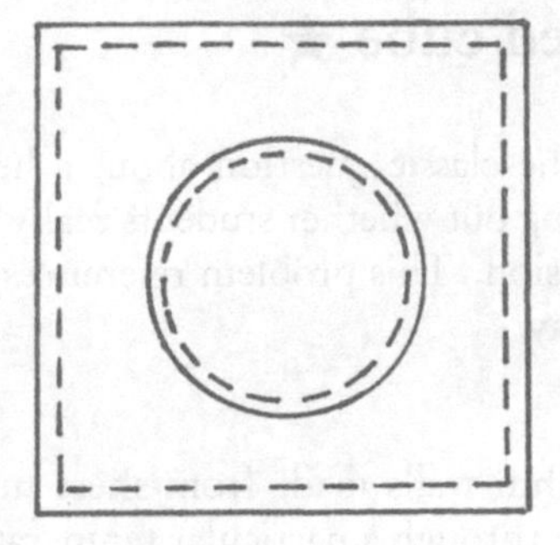

So all features on the sheet, including any holes, increase their linear dimensions at the same rate. The hole gets bigger.

When components are to be assembled with an *interference fit*[1] they are often heated or cooled to aid assembly. I remember spending three very cold

[1] This is an engineering term used to describe parts which, in their natural state, *interfere* with each other–that is, they have very slightly overlapping volumes. To assemble interference-fit parts, we need very large forces, press-fitting, or shrink-fitting. Normally in the latter process the female part (the rotor into which an axle is to be mounted, for example) is heated to increase its size, and the male part (the axle or shaft of a machine, for example) is cooled to reduce its size. The parts are then assembled. At a common temperature very large frictional forces are acheived at the interface, which allows large torques to be transmitted.

days in Swansea Marina waiting for a large team of engineers to recommission the Tawe Lock. We were meant to be sailing, but the lock gates had been taken off to be repaired, and our boat was stuck in the marina until the gates could swing again. The lock was quite an impressive piece of engineering. It was required to hold back a tidal range of about 10 m, one of the largest in the world. As you might imagine, the lock gates were quite substantial. They were a pair of quarter cylinders with an internal framework, which rotated on a vertical axis into huge recesses within the lock walls. The retaining pins alone were about 20 cm in diameter and 50 cm long, and made of solid steel. Each one must have weighed about 150 kg. With very little else to do I appointed myself chief engineer, and would take a twice-daily trip to the lock to see what progress was being made. The cranes were being hampered by high winds, so the real chief engineer had plenty of time to discuss the operation. I asked him how they were planning to get the pins in place. He explained that the pin and the hole were a size-for-size fit, and that the plan was to heat the female part (slightly) with a blow-torch, cool the pin with liquid nitrogen, and then in a tense and presumably fairly rapid operation, insert the pin. I can only presume that reversing the operation was essentially impossible. Back at the boat we had a cup of coffee and calculated the clearance they would achieve. According to our quick calculation,[2] the clearance on diameter[3] would be just over 0.6 mm, which seemed quite impressive. A day later they lowered the pins into position. The method worked perfectly.

12.2 The heated cube ★

This is a variation of the classic question about a heated plate with a hole, which is useful for finding out whether students really do understand the principles of thermal expansion. This problem references the same concepts, but in a slightly different way.

A HOLLOW CUBE with thin walls made from sheet metal, and a ruler of the same material, are taken through a particular temperature change. If the ruler changes length from x to x', write an expression for the relationship between V and V', the initial and final volumes enclosed by the walls of the cube. If the ruler and the side of the cube are both 100 mm in length, and the ruler expands by 1 mm, without using a calculator estimate the percentage change in volume of the cube. Assume that the sheet metal is an isotropic material.

[2]For steel, a typical value of α_L is 13×10^{-6} K^{-1}. Liquid nitrogen boils at -196 °C, which we assumed the pin was cooled to. We assumed the female part was heated to 50 °C. This gives $\Delta T = 246$ K. Thus the change in diameter (for $d = 200$ mm) is given by $\Delta d = d\alpha_L \Delta T \approx 200 \times 13 \times 10^{-6} \times 246 = 0.64$ mm.

[3]The difference in diameter between the heated female part and the cooled male part.

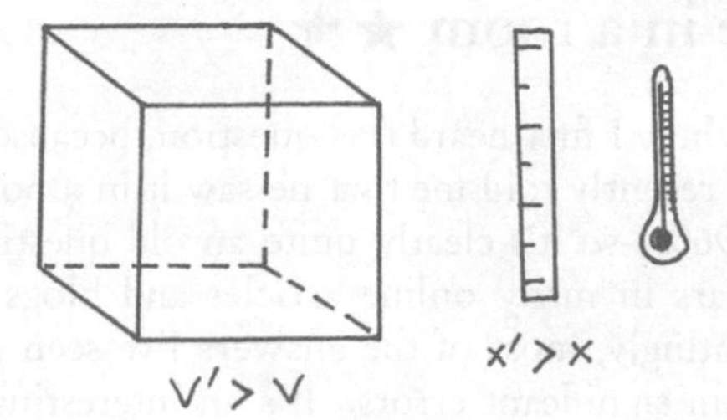

Answer

If the ruler and the cube are made from the same isotropic material, the fractional change in every linear dimension is x'/x. If the cube has initial side length L, then the new side length L' is given by $L' = L\,(x'/x)$. The ratio of volumes is therefore

$$\frac{V'}{V} = \left(\frac{L'}{L}\right)^3 = \left(\frac{x'}{x}\right)^3$$

If the ruler is initially 100 mm long and expands by 1 mm (or 1%), then taking the approximation for small changes the cube expands by approximately 3%. We can see this by expanding the equation above. We start with

$$\frac{V'}{V} = \left(\frac{x'}{x}\right)^3 = \left(\frac{x+\delta x}{x}\right)^3 = \left(1+\frac{\delta x}{x}\right)^3$$

We recognise this as a third-order *binomial expansion*,[4] which takes the general form

$$(x+y)^3 = x^3 + 3x^2y + 3xy^2 + y^3$$

Neglecting the terms $(\delta x)^n$ for $n \geq 2$ (which are negligibly small), we have

$$\frac{V'}{V} \approx 1 + 3\frac{\delta x}{x}$$

If $\delta x/x = 0.01$, it follows that $V'/V \approx 1.03$. We see that the increase in volume is almost exactly 3%. We would get exactly the same result, and with the same reasoning, if the cube were made of solid metal rather than sheet metal.

[4]The *binomial theorem* is a general method for expanding powers of a *binomial*. A binomial is the sum or difference of two terms, such as $(x+y)$ or $(x-y)$. The theorem states that $(x+y)^n$ can always be expanded into a sum of terms of the general form ax^by^c, where b and c are non-negative integers that obey $b + c = n$. The coefficient a depends on n and b, and is described by *Pascal's triangle*, or simply $a = n!/\,(b!c!)$. Other than its obvious application to expanding binomials, Pascal's triangle has applications in combinatorics, to describe the number of ways of choosing b elements from an n-element set. The triangular representation of the binomial coefficients (which is really a visual aid for remembering them) is attributed to Blaise Pascal (in the seventeenth century), but the generalised proof was first derived by Sir Isaac Newton (c. 1665).

12.3 Fridge in a room ★★

I can't remember where I first heard this question, because it was such a long time ago. A friend recently told me that he saw it in a popular science book published in the 1960s—so it's clearly quite an old question. Googling it, I found that it appears in many online articles and blogs related to physics puzzles. Disappointingly, most of the answers I've seen are at best unclear, and at worst contain significant errors. It's an interesting question which I hope you will enjoy. I find it very useful, because it tests students' fundamental understanding of a number of basic principles, and is ideally suited to a discussion. In the form I have asked it, it also tests students' ability to make sound estimates and to sketch graphs. I find that most students are able to make good headway with it if they are given enough prompts. Many are able to sketch representative graphs typically on the second or third attempt, with hints in between.

A NORMAL DOMESTIC refrigerator is placed inside a perfectly thermally insulated room and turned on. Sketch what happens to the temperature in the room $T(t)$ over the next few (say four) hours. The fridge door is then opened and left open. Sketch the temperature in the room during the following few (again, four, say) hours.

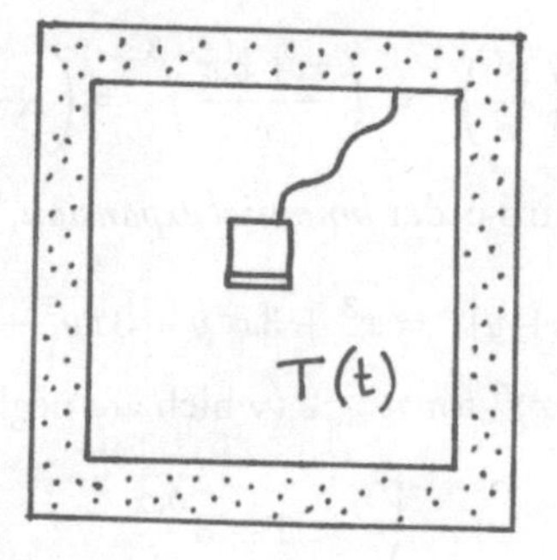

Answer

Where to begin? Here the cornerstone of understanding is simply to realise that energy conservation must hold. We could put a *control volume*[5] around the room and determine that the only way energy can enter or leave the system is through the electrical wire that passes through the surface of the control volume. Otherwise the system is isolated. Electrons are entering the room

[5] A control volume is the region contained within an imaginary surface that defines a *system* we wish to examine. We use the concept to allow us to analyse thermodynamic or fluid dynamic systems, by considering the flows of, say, mass, heat, and work across the control volume boundaries. The abstraction helps us reduce systems to a simple form in which their fundamental behaviour can be better revealed.

with a certain electrical potential and leaving with a lower electrical potential. Electrical energy is being *dissipated*[6] in the *load* that is the fridge. Over time the temperature in the room must rise.[7]

The rate at which the temperature in the room will rise is dictated primarily by the power dissipated by the fridge. There is a maximum power that the fridge can dissipate, but it may use less than this under certain conditions when the motor does not need to run so hard. There is a secondary consideration, which is the *heat pump effect* of the fridge. The difficult job a fridge has to do is to take heat from a cold place (the inside of the fridge) and transfer it to a hot place (the outside of the fridge). In the words of a student of thermodynamics, the heat is being moved *up* the temperature scale, and this takes energy (which is the power dissipated by the fridge). So we consume energy to run the *cycle*,[8] but the entire process is one in which no heat energy is created or destroyed, but simply moved. We say the heat is *pumped*.[9]

What's interesting about this question is considering the *qualitative* form of the temperature against time graph, which is all I tend to ask for when setting the problem. In this answer, we'll also estimate some crude numbers so we can sketch a quantitatively accurate graph.

Let the starting room temperature be, say, $T_a = 18°\text{C}$. Assume also that the fridge is designed to maintain a temperature of 3°C. We could estimate the power of a fridge according to how long it takes to cool things down (from experience), or we could make an educated guess about how much energy is dissipated from the back of the fridge. A good estimate might be a maximum power $P_{max} = 200$ W. At nominal conditions (when it has reached steady state with an internal temperature of 3°C and an external temperature of 18°C), the fridge might spend about 50% of its time on and 50% off,[10] so

---

[6]By which we mean it is being converted to a less useful form of energy, in this case heat. In the language of thermodynamics, we describe processes in which energy is converted to a less useful form as *irreversible*. Irreversible processes are associated with entropy generation, where entropy is the thermodynamic measure of how much the energy's usefulness has degraded. One measure of the usefulness of energy is the efficiency with which we can convert the energy to mechanical work in a realisable system.

[7]If this wasn't obvious I wouldn't worry too much—about a third of the people I have tried this question on thought the fridge would cool the room down.

[8]Here we mean a thermodynamic cycle. The fridge is a machine set up to take in power and, in a continuous cycle, pump heat from a cold location to a hot location.

[9]This curious terminology arose due to the early (and now superseded) *caloric theory* of heat, in which heat was thought to be transmitted via the flow of a fluid called caloric. This theory seems to have been initiated by the French chemist Antoine Lavoisier (in 1783), and it persisted until the end of the nineteenth century. As late as about 1880, it appears in one of the first spectacularly successful popular science books (first published around 1840, and republished 47 times in English alone)—the Rev. Dr. Brewer's A Guide to the Scientific Knowledge of Things Familiar. Brewer explains:

> Q. What is that "stream of heat" called, which flows thus, from one body, to another?
> A. CALORIC. Caloric, therefore, is the matter of heat, which passes from body to body; but heat is the sensation, of warmth, produced by the influx of Caloric.

[10]These numbers are fairly typical for an average-sized fridge, and most fridges run at a single

it would have an average *steady state* power consumption of $\bar{P}_{SS} = 100$ W. Remember that exact numbers are not important here, we're just estimating reasonable values.

How fast would the temperature in the room rise? For this we need to know the heat capacity of the room. Let's define the hypothetical room to have a volume $V = 5\text{ m} \times 4\text{ m} \times 3\text{ m} = 60\text{ m}^3$. We'll take the air density to be $\rho = 1.2\text{ kg m}^{-3}$, and the *specific*[11] heat capacity[12] to be $c = 1,000\text{ J kg}^{-1}\text{ K}^{-1}$. The *thermal mass*[13] or total *heat capacity* of air in the room is therefore given by

$$C = mc = V\rho c = 7.2 \times 10^4\text{ J K}^{-1}$$

We could complicate matters here by estimating the thermal mass of the fridge, and by contemplating whether the "perfectly thermally insulated" walls have any heat capacity. There is certainly nothing *wrong* with doing this, but in the spirit of the question these considerations were typically ignored. All we expected was an order of magnitude answer of the right *form*.

The rate of temperature increase with time in the *nominal* steady-state condition is

$$\left.\frac{dT}{dt}\right|_{SS} = \frac{\bar{P}_{SS}}{C} = \frac{100}{7.2 \times 10^4} = \frac{1}{720}\text{ K s}^{-1} = 5.0\text{ K h}^{-1}$$

At the maximum power condition, $P_{MP} = 200$ W, the rate of heating the room would be

$$\left.\frac{dT}{dt}\right|_{MP} = \frac{\bar{P}_{MP}}{C} = 10\text{ K h}^{-1}$$

which is double the heating rate of the nominal steady-state condition.

What about the heat pump effect? When the fridge is switched on, a certain amount of heat is pumped out of the fridge system[14] and released into the room, via the relatively warm heat exchanger at the back of the fridge. During this initial period the fridge works as hard as it can to establish the required temperature inside the fridge. The fridge consumes power at $P_{MP} = 200$ W, and the rate of heating the room due to this alone is 10 K h^{-1}. If we estimate

power setting, cycling in 15 minute bursts in typical domestic conditions.

[11]In this context *specific* means *per unit mass*.

[12]This point really deserves some discussion. As the question is set up, implying what we refer to thermodynamically as a *closed system*, we might more readily want to take the heat capacity at constant volume, c_v, rather than the heat capacity at constant pressure, c_p. Many first-year undergraduate students are confused about which heat capacity to use in such situations, however, so it seems an unhelpful distraction to be pedantic in the context of the question. As set, I think we would say that any estimate of either heat capacity would suffice. In the question I simply write c to indicate that we have not defined precisely which one we are taking.

[13]Here I think I would expect someone to be able to derive an appropriate formula with a little help, but I would certainly not expect them to be completely familiar with the concept (although I understand it is taught in every A-level syllabus).

[14]Here we mean the entire heat capacity of the parts of the fridge that cool down, including the enclosed air, the contents of the fridge, and some of the mass of the fridge itself.

that it takes half an hour for the fridge to achieve the required internal temperature, the temperature rise in the room due to power consumption alone would be 5 K. We could estimate that the additional temperature rise due to the heat pump effect[15] is an additional 2 K. In the first half hour we move from $T_a = 18°\text{C}$ to $T_b = 25°\text{C}$.

At T_b the room isn't far above its initial temperature, so we could say that in the next hour the room heats up at the nominal steady-state rate of 5 K per hour.[16] This takes us to $T_c = 30°\text{C}$. At this point the temperature difference driving heat into the fridge has increased substantially—that is, the room is now noticeably warm. The fridge has to work harder to maintain the cold internal temperature. We could estimate[17] that it will now take about 45 minutes for the room to increase in temperature by another 5 K. This takes us to $T_d = 35°\text{C}$.

At this point we see that the temperature difference across the fridge is almost double what it was in the nominal steady-state condition. Instead of working 50% of the time, the fridge's motor will be working 100% of the time. The fridge will be working at the maximum power condition and the rate of temperature increase will be $10\ \text{K h}^{-1}$. In half an hour we get to $T_e = 40°\text{C}$. From here we are on a linear trend, with the temperature increasing at $10\ \text{K h}^{-1}$. The fridge will now continue to work at maximum power but will fail to maintain its internal target (or *setpoint*) temperature, because heat is leaking into the fridge very rapidly from the warm room.

We can now plot all this interesting behaviour on a graph. If we were being really pedantic we might even superimpose an asymmetric triangular wave onto the trend (with increasing frequency), to show the effect of the fridge turning on and off during the period between a and d, when the motor is only running intermittently. From d onwards the motor runs continuously.

What happens if we open the door of the fridge after four hours? In the few minutes after we open the door, the warm air in the room equilibrates with the cool contents of the fridge (the contained air, most of the internal surfaces, and any contents), lowering the temperature in the room. The heat that was pumped out (in the time period between a and b) to achieve the internal temperature of 3°C flows back in. There is a small reduction in temperature from $T_f = 52.5°\text{C}$ to T_g. The temperature drop $T_f - T_g$ might be a few degrees (according to our calculations above). It is certainly rather larger than the additional increase in temperature between T_a and T_b due to the heat-pump effect. This is because the temperature difference between the room and the inside of the fridge is larger at f than it was at b. Once the internal and external temperatures have equilibrated, we return to an upward

[15]To justify this, we need to do further calculations of the type we've just performed, but I haven't included them here.

[16]The fridge has reached its nominal internal temperature, 3°C, so there is zero heat-pump effect, and the room is not very much warmer than the nominal temperature of 18°C. So as a crude estimate we can say it is at its steady-state power condition.

[17]It's relatively easy to calculate this precisely and you should do so if you're interested, perhaps using integration.

trend of 10 K h^{-1}. The fridge is again working as hard as possible to achieve the internal target temperature but, because the door is open, it is failing.

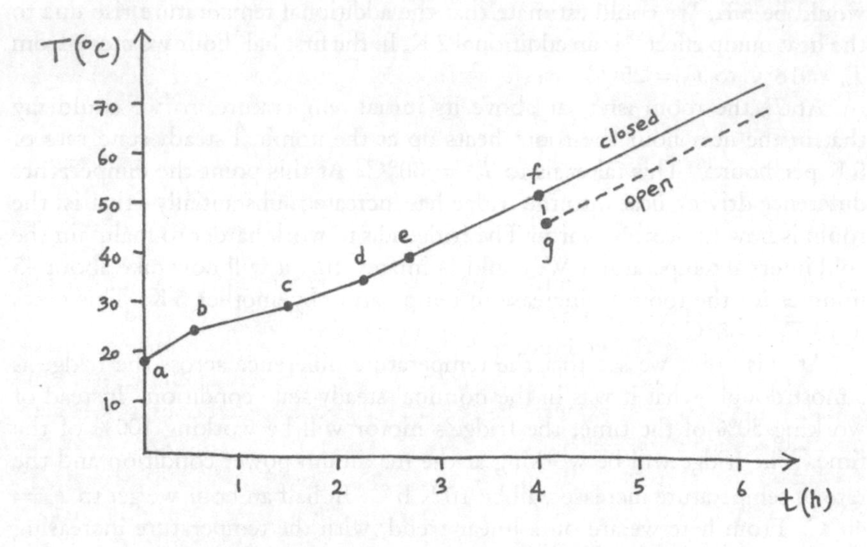

12.4 Ice in the desert ★★★

The history of ice in the desert is a long and fascinating one. Even in ancient times, those who could afford it had ice in the middle of summer, even in the hottest countries. Ice was used to preserve meats and perishable vegetables, and to provide refreshment on hot days. And, no doubt, to impress guests. The old technologies developed to provide ice have been somewhat forgotten in the modern age. It is believed that the Persian Yakhchāl, or ice house, dates back over two thousand years.[18] These huge cone-shaped buildings are rather reminiscent of massive up-turned wasp's nests. They were made of mud (in an adobe style), or the more sophisticated sārooj, a mortar of clay, lime, sand, ash, egg whites and goat hair. When dried in the sun, not only is sārooj very water-resistant, it also has extremely low thermal conductivity. Some Yakhchāls were equipped with a spiral runnel, like that of a helter-skelter, down which water was poured continuously during the day, to keep the outer surface cool. The ice houses were stocked with enough ice during the winter months to last the entire summer. Ice was cut in blocks from nearby mountains and then transported to the houses, or manufactured on site. The method of manufacture was rather ingenious and employed *radiative cooling*.

Radiative cooling was employed in both ancient India and Persia.[19] In

[18] Jorgensen, H., 2012, "Ice houses of Iran: where, how, why," Mazda Publishers, ISBN-10: 1568592698 ISBN-13: 978-1568592695.

[19] Chalom, M., Stickney, B., 2006, "Potentials of night sky radiation to save water and energy in the State of New Mexico," Governor Richardson's Water Innovation Fund, PSC #05-341-1000-0035.

India, shallow pottery dishes were placed in hay, which was used as insulation, and then exposed to the night sky. On clear nights, even when the ambient air temperature was well above freezing, ice would form around the edges of the dish, and it was harvested for storage in an ice house. This process would go on over many days or weeks until the ice house was full. This meant that the Mughal emperors of Delhi and Agra had a good supply of ice during the summer months. The Persian technique was similar, but involved flooding long shallow ditches with meltwater from ice within the ice house. The ditches were protected from the sun by carefully positioned walls which allowed the surrounding ground to cool towards the end of the day.

In the early 1800s an entrepreneurial American businessman called Frederic Tudor established what became known as the *ice trade*. Tudor started shipping ice (cut from the Hudson River in winter, and stored in an ice house all year) to Martinique and Cuba in the Caribbean, selling the ice for substantial sums to the European aristocracy. The ice trade became extremely profitable, and by the 1840s it had spread to India, South America, China and even Australia! Frederic Tudor, now extremely wealthy, was dubbed The Ice King.

But a technological heyday never lasts forever. At the end of the 1800s a huge industry was about to be destroyed by the development of technology that enabled more efficient refrigeration cycles. This led to compact plant machinery that could manufacture ice inexpensively. By the early 1900s more ice was being manufactured in refrigeration plants than was being harvested from lakes. By the 1920s the ice trade had collapsed, putting 90,000 people out of work. The history of making, transporting, and selling ice will never be quite as rich as it once was.

I was rather pleased to come up with a number of variants on the following question, which I have enjoyed asking students in the last few years. I recall that students struggled to set up the initial equations, and often needed considerable help. But they were pleasantly surprised by how easily the question yielded, and how neat the final solution was. This is a slightly harder variant (in terms of logic) than the one I asked, but the principle is identical.

THE OWNER OF End of the Earth Ice sells ice in the desert at a point A. The owner of End of the Universe Ice sells ice in the desert at a point B. From the city, which sells spheres of ice of any mass, it takes exactly twice as long to get to point B as to point A. If seller A leaves the city with an 800 kg sphere of ice and arrives with a 100 kg sphere, how much ice would seller B need to purchase to arrive at his stall with the same amount? The ice sellers make the journeys at night, so we can ignore the effect of radiation from the sun. We should also assume that the temperature of the desert is the same and constant during both journeys.

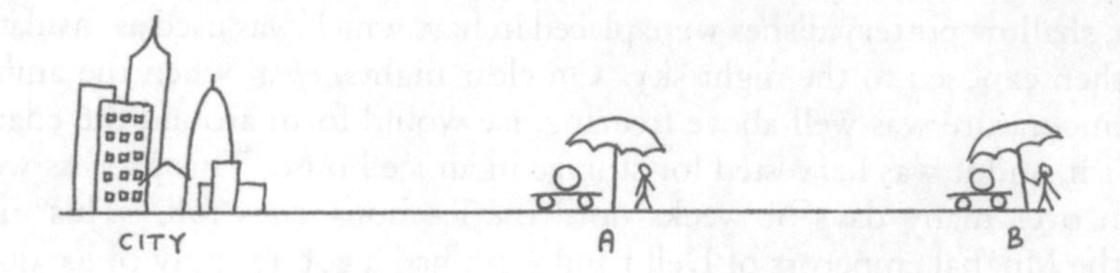

Answer

This rather open-sounding question encourages us to think first about the possible physical mechanisms, before writing equations and solving the problem. A tutor would normally want to discuss and agree the dominant mechanisms before proceeding to a solution. Although the question is rather open, it is not a riddle. In the generally very pleasant process of teaching clever students, one of the few things that mildly irritate me is the student who continually introduces extraneous ideas[20] to avoid tackling what would otherwise be a perfectly interesting question. What are the prevailing weather conditions? Do the men have fridges? Does one man have a car? And so on. You can see, I hope, that this could become rather wearing when we are all geed up and ready to help a student tackle a new idea. I think the sphere of ice is enough of a clue that we are dealing with an idealised problem, one that's amenable to what we hope will be fairly straightforward solution.

Heat is transferred to the sphere of ice from the surroundings, causing it to melt. But what does the rate of heat flow dQ/dt depend on? There are three mechanisms we might know something about: conduction, convection, and radiation (between the sphere of ice and the desert). Recalling the equations for each (which are given in the introduction to this section), we see that the rate of heat flow is proportional to area in each case: $dQ/dt \propto A$. Here, as we are told to do in the question, we assume that the temperature difference is the same for both A and B. Considering that $A = 4\pi r^2$, where r is the radius of the sphere, we have $dQ/dt \propto r^2$. The rate of reduction of volume of ice, $-dV/dt$, is proportional to the rate of heat flow, dQ/dt. The constants of proportionality are the density ρ and the latent heat of fusion L. We have

$$\frac{dQ}{dt} = -\rho\frac{dV}{dt}L \propto -\frac{dV}{dt}$$

The relationship between volume, V, and radius, r, is $V = (4/3)\,\pi r^3 \propto r^3$. Differentiating with respect to time we have $dV/dt \propto r^2\,(dr/dt)$. Noting that

[20] A "god from the machine" or, more commonly, *deus ex machina* (literary types like Latin), is a plot device in literature whereby intractable problems are suddenly and completely solved by the intervention of a completely new character or event. In good physics puzzles, as in literature, it is considered rather poor form to solve difficult problems by relying on some critical piece of information that has not been provided. An example in this case might be when a student introduces a car or a fridge to solve the problem. In contrast, it's perfectly all right for a student to need help with the value of a universal constant, or with the form of an equation—indeed this is expected.

$dQ/dt \propto -dV/dt$, and substituting for both dQ/dt and dV/dt, we have

$$\begin{aligned} \frac{dQ}{dt} &\propto -\frac{dV}{dt} \\ r^2 &\propto -r^2\frac{dr}{dt} \end{aligned}$$

So dr/dt is equal to a negative constant. We write $dr/dt = -c$, where c is a positive constant. We see that the radius of a sphere decreases at a constant rate. Let's use this to solve the problem.

Seller A sets out with 800 kg of ice and arrives at his stand with 100 kg. The volume and mass of ice are proportional to r^3. If the initial radius of seller A's ice sphere is r_{Ai}, and if the final radius is r_A, we have

$$\frac{r_{Ai}}{r_A} = \left(\frac{m_{Ai}}{m_A}\right)^{1/3} = \left(\frac{8}{1}\right)^{1/3} = 2$$

where m_{Ai} is the initial mass of the sphere, and m_A is the final mass of the sphere. So $r_{Ai} = 2r_A$. The reduction in radius is therefore given by $r_{Ai} - r_A = r_A$. We know that seller B travels twice as far as seller A, so the reduction in radius (which is proportional to the time spent travelling) is twice as large. That is, $r_{Bi} - r_B = 2r_A$. We also know that the sellers arrive with the same amount of ice, so $r_B = r_A$. Combining these facts we have $r_{Bi} = 3r_A$. The initial mass of sphere B is given by

$$m_{Bi} = m_A\left(\frac{r_{Bi}}{r_A}\right)^3 = 100\,(3)^3 = 2,700 \text{ kg}$$

It seems that seller B is going to have a very heavy load to pull when he sets out.

It's interesting to plot a graph of the radius of the spheres as a function of time. We have a downward-sloping linear trend. If the sellers arrive at their stalls at $t = 0$, we represent seller A setting off at $-t_A$ and seller B setting off at $-t_B$. At a common time t_f the sphere of ice is exhausted. We see that the sellers can only continue to sell ice for as long as it takes A to arrive at his stall. That is, $|t_f| = |t_A|$. The ice trade is a tough business.

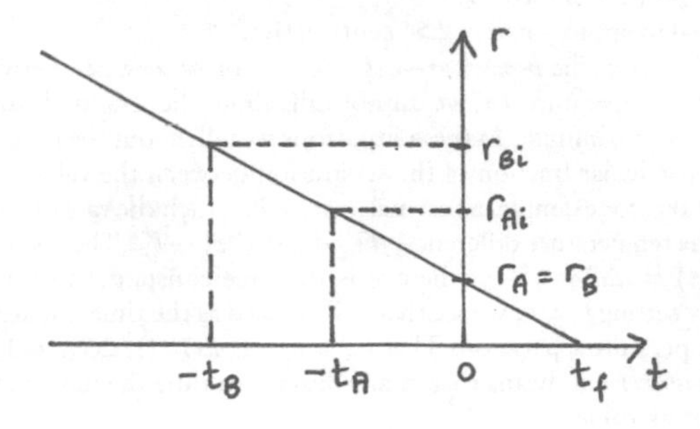

12.5 The cold end of the Earth ★★

In Newton's Principia,[21] he discusses the *ablation*[22] of a comet due to heat from the Sun, and presents a thought experiment designed to allow us to calculate the rate of cooling of the Earth. The thought experiment considers the ratio between the times taken to cool spheres of different radii.

> ...a globe of iron of an inch in diameter, exposed red-hot to the open air, will scarcely lose all its heat in an hour's time; but a greater globe would retain its heat longer in the ratio of its diameter, because the surface (in proportion to which it is cooled by the contact of the ambient air) is in that ratio less in respect of the quantity of the included hot matter; and therefore a globe of red-hot iron equal to our Earth, that is, about 40,000,000 feet in diameter, would scarcely cool in an equal number of days, or in above 50,000 years.

CONSIDER TWO HOT solid spheres of iron in the vacuum of space, cooling only due to radiation. Let one solid sphere be an inch[23] in diameter, and let the other have the diameter of the Earth, $d_E \approx 12.8 \times 10^6$ m. If the *time constant*[24] of the one-inch solid sphere is an hour, what is the time constant of the solid sphere with the diameter of the Earth? Assume that the spheres are *instantaneously isothermal*[25] due to high internal heat transfer.

[21] Newton, I., 1687, "Philosophiæ naturalis principia mathematica." Translated from Latin the title would be Mathematical Principles of Natural Philosophy. It is often simply referred to as The Principia.

[22] *Ablation* generally refers to the removal of material from the surface of an object due to heating, though it can also refer to material removal by mechanical means. It is a word commonly used in aerospace contexts, to refer, for example, to the erosion of the *ablative heat shield* (or, simply, ablator) on a rocket nose-cone or atmospheric re-entry vehicle. Good ablative materials should have low density and low conductivity, and produce gases (through both sublimation and pyrolysis) which protect the heat shield by reducing both the convective (flows of hot gas to the surface) and radiative heat loads. One such material is Phenolic Impregnated Carbon Ablator, or PICA. This is a carbon fibre structure impregnated with phenolic resin. Another material with comparable ablative properties is cork!

[23] One inch is equal to approximately 2.54 centimetres.

[24] In systems with *asymptotic behaviour*—exponential or otherwise—between, for example, a temperature T_1 and a temperature T_2, we cannot talk about the length of time it takes to cool to T_2. That time is, of course, infinite. In these situations we talk about the *time constant*, or the time taken to decay to a particular fraction of the separation between the values the function takes at $t = 0$ and $t = \infty$. Take, for example, a thermal system in which the rate of change of temperature is proportional to the temperature difference, $dT/dt \propto T(t) - T_2$. The solution is of exponential form, namely $\Delta T(t) = \Delta T_0 e^{-t/\tau}$, where τ is the time constant, $\Delta T(t) = T(t) - T_2$, and $\Delta T_0 = T_1 - T_2$. By setting $t = \tau$, we see that τ is defined as the time it takes the system to reach $1/e$ of its initial temperature separation. That is, at $t = \tau$, $\Delta T(t)/\Delta T_0 = 1/e \approx 0.3678$.

[25] *Instantaneously isothermal* means that at any instant in time the temperature of the sphere is the same throughout its volume.

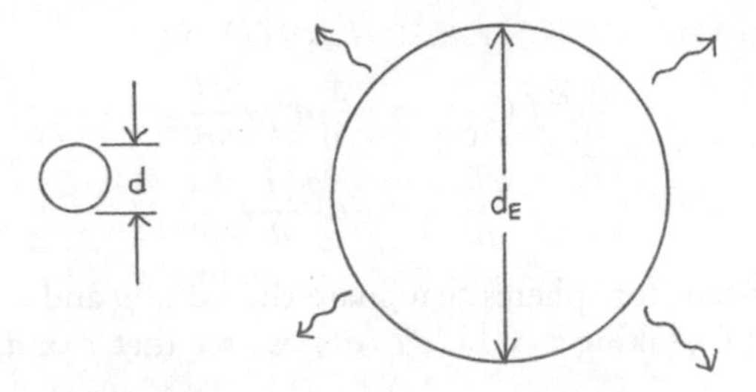

Answer

Here we are being asked to consider the *scaling* of the problem. In other words, we're not being asked to solve the problem directly, but to relate one time constant to another. This technique, which is part of a wider group of techniques referred to as *dimensional analysis*, can prove quite powerful in understanding physical systems.

The *thermal mass* C, or overall heat capacity, of an object is the product of its mass m and the specific heat capacity c. For a sphere, the mass is given by $m = (\pi/6)\, d^3 \rho$, where d is the diameter and ρ is the density. The thermal mass is then

$$C = \frac{\pi}{6} d^3 \rho c$$

The thermal mass, C, relates the heat lost due to radiation, ΔQ, and the temperature change of the object, ΔT, by the equation $\Delta Q = C \Delta T$. In differential form

$$\frac{dQ}{dt} = C \frac{dT}{dt}$$

The rate at which heat is lost due to radiation is proportional to the surface area of the sphere, $A = \pi d^2$, and some unknown function of temperature, $f(T)$.[26] We have $dQ/dt = \pi d^2 f(T)$. Putting this information together we have

[26] In fact, although we do not need to know this to solve the question, the rate of heat loss is simply given by the Stefan-Boltzmann Law.

$$\frac{dQ}{dt} = \varepsilon \sigma A \left(T_1^4 - T_2^4\right)$$

where the instantaneous temperature of the body is T_1, and, for a solid sphere floating in space (assuming it to be far from radiation sources), T_2 is the background temperature of the Universe, $T_2 \approx 2.735$ K. Here A is the surface area of the sphere, ε is the emissivity of iron (which has a wide range of values, between about $\varepsilon = 0.2$ and $\varepsilon = 0.8$, depending on the level of surface oxidation), and σ is the Stefan-Boltzmann constant, which has the approximate value $\sigma = 5.7 \times 10^{-8}$ W m^{-2}K^{-4}.

$$\pi d^2 f(T) = \frac{\pi}{6} d^3 \rho c \frac{dT}{dt}$$
$$\frac{dT}{dt} = a \frac{f(T)}{d}$$

where a is a constant for spheres that share the same ρ and c. If the spheres have the same $f(T)$, taking $\tau \propto 1/(dT/dt)$, we see that $\tau \propto d$, so

$$\tau_E = \tau_{1"} \frac{d_E}{d_{1"}}$$

Putting in numbers, we get $\tau_E = 5.0 \times 10^8$ hours, or approximately 57,000 years. Our result is in good agreement with Newton's calculation of 50,000 years.

Further discussion

There are two significant differences between the real problem of the cooling Earth and the problem as set. The first is that for large bodies the temperature gradient within the body becomes significant, and the conduction through the body limits the rate at which cooling can occur. This significantly increases the time constant of cooling for large bodies. The calculations for this situation were performed by Lord Kelvin,[27] who used the more sophisticated mathematical methods available at the end of the 1800s (specifically Fourier analysis) to estimate that if a conductivity model is taken into account, the age of the Earth could be very much greater than Newton's estimate. In his 1864 paper, On the Secular Cooling of the Earth,[28] Kelvin discusses the problem of the temperature gradient.

> In all parts of the world in which the Earth's crust has been examined, at sufficiently great depths to escape influence of the irregular and of the annual variations of the superficial temperature, a gradually increasing temperature has been found in going deeper.

[27] Lord Kelvin (1824–1907) was a pre-eminent British mathematician, physicist and engineer. Among his many and varied contributions to science was the absolute temperature scale that bears his name. He is also credited with being the first to determine the true value of absolute zero to a high degree of accuracy. Every good high-school student knows the value of absolute zero (−273.15 °C), but I find it especially hard to forget because my physics tutor had office 273 in the Clarendon Laboratory in Oxford. I think I only visited that office once, but almost twenty years later I remember the instructions regarding its location. Dr Leask had cocked his head to one side, as though lost in thought, and said "273... absolute zero, my dear boy. Absolute zero." It has stuck with me ever since.

Office 273 at the Clarendon was significant because that laboratory specialised in low-temperature physics. Between 1960 and about 1980 the Clarendon was known as "the coldest place on Earth" due to experiments by the Hungarian physicist Professor Nicholas Kurti, who achieved temperatures as low as one millionth of a degree Kelvin. Dr Leask had to fight to get that room!

[28] Kelvin, W.T., 1890, "On the secular cooling of the Earth," Transactions of the Royal Society of Edinburgh, Vol. XXIII, pp. 167–169, 1864.

He then comments on the application of Fourier analysis to the problem of conduction in the Earth, and finally on his own solutions using that analysis.

> I think we may with much probability say that the consolidation[29] cannot have taken place less than 20,000,000 years ago, or we should have more underground heat than we actually have, nor more than 400,000,000 years ago, or we should not have so much as the least observed underground increment of temperature.

Some scientists believed those predicted ages to be incompatible with their understanding of the solar system. It was not until the very early 1900s that the mystery finally began to unravel, however, with the advent of the science of radioactivity. Using very accurate radiometric dating, we now believe that the Earth is about 4.5 billion years old—a little older than Newton's estimate. The solution to the mystery, of course, was that the core of the Earth has a vast store of nuclear energy. It is interesting how science is so often a gradual unravelling, with future steps impossible to foresee. Arthur Stinner gives an enjoyable historical review of this topic in his paper Calculating the Age of the Earth and the Sun.[30]

---

[29] Formation of the Earth.

[30] Stinner, A., 2002, "Calculating the age of the Earth and the Sun," Physics Education, 37 (4), pp. 296–305.

Chapter 13

Buoyancy and hydrostatics

In this chapter we look at questions involving buoyancy and very simple hydrostatics. This is a science that goes back to antiquity, to the time of Archimedes. The rediscovery of the science of Archimedes is rather interesting, as I will explain, but first I should say a few words about the context.

Numerous well-known physics puzzles have explored Archimedes' Principle. When I was interviewed for physics, I was asked a classic question about a man in a boat, on a lake, with a brick. He throws the brick into the water and we are asked whether the level of the lake goes up, down, or stays the same.

There is actually relatively little we need to know to solve problems in this area, but we do need a very firm grasp of a few ideas, and this can take time. In fact, at an introductory level both hydrostatics and fluid dynamics are deceptively simple. The lack of advanced mathematics lulls people into a false sense of security and they forget that they need to think hard about the application of principles. At least, that's *my* experience.

I think for most problems in this area we need to understand only two things.

- ARCHIMEDES' PRINCIPLE. Archimedes was one of the greatest mathematicians and physicists of antiquity. He lived in Syracuse (modern-day Sicily) between (according to some sources) 287 BC and 212 BC. He made many contributions to mathematics, but also pioneered the laws of buoyancy, and invented the screw pump and a number of weapons of war. The *heat ray* as a means of setting fire to invading ships using parabolic mirrors[1] is attributed to him, for example. Remarkably, the only surviving record of many of his works is the Archimedes Palimpsest,[2] a thirteenth-century goatskin parchment found in Constantinople about

[1]It seems that most modern scholars regard the device as apocryphal.

[2]The word *palimpsest* refers to a page from a book (normally of *parchment* or animal skin) which has been re-used by scraping off the original text and writing over the top. This was fairly common in antiquity, because preparing animal skin was labour-intensive.

one hundred years ago by the Danish professor Johan Heiberg. Underneath a Christian text written by monks, he found a tenth-century copy of several of Archimedes' works, including the only known copy of his treatise on buoyancy On Floating Bodies. The text was translated[3] from the Greek into English by Sir Thomas Little Heath, making Archimedes' work accessible to modern day mathematicians. In Sir Thomas Heath's translation, the theorems of buoyancy are advanced by way of a series of nine propositions. Propositions five to seven are the ones most commonly associated with what is now known as Archimedes' Principle.

> PROPOSITION 5. Any solid lighter than a fluid[4] will, if placed in the fluid, be so far immersed that the weight of the solid will be equal to the weight of the fluid displaced.
>
> PROPOSITION 6. If a solid lighter than a fluid be forcibly immersed in it, the solid will be driven upwards by a force equal to the difference between its weight and the weight of the fluid displaced.
>
> PROPOSITION 7. A solid heavier than a fluid will, if placed in it, descend to the bottom of the fluid, and the solid will, when weighed in the fluid, be lighter than its true weight by the weight of the fluid displaced.

- PRESSURE AT DEPTH. The pressure at depth in a body of fluid, or the *hydrostatic pressure*, is given by $\rho g h$, where ρ is the fluid density, g the acceleration due to gravity, and h the depth below the *free surface*. Pressure has units equal to force per unit area, and acts equally on all immersed surfaces normal to the surface.

Hopefully many of these questions will be unfamiliar and enjoyable. I have selected and invented slightly more difficult and unusual problems, so that you can test your understanding of the material. However, we start this section with two straightforward—and well-known—puzzles.

13.1 Archimedes' crown and Galileo's balance ★

In his ten-volume work On Architecture,[5] the Roman author Marcus Vitruvius Pollio (c. 80 BC to 15 BC) tells a story about how Hiero, then king of Syracuse, hired Archimedes to determine whether his crown was made of pure gold. The story goes that Hiero had given a jeweller a quantity of pure gold, and had received his crown in return. The crown was the same mass as the

[3] Heath, T.L., 1897, "The works of Archimedes," Cambridge University Press.

[4] By "lighter than a fluid", Archimedes meant *less dense than*.

[5] Translated by Morgan, M.H., 1914, "Vitruvius: the ten books on architecture," Harvard University Press, Cambridge.

original metal but slightly more yellow in colour. He suspected the jeweller of being crooked, and of having mixed his gold with silver to form electrum.[6] Vitruvius (as he is commonly known) writes:

> In the case of Archimedes, although he made many wonderful discoveries of diverse kinds, yet of them all, the following, which I shall relate, seems to have been the result of a boundless ingenuity. Hiero, after gaining the royal power in Syracuse, resolved, as a consequence of his successful exploits, to place in a certain temple a golden crown which he had vowed to the immortal gods. He contracted for its making at a fixed price, and weighed out a precise amount of gold to the contractor. At the appointed time the latter delivered to the king's satisfaction an exquisitely finished piece of handiwork, and it appeared that in weight the crown corresponded precisely to what the gold had weighed.
>
> But afterwards a charge was made that gold had been abstracted and an equivalent weight of silver had been added in the manufacture of the crown. Hiero, thinking it an outrage that he had been tricked, and yet not knowing how to detect the theft, requested Archimedes to consider the matter. The latter, while the case was still on his mind, happened to go to the bath, and on getting into a tub observed that the more his body sank into it the more water ran out over the tub. As this pointed out the way to explain the case in question, without a moment's delay, and transported with joy, he jumped out of the tub and rushed home naked, crying with a loud voice that he had found what he was seeking; for as he ran he shouted repeatedly in Greek, "Eνρηκα, ενρηκα."

If your eyes are not watering too much at the thought of the naked Greek running about shouting in the street, you'll realise that this story is the origin of the most famous interjection in the history of science: 'Eureka!'

Once reclothed, Archimedes went to the king. Armed in one hand with the crown, and in the other with a mass of pure gold equal to the mass of the crown, he placed them one after the other in a vessel full to the brim with water. The crown was, of course, less dense, which Archimedes detected by noting that more water overflowed when it was placed in the vessel. So Vitruvius's story goes, anyway.

Writing in the year 1586, a mere sixteen centuries later, Galileo cast doubt on Vitruvius' story. Galileo was a keen student of Archimedes and thought that the method outlined by Vitruvius was from a mind far inferior to that of Archimedes, and would have been neither elegant nor precise enough to satisfy the "divine man". Galileo also knew that Archimedes was familiar with

[6]Electrum is a naturally occurring alloy of gold and silver, often mixed with traces of other metals such as platinum. It has been used extensively throughout history for coinage. There is much historical evidence that the percentage of gold in many trade currencies was significantly lower than the percentage of gold in the electrum mined from the same regions, suggesting that the ore was deliberately diluted with silver for profit.

both the lever and the principles of buoyancy (the only principle required in Vitruvius' account is that of equivalent volumetric displacement for immersed objects). This resulted in Galileo's treatise The Little Balance,[7] which describes a far more elegant way to determine the density of two objects. Galileo writes:

> Some authors have written that he proceeded by immersing the crown in water...But this seems, so to say, a crude thing, far from scientific precision; and it will seem even more so to those who have read and understood the very subtle inventions of this divine man in his own writings...

He continues:

> And at last, after having carefully gone over all that Archimedes demonstrates in his books On Floating Bodies and Equilibrium, a method came to my mind which very accurately solves our problem. I think it probable that this method is the same that Archimedes followed, since, besides being very accurate, it is based on demonstrations found by Archimedes himself.

Consider the modified version of a drawing by Galileo describing what is now known as Galileo's balance. I have heard a number of puzzles based on this principle, all of which ask students to explain the physics underpinning its operation. This quantitative question is my own, although I'm sure similar problems have been posed time and again over the years.

A SOLID LUMP of metal B is suspended from a balance arm from a point b. On the other side of the balance a solid counterpoise A is suspended from a point a. The distance $|ac|$ is chosen on a finely graduated scale, so that A balances B exactly. B is then immersed in water, and A moved to a new position a' so that the arm is again balanced. Write an equation to describe the ratio $|a'c| / |ac|$ in terms of the densities of air, water, and substance B. Estimate the value of the ratio for the cases in which B is pure silver or pure gold.

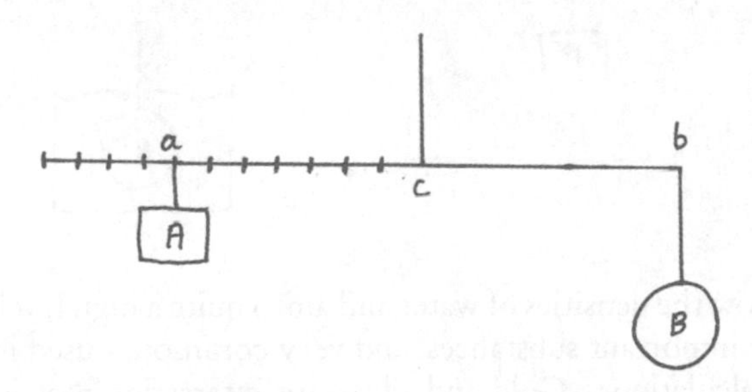

[7]Full translation in: Fermi, L., Bernardini, G., 2003, "Galileo and the Scientific Revolution," Dover Publications, ISBN-10: 0486432262.

Answer

The question is really asking us to consider the change in the downward force at b when object B goes from being immersed in air to being immersed in water.

When B is immersed in air, the downward force at b is given by $F_B = (\rho_B - \rho_a) V_B g$, where ρ_B and ρ_a are the densities of object B and the air respectively, V_B is the volume of B, and g is the force per unit mass due to gravity. We take the difference in densities because there is an upthrust equivalent to the weight of displaced air, according to the principles of Archimedes. When B is immersed in water, the downward force at b becomes $F'_B = (\rho_B - \rho_w) V_B g$. Here we have exchanged ρ_a for ρ_w, the density of water.

Consider now the counterpoise on the other side of the balance. When the lump of metal B is in air, the counterpoise A balances the scales in position a. When B is in water, the counterpoise balances the scales in position a'. In both situations the downward force due to A is $F_A = m_A g - \rho_a V_A g$, where m_A is the mass of the counterpoise, and V_A its volume. The buoyancy force in air of A is $\rho_a V_A g$.

Equating torques (moments) about c in both cases we have, for air

$$F_A\,|ac| = F_B\,|bc| = (\rho_B - \rho_a)\,V_B g\,|bc|$$

and for water

$$F_A\,|a'c| = F'_B\,|bc| = (\rho_B - \rho_w)\,V_B g\,|bc|$$

Taking the ratio $|a'c|\,/\,|ac|$, we get

$$\frac{|a'c|}{|ac|} = \frac{\rho_B - \rho_w}{\rho_B - \rho_a}$$

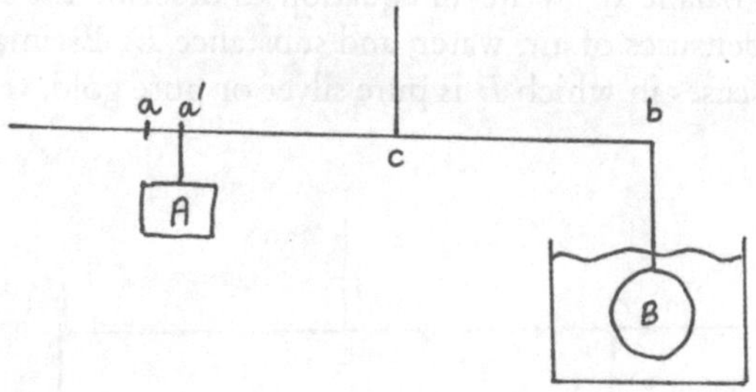

We should know the densities of water and air to quite a high level of precision. They are very important substances, and very commonly used in physics and engineering calculations. Gold and silver are interesting metals for all sorts of reasons. I think it's reasonable to expect students (especially those who have studied chemistry at A-level) to be able to estimate the densities of gold and silver to better than just order of magnitude approximations. We'd expect them to know that gold has a density that's very roughly twenty times that of water (one of its interesting properties), and that silver has a density that's very roughly ten times that of water. In the case of silver I think a slightly higher or lower value is reasonable, because its density is not unusual enough to be particularly memorable. Using these very rough estimates we get

$$\text{Silver (estimate): } \frac{|a'c|}{|ac|} = \frac{10 - 1}{10 - 0.001} \approx 0.900$$

$$\text{Gold (estimate): } \frac{|a'c|}{|ac|} = \frac{20 - 1}{20 - 0.001} \approx 0.950$$

For an initial length $|ac|$ equal to $1,000$ mm, for silver we get $|a'c| = 900$ mm, and for gold we get $|a'c| = 950$ mm. The difference between these values is 50 mm—a significant distance. This demonstrates the practicality of using a Galileo balance to determine, say, the ratio of gold and silver in a crown to quite a high degree of accuracy (probably to within 1% or so with a well-made balance - even in Galileo's time).

Further discussion

We now repeat the same calculation with precise values for the densities of air, water, gold and silver. This is the sort of thing engineers and physicists look up in data tables. We tend to only remember most physical properties (density, specific heat, boiling point, and so on) to orders of magnitude. The actual values[8] are as follows: $\rho_{\text{silver}} \approx 1.049 \times 10^4$ kg m^{-3}, $\rho_{\text{gold}} \approx 1.930 \times 10^4$ kg m^{-3}, $\rho_{\text{w}} \approx 1.000 \times 10^3$ kg m^{-3}, and $\rho_{\text{a}} \approx 1.28$ kg m^{-3}.

$$\text{Silver (precise): } \frac{|a'c|}{|ac|} = \frac{10.49 - 1.00}{10.49 - 0.001} \approx 0.905$$

$$\text{Gold (precise): } \frac{|a'c|}{|ac|} = \frac{19.30 - 1.00}{19.30 - 0.001} \approx 0.948$$

As we can see, this modifies the result only slightly. It's also worth noting, if it weren't already obvious, that the influence of including air in even this more precise calculation is entirely negligible and could have been ignored. If we were measuring the density of silver, we could use the approximation.

$$\frac{|a'c|}{|ac|} = \frac{\rho_B - \rho_w}{\rho_B - \rho_a} \approx \frac{\rho_B - \rho_w}{\rho_B}$$

to an accuracy of approximately 1 part in 1×10^4, or 0.01%.

[8] At *standard temperature and pressure,* or STP, which is 0° C and 100 kPa.

13.2 Another Galileo's balance puzzle ★★

The Galileo balance operates on the following principle. It has an equal mass on either side of its arm. On one side of the balance there is a change in torque (moment) due to a buoyancy force caused by submerging a mass in water. On the other side of the balance there is a change in torque caused by altering the distance to a counterpoise so that it matches the torque caused by the additional buoyancy force. The amount by which the counterpoise must be moved is a measure of the relative density of the submerged object.

Consider two solid objects A and B, each of unknown mass and unknown but uniform density. Is there a way of arranging a Galileo balance to accurately determine which of the two objects is more dense? If there is, prove that the arrangement could work. Assume that no other masses are available, and that we have no precise distance-measuring equipment.

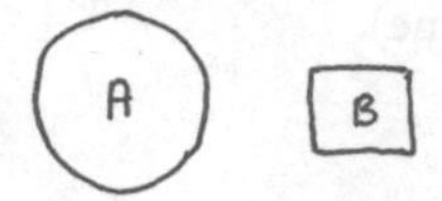

Answer

When thinking about the Galileo balance, I was struck by something that's perhaps rather obvious. The counterpoise doesn't need to be either the same mass or the same density as the other object. (In this case we can think of either object as a counterpoise, but we arbitrarily make it A.) With arbitrary mass for A and B we can still make a comparative measurement of density.

Consider that we have two solid objects A and B with volumes V_A and V_B and uniform densities ρ_A and ρ_B. To make a precise measurement of relative density, the apparatus must first be balanced. The downward force due to object A at point a is $F_A = V_A(\rho_A - \rho_a)\,g$, where ρ_a is the density of air. The downward force due to object B at point b is $F_B = V_B(\rho_A - \rho_a)\,g$. So by taking the moment about c, we get

$$V_A(\rho_A - \rho_a)\,|ac| = V_B(\rho_B - \rho_a)\,|cb|$$

or

$$\frac{V_A\,|ac|}{V_B\,|cb|} = \frac{\rho_B - \rho_a}{\rho_A - \rho_a}$$

We'll return to this equation in a moment.

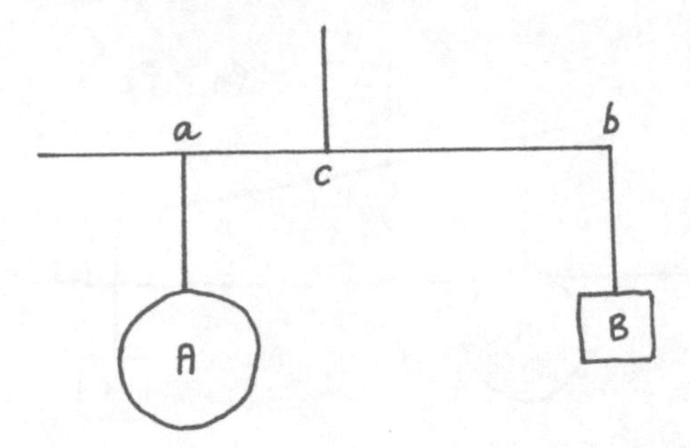

Now we immerse both A and B in water. If the objects were *fully immersed* there would be upthrust forces on A and B equal to the weights of displaced water. The upthrust forces reduce the net downward forces at a and b. When A is immersed in water, the downward force at a becomes $F'_A = (\rho_A - \rho_w) V_A g$. The change in force at a is

$$\Delta F_A = F'_A - F_A = (\rho_a - \rho_w) V_A g = kV_A$$

Here we set $k = (\rho_a - \rho_w) g$, to note that this term appears only as a constant for this experiment. Similarly, when B is immersed in water, the downward force at b becomes $F'_B = (\rho_B - \rho_w) V_B g$. The change in force at b is

$$\Delta F_B = F'_B - F_B = (\rho_a - \rho_w) V_B g = kV_B$$

The changes in torque are given by $kV_A |ac|$ and $kV_B |bc|$. We showed earlier that the condition for initial balance was

$$\frac{V_A |ac|}{V_B |cb|} = \frac{\rho_B - \rho_a}{\rho_A - \rho_a}$$

Thus we see that the ratio of the changes in torque about the axis when the objects are fully immersed is equal to the ratio of density differences

$$(\rho_B - \rho_a) / (\rho_A - \rho_a)$$

Now consider three cases of interest.

- CASE 1: $\rho_B > \rho_A$. In this case, $(\rho_B - \rho_a) / (\rho_A - \rho_a) > 1$, and the reduction in torque due to fully immersing A would be less than the reduction in torque due to fully immersing B. The balance can no longer be level, and no matter how far it is immersed (within the range of motion of the arm), a portion of A remains above the surface.

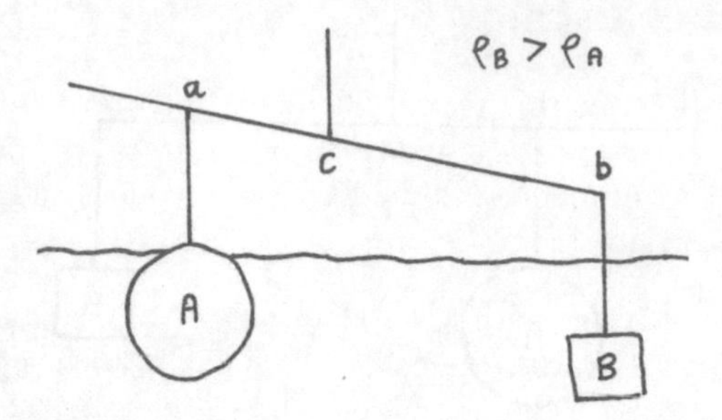

- Case 2: $\rho_B < \rho_A$. In this case, $(\rho_B - \rho_a) / (\rho_A - \rho_a) < 1$, and the reduction in torque due to fully immersing B would be less than the reduction in torque due to fully immersing A. So a portion of B remains above the surface.

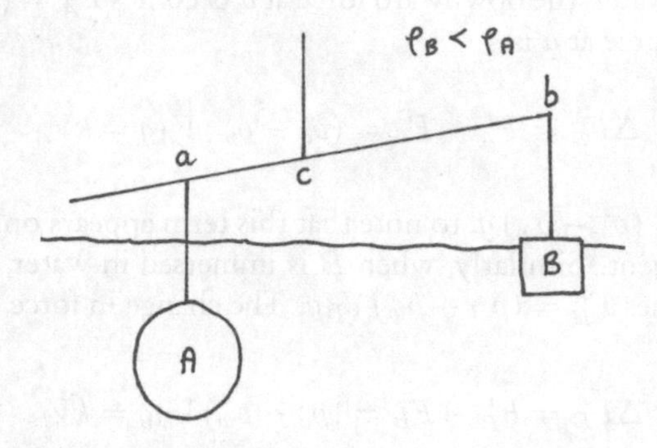

- Case 3: $\rho_B = \rho_A$. In this case, $(\rho_B - \rho_a) / (\rho_A - \rho_a) = 1$, and the reductions in torques due to fully immersing A and B are the same. The balance remains level.

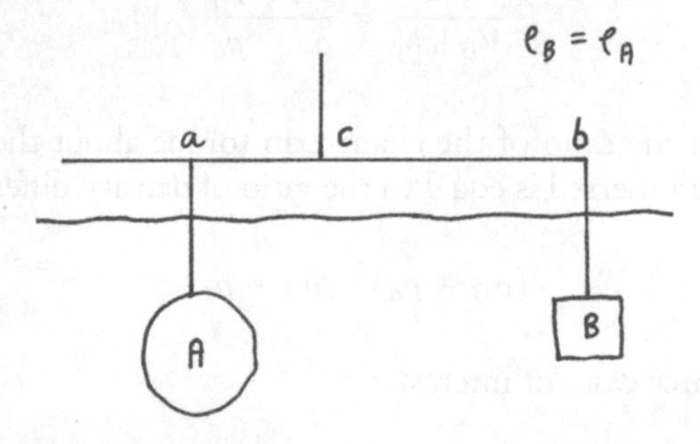

It's worth restating that this result is entirely *independent* of the initial masses and densities of A and B. We have proven that a balance for distinguishing density differences can work even when the masses are unknown. Take any solid lump of unknown metal and we can compare its density accurately to any other solid lump of gold or silver.

13.3 Balanced scales ★

I really like questions of this type. There are an almost infinite variety of them, designed to test the basic principles of hydrostatics and the principles of Archimedes. If you understand these properly you should be able to solve these questions very quickly.

YOU HAVE A pair of scales with a vessel of water on one side that is exactly balanced by a mass M on the other side. A solid cylinder is introduced to the vessel containing water. It displaces some of the liquid, but it is arranged so it does not touch either the base or the side of the vessel. The water level rises but the mass of water is unchanged. Determine whether the mass M' required to achieve balance in the second configuration is greater, smaller, or the same as mass M.

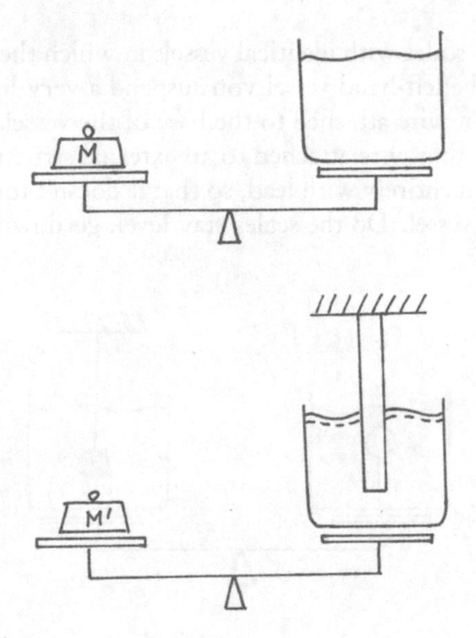

Answer

This is a simple question so we'll keep the answer correspondingly brief. When the solid cylinder is introduced, the level of water in the vessel rises. This increases the hydrostatic pressure on the base of the vessel accordingly.[9] For a cylindrical vessel, the total downward force on the base of the vessel is the product of the pressure and the base area.[10] Whatever the shape of the vessel, in this example the rising water level causes an increase in the downward

[9]If we were asked to quantify the change, the relevant equation would be $p = \rho_w gh$, where ρ_w is the density of water, h is the height of water above the base of the vessel, and g is the force per unit mass due to gravity.

[10]Of course, this is only true if the pressure is constant over the entire base area of the vessel (that is, for a flat-bottomed vessel). In which case, we can write $F = pA_{base} = \rho_w ghA_{base}$. In

force on the base. We need a larger mass to balance the vessel in the second configuration. So $M' > M$.

If you need further physical arguments to convince you, consider that there is an upward force (the buoyancy or upthrust force) on the cylinder. This must be supplied from somewhere, and it comes from the water in contact with the base of the cylinder. There must be an equal and opposite force on the water in the vessel.

13.4 The floating ball and the sinking ball ★★

This is my favourite question on buoyancy. It is extremely simple to phrase, doesn't need technical language, and combines quite a number of important concepts. Some people find it surprisingly hard!

YOU HAVE A pair of scales with identical vessels in which there are equal quantities of water. In the left-hand vessel you suspend a very light ping-pong ball on a very thin, light wire attached to the base of the vessel. In the right-hand vessel, from a very thin wire attached to an external structure, you suspend a ping-pong ball filled entirely with lead, so that it doesn't touch either the base or the sides of the vessel. Do the scales stay level, go down on the left, or go down on the right?

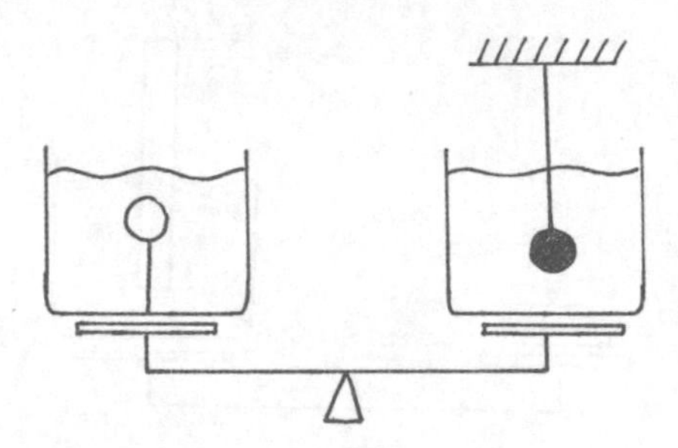

Answer

This is another problem for which we can give a very brief answer. There is much here to confuse us and potentially trip us up, but the basic principles are very simple indeed.

Consider first the right-hand vessel. We introduce the ping-pong ball full of lead to the water and the ball sinks. There is an upthrust on the ball equal to the weight of displaced water, and an equal and opposite (downward) force on the vessel. The downward force on the right is larger than it would have been had no lead-filled ping-pong ball been present. Another way to visualise the increase in the downward force (when the lead-filled ball is introduced)

the more general case of a curved base, we would need to perform a *surface integral* (over the area of the base) of the component of the pressure force that acts in the downward direction.

is to note that there is an increase in hydrostatic pressure on the base of the vessel.

Consider now the left-hand vessel. We imagine introducing a ping-pong ball which is suspended in the water, restrained by a wire from the base of the vessel. The ping-pong ball floats, and the level of liquid rises. There is an increase in hydrostatic pressure on the base of the vessel. However, there is also an upward force on the base of the vessel where the wire is connected. These two forces cancel. Another way to visualise this is to note that there is no change in any external force acting on the vessel when we introduce the ping-pong ball.[11] The downward force on the left is the same as it would have been if the floating ping-pong ball had not been introduced.

In summary, there is an increase in the downward force on the right, and the downward force on the left is unchanged. The scales tip down on the right.

13.5 Floating cylinders ★★★

I find that even the simplest buoyancy questions are apt to cause confusion in high-school and even university students, but they are good discussion points, and a good way of testing students' physical insight. I never got round to asking this one, but had it filed away with another ten or so questions which provide relatively simple tests of the basic principles. I think students would have found it quite hard, and I explain why in the answer.

THREE SOLID SHUTTLES A, B, and C are manufactured from Styrofoam with uniform density in the shapes shown. A is a narrow cylinder, B is a wide cylinder surmounted by a narrow cylinder, and C is a narrow cylinder surmounted by a wide cylinder (that is, an upside-down version of B). The shuttles are suspended from wires which carry forces F_A, F_B and F_C respectively.

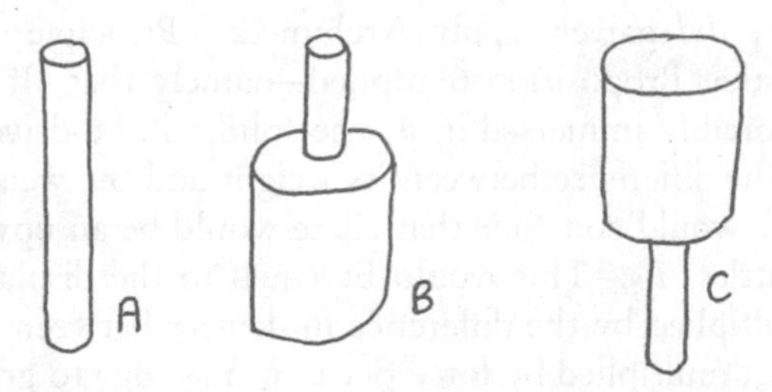

[11]Here we are neglecting the weight of the ball and the wire. We are told they are *light*, so we are right to do this. If we took realistic values instead, the result would be the same. We are also neglecting the insignificantly small changes in atmospheric pressure with height.

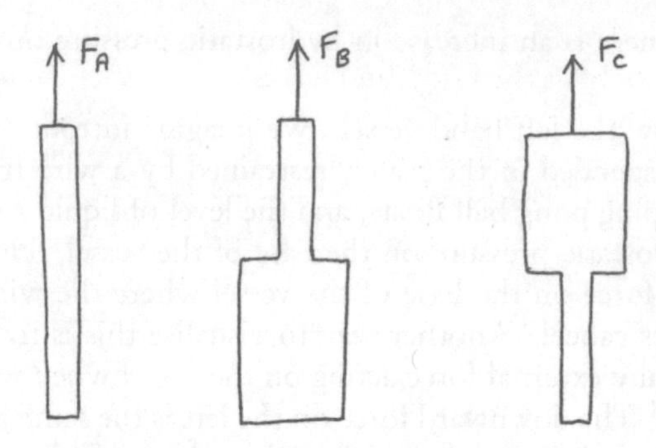

Without moving the shuttles, they are enclosed in purpose-made vessels which make a frictionless seal against the shuttles on both the upper and lower cylindrical surfaces. The vessels are then filled with water through vent holes at the top, which are left open to the atmosphere. Are the new forces F'_A, F'_B and F'_C greater or lesser in magnitude than the original forces?

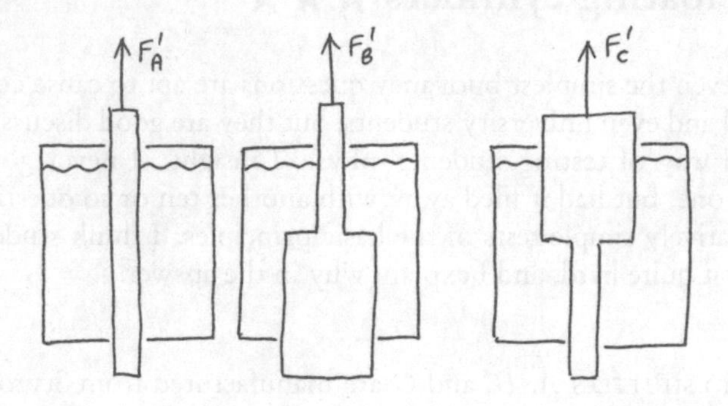

Answer

The reason I think most students would have struggled with this is that it looks like we can apply Archimedes' Principle. Unfortunately it's not that simple.

So what if we *did* naively apply Archimedes' Principle? I suppose we would conclude that Proposition 6 applied—namely that "If a solid lighter than a fluid be forcibly immersed in it, the solid will be driven upwards by a force equal to the difference between its weight and the weight of the fluid displaced." So we would conclude that there would be an upward buoyancy force on the shuttles, F_b. This would be equal to the displaced volume in each case, V, multiplied by the difference in density between water, ρ_w, and the Styrofoam, ρ_s, multiplied by force per unit mass due to gravity, g. So we would get something with the form $F_b = V(\rho_w - \rho_s)\,g$. We would then say that $F'_A < F_A$, $F'_B < F_B$ and $F'_C < F_C$. And we would be wrong!

The problem is that Archimedes' Principle does not directly apply in this situation. The various statements of the principle apply only to floating bodies and fully immersed bodies. In the first case, a body displaces a volume of water with the same weight as the body, and has an upthrust equal to the weight. In the second case, the displaced volume is equal to the volume of the body, and the upthrust is equal to the weight of displaced fluid, which is less than or equal to the weight of the body. Unfortunately the shuttles in this problem are neither floating nor fully immersed. I'm not referring to the possibility that there may be forces due to the wires from which they are suspended. I'm referring to the fact that the pressure boundary conditions on the uppermost—and particularly the lowermost—horizontal surfaces of the shuttles are *not* set by the hydrostatic pressure of the liquid. In fact, both surfaces are exposed to atmospheric pressure.

The easiest way to think about this problem is to simply consider the change in pressure on the surfaces of the shuttles when the water is introduced to their containing vessels. We start with the observation that changes in pressure on the vertical surfaces of the shuttles do not give rise to forces in the vertical direction (they only act to cause compression of the Styrofoam). We can therefore neglect the forces due to pressure changes on these surfaces.

Consider the change in pressure on the horizontal surfaces of the shuttles. For all the shuttles, the upper and lower surfaces are exposed to atmospheric pressure both before and after water is introduced, so experience no change in pressure. Now consider the shuttles in turn.

- SHUTTLE A. There is no intermediate[12] horizontal surface. There is therefore no additional vertical force that arises when water is introduced into the containing vessel. So $F'_A = m_A g = F_A$.

- SHUTTLE B. The intermediate horizontal surface faces upwards. The additional pressure acting on this surface is equal to the hydrostatic head, $\rho_w g h$, where h is the height of the *free surface*[13] above the intermediate horizontal face. The total force due to the action of this pressure is equal to the hydrostatic head multiplied by the surface area of the intermediate horizontal surface, A. This gives $\rho_w g h A$. The force acts downwards. Therefore $F'_B = m_B g + \rho_w g h A > F_B = m_B g$.

- SHUTTLE C. The intermediate horizontal surface faces downwards. The total force on this surface is again $\rho_w g h A$, but this time it acts upwards. Therefore $F'_C = m_C g - \rho_w g h A < F_C = m_C g$.

[12]Between the uppermost horizontal surface and the lowermost horizontal surface.

[13]Where the air is in contact with the atmosphere and at atmospheric pressure.

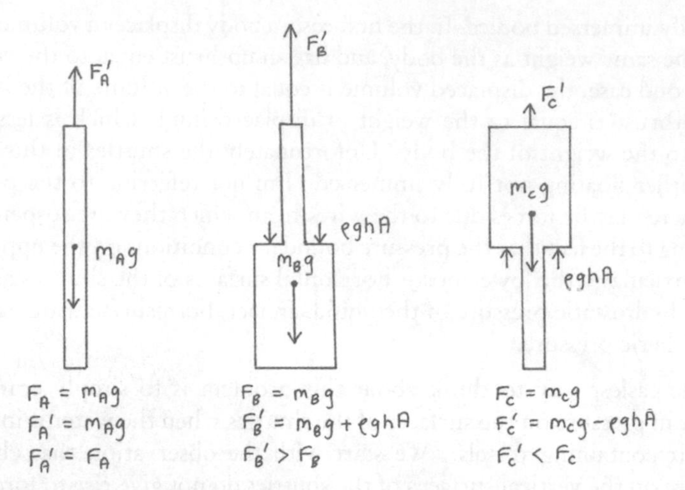

You may be surprised by the different result for each cylinder. Of course, many of the principles of Archimedes do apply to this situation, but we may come unstuck if we've memorised simple rules without understanding the underlying principles. Even for simple questions of this sort we often need to think.

13.6 The hydrostatic paradox ★★

The hydrostatic paradox is attributed to the Flemish scientist Simon Stevin, who in 1586 wrote De Beghinselen des Waterwichts (Elements of Hydrostatics). Over the years the supposed paradox has been posed in many ways,[14] but essentially it relates to the principle that the pressure exerted at a given depth in a fluid is independent of the shape of the vessel. The archetypal Stevin paradox is that by this logic we could exert a force on the inner base of a vessel that is greater than the weight of the enclosed fluid (by tapering the vessel inwards towards the top).[15]

I have heard of numerous physics puzzles that test one or other of these supposed paradoxes. The following problem, which is also a classroom demonstration, is rather elegant.

A PAIR OF scales has integral vessels (that is, vessels that are part of the scales themselves, rather than sitting on top of the scales) of the same base area. The left-hand vessel tapers outwards towards the top. The right-hand vessel tapers inwards towards the top. The vessels are filled to the same height with water. By the laws of hydrostatics the force on the base of the vessels must be identical

[14] Wilson, A.E., 1995, "The hydrostatic paradox," The Physics Teacher, Vol. 33, pp. 538–539.

[15] A related paradox, the *Archimedes paradox*, states that with a suitably shaped vessel we can float an object in less than its own weight of water.

because they have the same area and are exposed to the same pressure. Thus the scales should remain balanced. Is this logic correct?

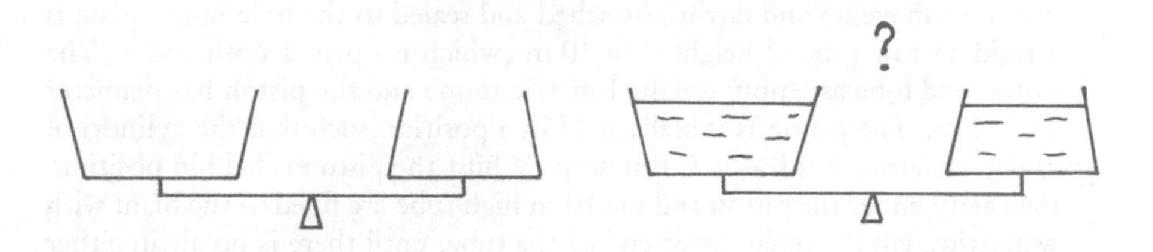

Answer

No. The logic is incorrect, and the scales tip down on the left because the weight of water on that side is greater. We are encouraged to examine the problem in terms of forces, however, which is instructive.

It *is* true that the force on the base of the vessels is the same. They have the same area, and are exposed to the same hydrostatic pressure ρgh, where ρ is the fluid density, g the force per unit mass due to gravity, and h is the height of water in the vessel. If we stop here we are led to the erroneous conclusion that the scales remain in balance. We have neglected the forces on the inclined walls of the vessels, however. The hydrostatic pressure varies linearly with depth. At the free surface the hydrostatic pressure is zero, and at the base of the vessels the hydrostatic pressure is ρgh. Along the inclined walls of the vessels the *pressure distribution* is represented by the triangular form shown in the diagrams below. The pressure force is always normal to the surface of the vessel. The horizontal components of the pressure force act in all angular directions around the vessel and cancel out. There is, however, a net vertical component of force. On the left vessel this net force acts downwards. On the right vessel this net force acts upwards. When analysed in terms of forces, it is these two terms (which are often neglected by those who have not seen a problem like this before) that give rise to the imbalance and cause the scales to tip in the direction we should expect on the basis of the mass imbalance. Of course, if we perform an integral of the pressure forces on both vessels, we get exactly the same imbalance force as if we take the two weights of water. The methods are equivalent.

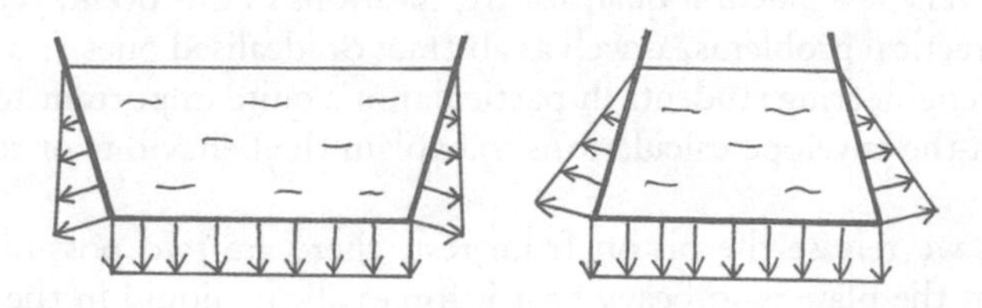

13.7 A quantitative piston puzzle ★

A THICK STEEL plate with a hole in the middle (drilled through the plate) is housed in a close-fitting cylinder, sealed with a frictionless seal that allows the piston to move up and down. Attached and sealed to the hole in the plate is a rigid narrow tube of height $h = 10$ m (which is open at both ends). The piston and tube assembly weighs 1 metric tonne and the piston has diameter $D = 1$ m. The piston is initially held in a position such that the cylindrical cavity underneath it is $d = 0.1$ m deep. Whilst the piston is held in position, the cavity under the piston and the 10 m high tube are filled to the brim with water through the open upper end of the tube, until there is no air in either the tube or the cavity. The system is then released from rest. What happens?

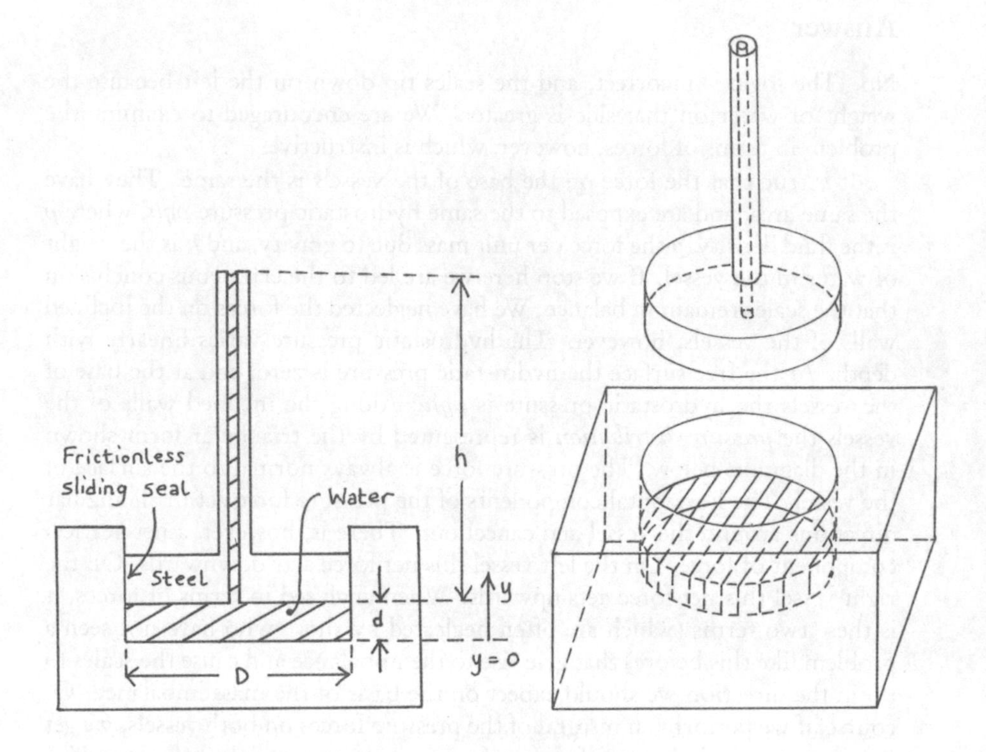

Answer

There are very few *practical* quantitative questions in this book, yet being able to solve practical problems, as well as abstract or idealised ones, is a very useful skill. For engineering students in particular, it's quite important to be able to do back-of-the-envelope calculations to explain the behaviour of real physical systems.

When we release the piston from rest, there are two possibilities. The first is that the plate is so heavy that it forces all the liquid in the cylindrical reservoir out of the top of the tube, closing the gap beneath it from a height d to zero height, and creating a fountain of liquid. The second possibility is that the water level in the tube falls from a height h to a lower height h', increasing the gap beneath the piston from d to $d + \delta d$, where δd is a very small increase in height. It's the second possibility that occurs in this case, as we can show easily with the following argument.

Consider the diagram below. We define h' as being the equilibrium position of the water in the tube. That is, the position it settles to after being released from rest.

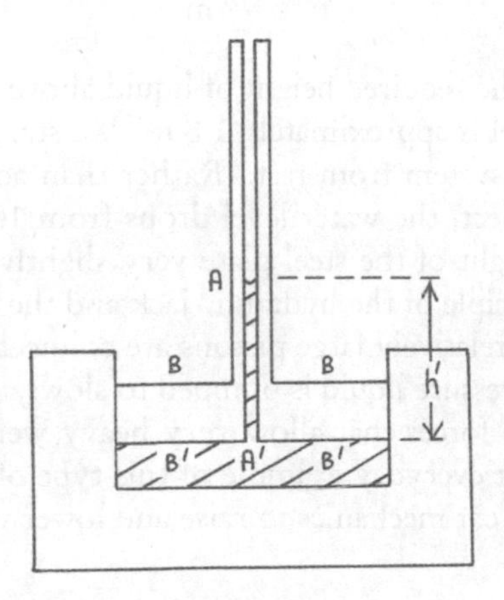

In this equilibrium position we note that A and B are both exposed to the atmosphere. At both points, therefore, the pressure is equal to atmospheric pressure: $p_A = p_B = p_{atm}$. The pressure at point A' is greater than the pressure at A because of the hydrostatic pressure acting on it due to the column of water it supports. We write

$$p_{A'} = p_A + \rho g h' = p_{atm} + \rho g h'$$

Considering point B', in addition to resisting the atmosphere, the pressure at B' must be high enough to support the weight of the steel plate. We write

$$p_{B'} = p_B + \frac{Mg}{A} = p_{atm} + \frac{Mg}{A}$$

where M is the mass of the plate, A the area of the plate, and g the force per unit mass due to gravity.

It is a condition of equilibrium that $p_{A'} = p_{B'}$, otherwise water would accelerate from A' to B' or vice versa. This leads to the equilibrium condition

$$
\begin{aligned}
p_{A'} &= p_{B'} \\
p_{atm} + \rho g h' &= p_{atm} + \frac{Mg}{A} \\
h' &= \frac{M}{A\rho}
\end{aligned}
$$

Putting in numbers we have $M = 1,000$ kg, $A = \pi D^2/4 = \pi/4$ m^2, and $\rho \approx 1,000$ kg m^{-3}. This gives

$$h' \approx \frac{4}{\pi} \text{ m}$$

So, for equilibrium, the required height of liquid above the interface between the water and the steel is approximately 1.3 m. We start with a head equal to 10 m and release the system from rest. Rather than acting as a fountain, as we might naively expect, the water level drops from 10 m to approximately 1.3 m, raising the height of the steel plate very slightly in the process. This is, of course, the principle of the hydraulic jack and the hydraulic ram. These are systems in which relatively large pistons are connected to small feed pipes though which high pressure liquid is pumped to slowly move the piston. The piston generates huge forces that allow very heavy weights to be raised and lowered. The simplest everyday example of this type of jack can be seen in a garage, and is used by car mechanics to raise and lower vehicles.

13.8 The floating bar ★★★★

I really like this question, and have enjoyed asking students to solve it over the years. I invented it over coffee with a friend. As with all good questions, it must have been asked many times before in various forms. I tried it out on some colleagues and found that most could intuitively spot the method, but that it took them a little time to rigorously work towards a solution. The mathematics is relatively straightforward, but also long-winded. I found that the question was amenable to an algebraic solution for some particular cases—the ones that I introduced in asking the question—but only if I gave students a lot of hints on methods and standard results to speed up the solution process. The results are slightly surprising, which students seem to enjoy. I discovered later that the question has a long pedigree. There is a very nice article on the topic by Walter Reid[16] from, appropriately, the US Naval Ordnance Laboratory.[17] His 1963 phrasing of the question is a little different from mine, but the principle is the same. At the beginning of the paper he writes:

---

[16] Reid, W.P., 1963, "Floating of a long square bar," American Journal of Physics, Volume 31, Number 8, August 1963, pp. 565–568.

[17] A lab built in Silver Spring, Maryland, on a remote tract of farmland, to combine the US Mine Laboratory and the US Experimental Ammunition Station. The lab was built towards the end of the Second World War to develop, among other things, mine and anti-mine technology.

> Physics students are sometimes asked to calculate how a block of a certain density will float. There is more to this problem than one might first expect.

And at the end of the paper:

> This problem serves as a nice example to illustrate how easy it is for one's intuition to be wrong.

Reid looks at the problem from the point of view of energy minimisation, which is what I had in mind when I dreamt it up. Relatively recent papers show that it is still a topic of interest. In the 1987 paper entitled The Floating Plank,[18] the author Robert Delbourgo considers a more general situation than Reid, but phrases it as a *metacentric problem.* Even Delbourgo tries to impress us, by writing:

> Although hydrostatics is considered to be an easy topic, there is one part that certainly is not, namely the determination of the equilibrium configuration of floating bodies.

A very nice description of the method of using metacentres[19] is given by Benny Lautrup, in his book Physics of Continuous Matter.[20] The method isn't exactly trivial to apply, so we won't attempt it here.

But I shouldn't put you off any further. The question is as follows.

A LONG SQUARE-SECTIONED bar of material of uniform density ρ is placed in water of density ρ_w. A stable equilibrium point is achieved when the potential energy of the system is minimised. This condition is met when the centre of gravity, G,[21] assumes the position with the minimum height above the centre of buoyancy, B_T.[22] Two solutions are proposed:

- SOLUTION A: with the top surface of the bar parallel to the surface of the fluid.
- SOLUTION B: with the top surfaces of the bar at 45°to the surface of the fluid.

---

[18]Delbourgo, R., 1987, "The floating plank," American Journal of Physics, Volume 55, pp. 799–802.

[19]The metacentre is an imaginary point about which a floating object can be thought to be suspended, like a pendulum, under conditions of displacement from equilibrium. Clearly, stable equilibria exist for situations where the centre of gravity is above the centre of buoyancy, a condition that applies to most vessels. This is because the centre of gravity is below the metacentre, on account of the hull being shaped in such a way as to cause a large restoring moment under conditions of roll.

[20]Lautrup, B., 2005, "Physics of continuous matter: exotic and everyday phenomena in the macroscopic world," Institute of Physics Publishing Ltd., ISBN 0 7503 0752 8.

[21]The *centre of gravity* is the point (normally, but not always, inside a body) from which the weight of an object may be considered to act. In a uniform gravitational field, the centre of gravity and the centre of mass are the same.

[22]The *centre of buoyancy* is the centre of gravity of the water that a body has displaced.

For any one (or more) of the cases $\rho/\rho_w = 0.2, 0.5$ and 0.8, consider which situation, A or B, has the lowest potential energy, and is therefore more likely.[23] Assume that the axis of the bar remains horizontal.

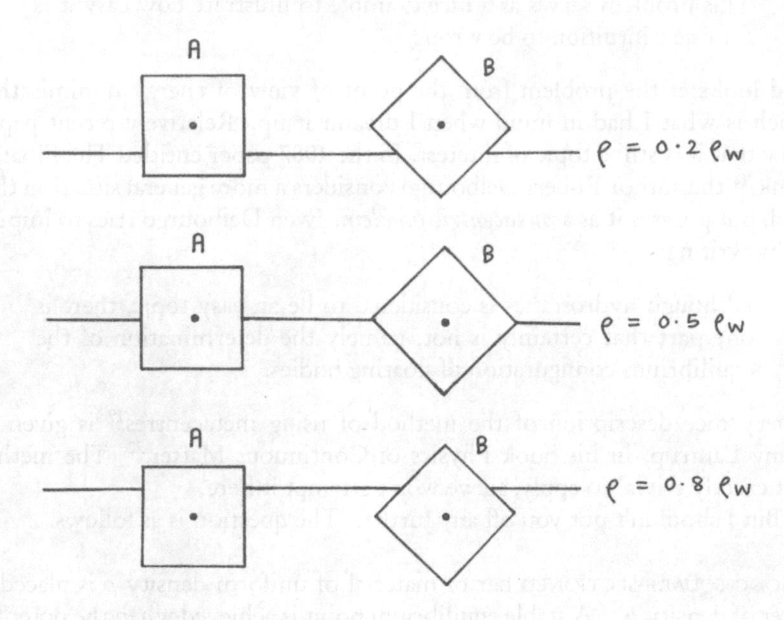

Answer

First consider the case where $\rho/\rho_w = 0.8$. The fraction of the bar above the surface of the water is $1/5$. So, taking the square section of the bar to be of unit length, and initially assuming that situation A is more stable, we see that the projection above the surface of the water is $y = 1/5$. The distance of the centre of gravity, G, below the highest point of the square section is $Y = 1/2$. So the distance of G below the waterline is $Y - y = (1/2) - (1/5) = 3/10$.

[23]This choice of wording is deliberate. It's not obvious that simply because one solution has lower potential energy than the other that it is the *only* solution. It is perfectly possible (indeed quite common in problems of this type) for more than one *metastable equlibrium* point to exist. In fact, yachts often have two metastable orientations, one the correct way up, and the other exactly upside down. If a yacht is rolled beyond the *angle of vanishing stability* (which is typically 120° away from the vertical; that is, with the mast underwater), it often stays the wrong way up!

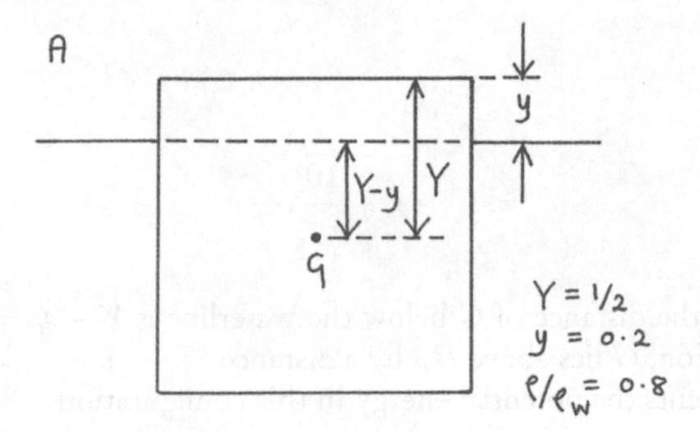

Now consider the position of the centre of buoyancy of the square section, B_T. The centre of buoyancy is at the effective centre of gravity of the liquid that has been displaced. We therefore consider only the part of the square section below the waterline. We split the body into two parts. The first part is symmetric about G, and therefore has a centre of buoyancy B_1 in the same position as G, at a distance $Y - y$ below the waterline. The area of this part (or the volume per unit thickness of the bar) is $2(Y - y)$. The second part has area y (and represents the image of the part that exists above the waterline), and has a centre of buoyancy B_2 at a distance $2(Y - y) + \frac{1}{2}y$ below the surface of the water.

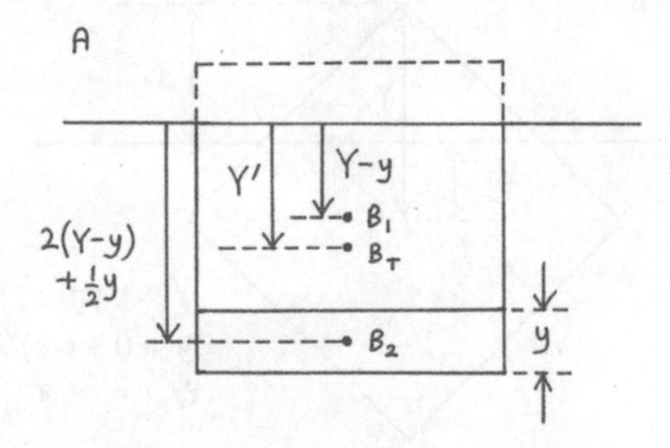

We represent the distance below the waterline of the overall (combined) centre of buoyancy, B_T, as Y'. We calculate this distance by *area-weighting* the distances of the individual centres of buoyancy B_1 and B_2 below the waterline. The symmetry of the problem (about the vertical axis) means that no further complexity is required. We write

$$Y'\left[2(Y - y) + y\right] = 2(Y - y)(Y - y) + y\left[2(Y - y) + \frac{1}{2}y\right]$$

Rearranging, and putting in values for Y and y, we obtain the distance below the waterline of the centre of buoyancy:

$$Y' = \frac{4}{10}$$

We recall that the distance of G below the waterline is $Y - y = 3/10$. So in this configuration, G lies above B_T by a distance $Y' - (Y - y) = 1/10$. This distance represents the potential energy in this configuration.

Now consider configuration B, in which the surfaces of the bar are at 45° to the surface of the water. The same proportion of the square section is above the surface of the water. By considering the area of the triangular section above the waterline we require $y^2 = 2/10$, or $y = 1/\sqrt{5}$. From the geometry of this configuration, we see that G is below the highest point on the square section by a distance $Y = 1/\sqrt{2}$. Thus, the distance of G below the waterline is $Y - y = (1/\sqrt{2}) - (1/\sqrt{5})$.

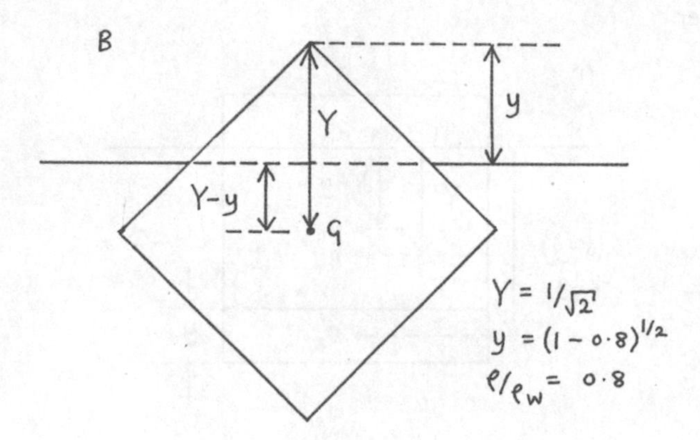

Consider again the position of the centre of buoyancy of the square, B_T. We look at the part of the square section below the waterline by splitting it into two parts. The first is symmetric about G, and has a centre of buoyancy B_1 in the same position as G, at a distance $Y - y$ below the waterline. The area of this part (or the volume per unit thickness of the bar) is $1 - 2y^2$. The second part has area y^2 (and represents the image of the part that's above the waterline), and has a centre of buoyancy B_2 at a distance $2(Y - y) + (1/3)\,y$ below the surface of the water. Here we have used the fact that a triangle's centre of buoyancy is $1/3$ of the triangle's height away from its base.

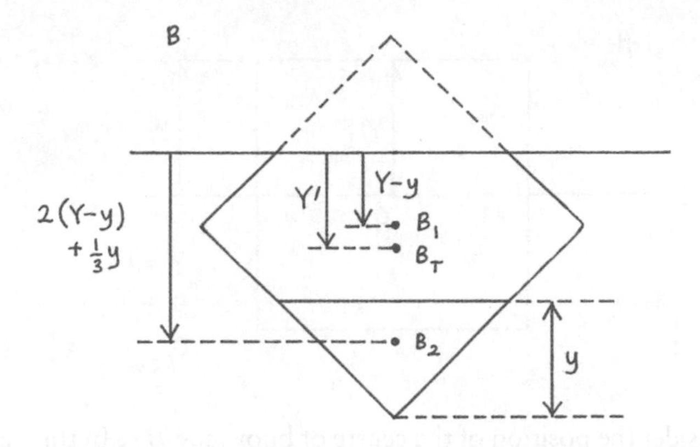

We again represent the distance below the waterline of the overall (combined) centre of buoyancy, B_T, as Y'. We calculate it by area-weighting the distances below the waterline of the individual centres of buoyancy B_1 and B_2. This gives

$$Y'\left[1-y^2\right]=(1-2y^2)(Y-y)+y^2\left[2(Y-y)+\frac{1}{3}y\right]$$

Putting in values for Y and y we get

$$Y'=\frac{5}{4\sqrt{2}}-\frac{7}{6\sqrt{5}}\approx 0.3621$$

We recall that the distance of G below the waterline is $Y-y=(1/\sqrt{2})-(1/\sqrt{5})\approx 0.2599$. So, in this configuration, G lies above B_T by a distance $Y'-(Y-y)=0.3621-0.2599=0.1022$. This distance represents the potential energy in this configuration.

In summary, in configuration A, G is above B_T by a distance exactly equal to $1/10$. In configuration B the distance is increased to approximately 0.1022. For a density ratio $\rho/\rho_w=0.8$, this shows that configuration A has lower potential energy than configuration B, so is more stable.

Let's now turn our attention to the case where $\rho/\rho_w=0.5$. We'll repeat the same analysis as above for this new density ratio. The fraction of the square section above the surface of the water is $1/2$. We first assume that situation A is more stable. In this configuration the projection above the surface of the water is $y=1/2$. The distance of the centre of gravity, G, below the highest point of the square section is $Y=1/2$. So the distance of G below the waterline is $Y-y=(1/2)-(1/2)=0$. As expected, the centre of gravity lies on the waterline.

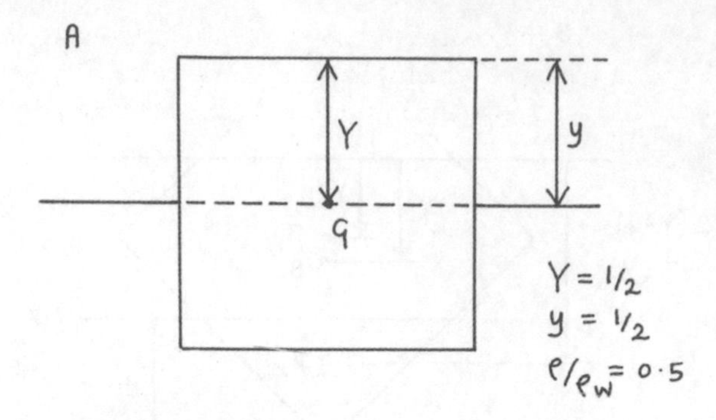

Now consider the position of the centre of buoyancy B_T. In this case, the calculation is simple. B_T lies at a distance $Y' = y/2 = 1/4$ below the waterline. So G lies above B_T by a distance $y/2$.

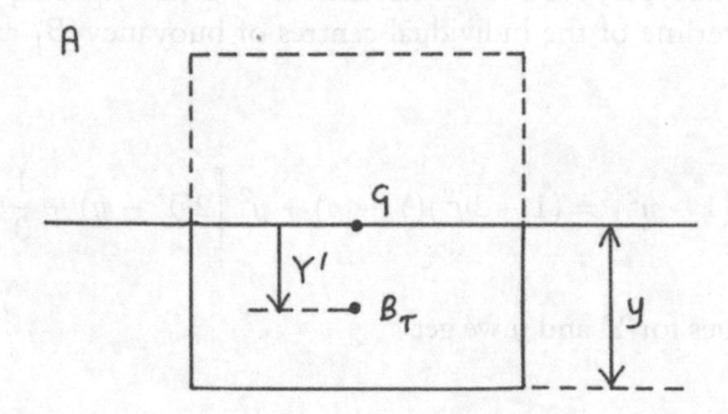

Now assume that situation B might be more stable. The fraction of the square section that is above the water is again $1/2$. The projection of the square section above the water is therefore $y = 1/\sqrt{2}$. The distance of the centre of gravity, G, below the highest point of the square is $Y = 1/\sqrt{2}$. So the distance of G below the waterline is $Y - y = (1/\sqrt{2}) - (1/\sqrt{2}) = 0$. Once again, of course, the centre of gravity lies on the waterline.

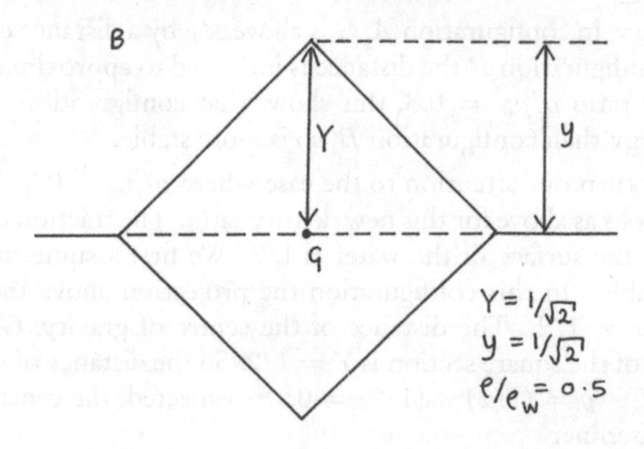

Consider the position of the centre of buoyancy B_T. By using the relation for the centroid of a triangle (which can easily be proven, if you wish), we discover that $Y' = y/3 = 1/(3\sqrt{2})$. In this case G lies above B_T by a distance $1/(3\sqrt{2})$.

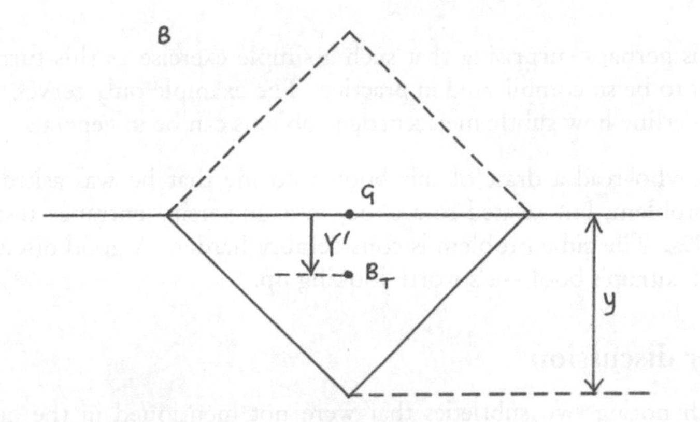

To summarise, in configuration A, G is above B_T by a distance exactly equal to $1/4$. In configuration B the distance is reduced to $1/(3\sqrt{2})$; approximately 0.2357. For a density ratio $\rho/\rho_w = 0.5$, this shows that configuration B has lower potential energy than configuration A and is therefore more stable. The result is different to that for $\rho/\rho_w = 0.8$, for which configuration A was more stable. It can easily be shown that for $\rho/\rho_w = 0.2$ the result is the same as for $\rho/\rho_w = 0.8$. We see that A is the most stable configuration.

The most stable configurations for the three density ratios are shown in the drawing below.

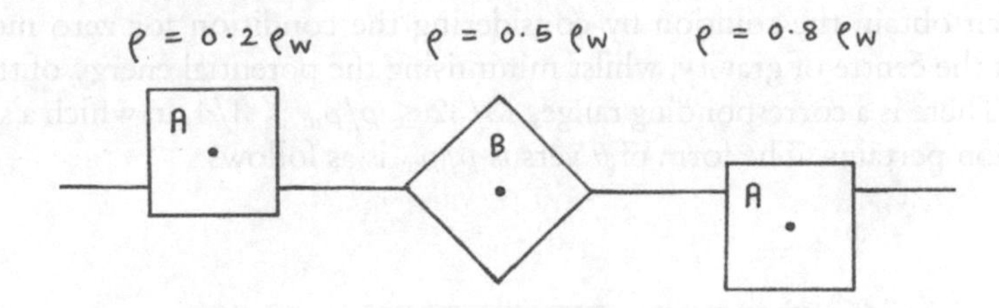

Are you surprised? I think you should be, unless your intuition is very, very good indeed. Let's return to Reid, who ends his paper with the words:

> This problem serves as a nice example to illustrate how easy it is for one's intuition to be wrong. If the reader will look through a number of physics texts, he will find that some good physicists have assumed that a block will always float with a side parallel to the fluid surface. The determination of the height to which the block will extend has been considered simple enough to be included in a list of problems at the end of a chapter.

And he's right. If you look at basic introductory texts, of the sort encountered in high school, and also undergraduate texts on continuum mechanics, you'll find examples of problems that neglect this energy minimisation effect. Delbourgo solved the same problem by considering metacentres, and comments as follows:

> It is perhaps surprising that such a simple exercise as this turns out to be so complicated in practice. The example only serves to underline how subtle metacentric problems can be in general.

A friend who read a draft of this book told me that he was asked a very similar problem, but related to a cube, in a university entrance test in the early 1990s. The cube problem is considerably harder. A good discussion is given in Lautrup's book—it's worth looking up.

Further discussion

It's worth noting two subtleties that were not mentioned in the question. In fact, to simplify the algebra, the question was posed a very specific way: "which situation, A or B...is therefore more likely?" Had we asked for a proof of the *most stable position* the analysis would have been more complex. You will probably be slightly appalled to learn that minimum solutions between 0° and 45°exist for density ratios in the range $1/4 \leq \rho/\rho_w \leq 9/32$. In this region the form of solution for θ, the angle away from the horizontal, is as follows.

$$\sin 2\theta = \frac{16\,(\rho/\rho_{\mathrm{w}})}{9 - 16\,(\rho/\rho_{\mathrm{w}})}$$

We can obtain the solution by considering the condition for zero moment about the centre of gravity, whilst minimising the potential energy of the system. There is a corresponding range, $23/32 \leq \rho/\rho_w \leq 3/4$, in which a similar solution pertains. The form of θ versus ρ/ρ_w is as follows.

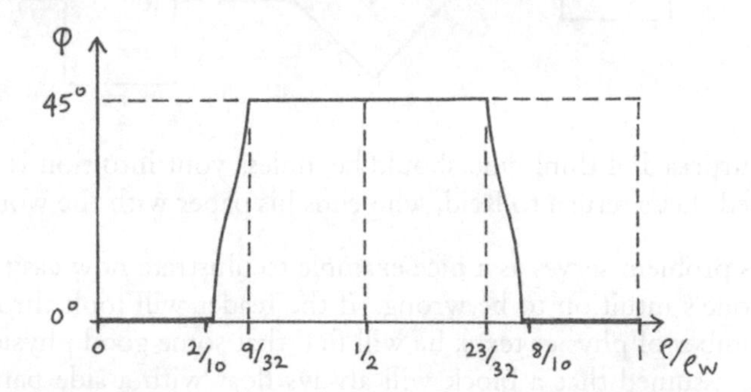

I wish to pose a final question to check that you really do have a good *physical* feel for what's going on. The question is posed in relation to the example given above for $\rho/\rho_w = 0.5$. In that example, the distance of G below the waterline was $Y - y = 0$ for both situations A and B. We argued that situation A was favoured over situation B because we minimised the potential energy $Y' - (Y - y)$. But why is that the best metric to use? We could have argued that because the surface of the water does not change height when the bar is introduced, we should minimise the potential energy of the bar (with respect to an arbitrary reference point). To minimise the energy of the bar we wish to maximise the value $Y - y$, the distance of G below the waterline. But for both situations A and B the value of this energy is zero with respect to the surface of the water. Neither situation is favoured. Following this logic we could come up with other examples with *different* results from the one we would get using the analysis employed earlier in this question. For $\rho/\rho_w = 0.2$, for example, the solution which minimises $Y' - (Y - y)$ has a *higher* potential energy for the bar. Try it. How do we reconcile these two arguments? The problem is that we cannot simply minimise the potential energy of the bar—we need to minimise the potential energy of the system that comprises water and bar.

Consider the following argument. The bar is initially at an arbitrary height h and will be lowered until it comes to rest in the water at the lowest energy condition for the *system*. We state that this condition is met when the centre of gravity of the bar is a distance Y'' below the water. The change in potential energy of the bar is $\Delta\text{PE}_1 = (-h - Y'')Mg$, where Mg is the weight of the bar. The change in potential energy is negative, indicating that the bar loses energy. What about the water? The weight of water displaced is the same as the weight of the bar. The water is displaced to the free surface. The mean change in height of the displaced water is therefore the distance from the centre of buoyancy, B_T, to the free surface, Y'. The gain in potential energy is $\Delta\text{PE}_2 = Y'Mg$.

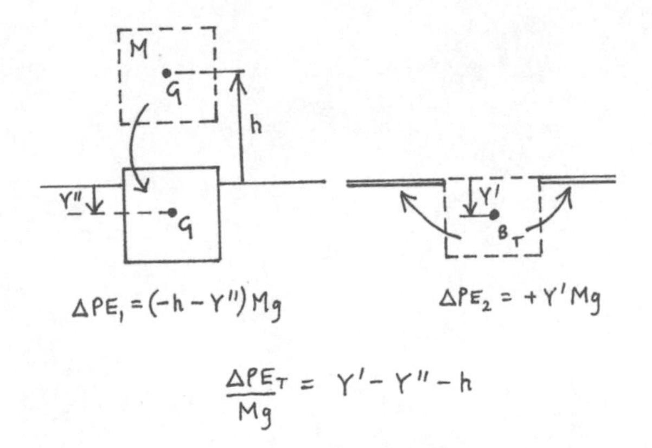

The total change in potential energy of the system, normalised by the weight, is therefore

$$\frac{\Delta\text{PE}_\text{T}}{Mg} = \frac{\Delta\text{PE}_1 + \Delta\text{PE}_2}{Mg} = Y' - h - Y''$$

The system is at the minimum energy point when we minimise this function. We can disregard the arbitrary starting height h, which is a constant. Thus we wish to minimise $Y' - Y''$. Stating this in words, we wish to minimise the distance between the centre of buoyancy and the centre of gravity. Our original constraint was correct.

Chapter 14

Estimation

I really enjoy questions on *estimation*, or back-of-the-envelope calculations. They are popular questions for a whole range of numerate disciplines, and are certainly very common elements in physics and engineering tests. Strangely, they are also very common in job interviews, particularly with companies that consider themselves to be "free thinking".

I remember several summer evenings when we were physics undergraduates, sitting in a friend's room, staring out over the cricket lawn, drinking coffee, and estimating things simply for the *fun* of it. The game had two stages. First, we would *guess* the answer to the problem at hand, going on instinct alone, and then compare our guesses. At this stage, to raise the stakes, you had to pretend to be very confident about your own guess and very sceptical about everyone else's. It didn't matter that we all made the least accurate guess about the same number of times. Then we would independently *estimate* the answer, usually without recourse to pen and paper. It could take a while, because we often had to devise strategies to do the calculation at hand. This was just before the internet changed everything. So if we needed some important but esoteric value we couldn't estimate easily, one of us would be sent to the library to consult a (printed) encyclopaedia, or to look at a set of data tables, and would reappear with the population of Bolivia, say, or the density of uranium, scribbled on a scrap of paper. Once we'd made our estimates we would compare them. It was often surprising how close the estimates were, even when we had taken diverging approaches to the solution.

We worked out that there were fewer blades of grass on the Merton College cricket ground than there were people in the world,[1] and that if we shared the books in our college library equally across the population of the United

[1] Merton College cricket ground is about 200 yards on a side, and we estimated about 50 blades of grass per square inch. This gave $50 \times 12^2 \times 3^2 \times 200^2 = 2.6 \times 10^9$ blades. Back in the mid-1990s this was roughly half the population of the world, which was 5.7 bn (it is now over 7 bn). I still remember gazing out over the brown field at the end of that summer and thinking what an *awful* lot of people that represented, and how insignificant we would each seem if we all stood there side by side.

Kingdom (something which we felt like doing more than once) we could offer each citizen a paltry third of a page.[2] We worked out how many college rooms it would take to store all the gold in the world—a question that appears in this book.

Wherever you look there is something to estimate, and it's useful to know how to set about it. Just recently someone told me that work had started on the Kingdom Tower in Saudi Arabia. If it's completed it will be over a kilometre high. The energy cost of pumping water up the tower, she speculated, will be terrific. I thought for a moment, and with a little mental arithmetic worked out the annual cost per person. When I told her the number she looked at me in surprise. "How did you know that?" she asked. When I told her I had just estimated it, but fairly accurately, she looked at me in disbelief. I had to go through the steps of the calculation to convince her that I really had just calculated it in my head.

Most everyday things can be estimated easily and relatively accurately. It's an essential skill for anyone who is serious about being a scientist. I shouldn't really admit to it here, but I'm always a little shocked when someone who has been through high-school science can't estimate—roughly—how big, long, heavy etc. some object of interest is. Often people simply shrug their shoulders and say, "I have no idea." But we should always have *some* idea.

I hope you will enjoy this selection of problems. They are based on the more unusual questions I have heard over the years, or are favourites of my own. Some of them require the odd value which you might not be able to estimate accurately enough to satisfy yourself. In which case you might need to look it up in a data table, or ask your tutor for it. Not knowing values is fine. It's the ability to come up with a *method* for solving a problem that's important.

14.1 Mile-high tower ★

Frank Lloyd Wright, the great American architect, was perhaps not the first to conceive of the notion of a mile-high building, but he was probably the first to work seriously on plans for one. His 1956 skyscraper "Illinois" would have been 528 stories high, making it four times taller than the Empire State Building. To get up and down it you'd have had to use one of 76 "atomic-powered" elevators (gulp!). The Illinois was never built, and it's unlikely that the technology to do so even existed at that time.

Almost sixty years later, the tallest structure ever constructed by man is the Burj Khalifa in Dubai, in the United Arab Emirates. Finished in 2010, it stands 829.8 m tall. Huge as it is, it is still only half the height of the magic mile.

---

[2] The library walls totalled a good 300 m, and were stacked six shelves high. A *ream* of 500 pages of paper is about 5 cm thick. This gives $(300 \times 6 \times 500)/0.05 = 18 \times 10^6$ pages, or a third of a page for each of the then 58 million inhabitants of the UK (the population is now over 63 million).

The Kingdom Tower, which is currently being built in Jeddah, Saudi Arabia, was originally designed to be one mile high. Plans were drawn up by the architect who designed the Burj Khalifa. But the geology of the area proved unsuitable, and the tower had to be scaled down. The current intended height is still a closely guarded secret, but it has been announced that it will be over 1 km, setting a new record for super-tall buildings.

In the architectural world there now seems to be little debate over the technical feasibility of the mile-high building—the question is purely one of finance. Given the sudden explosion in super-tall buildings in the last ten years, the near future will be very interesting.

I remarked in the introduction that someone I know recently speculated that the cost of pumping water to the top of a mile-high building would be fantastic. The purpose of this question is to calculate whether or not that is so.

ESTIMATE THE ANNUAL financial cost of the energy required per person to supply water to living accommodation at the very top of a mile-high building.

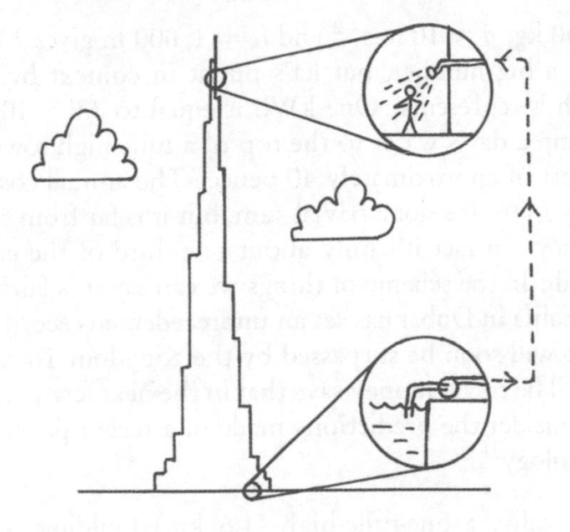

Answer

Of course, we have to assume that we're talking about the current economic and energy climate. The current cost of electricity in the UK is roughly 15 pence per kWh. It's reasonable to expect students to be able to estimate this to within, say, a factor of two. Energy is a key household expense, and one of the most significant drivers of the economy. Now let's calculate the annual energy required to pump water for a single person.

First we need to calculate the amount of water used by a person in a day. We can create a rough table to show this, and each of us might fill the table slightly differently. But we should certainly agree to within a factor of two or three, say. My rough estimates are as follows.

	Number per day	Litres	Total litres
Toilet flushes	5	10	50
Showers	2	25	50
Washing clothes	1/3	60	20
Washing dishes	1	20	20
Sundries	1	10	10
		Total	**150**

We estimate the total amount of water used at 150 l (or 150 kg) per day.

Now we can estimate the energy required to pump that water to the top of a mile-high tower. A mile is approximately 1.6 km, or 1,600 m. The energy cost is simply the gravitational potential energy gained, mgh, divided by the total efficiency of the pump. Let's take a very conservative electrical efficiency value of $\eta_E = 0.5$, and a very conservative mechanical efficiency value of $\eta_M = 0.5$. The total energy cost per day is

$$E = \frac{mgh}{\eta_E \eta_M}$$

Taking $m = 150$ kg, $g \approx 10$ m s^{-2} and $h = 1,600$ m gives $E = 9.6 \times 10^6$ J. This looks like a big number, but let's put it in context by calculating the number of kWh it represents. One kWh is equal to 3.6×10^6 J, so we need 2.7 kWh to pump a day's water to the top of a mile-high tower. That works out at a daily cost of approximately 40 pence. The annual cost of the energy is approximately £146. It's not a trivial sum, but it is far from being a fantastic amount of money. In fact it's only about one-third of the cost of a typical Oxford water bill. In the scheme of things we can say it is fairly insignificant.

The Burj Khalifa in Dubai has set an unprecedented record for height, and it's likely that it will soon be surpassed by the Kingdom Tower in Jeddah, at over 1 km high. The smart money says that in the next few years we will build even higher. Consider the predictions made in a recent paper in the Journal of Urban Technology:[3]

> Technologically, a one-mile-high (1.6 km) building is possible today—although not without the associated challenges of required high-strength materials, fire safety provisions, elevator systems, and, above all, constructability. As the building gets taller, the slenderness (height-to-width) ratio increases and the tower becomes dynamically extra sensitive to wind so it cannot be inhabited comfortably. To overcome this, the slenderness effect of the tower is offset by making it wide at the base, or connecting several towers with bridges at different levels so they will enforce and stiffen each other. Will it be possible someday to build a skyscraper even taller than a mile-high building? Probably yes. Since

[3] Al-Kodmany, K., 2011, "Tall buildings, design, and technology: visions for the twenty-first century city," Journal of Urban Technology, Vol. 18, No. 3, pp. 115–140.

> 1885 when the 10-storey Home Insurance Building was built in Chicago, we have come a long way to the 2,717ft (828m) tall Burj Khalifa in Dubai, 2010. Will building height increase at this rate in the future? We may never know and may not contemplate this at present but we cannot guarantee that this will not definitely happen. One thing we know is that an inborn human instinct is to overcome obstacles through creativity and dreaming. This raises fundamental philosophical and moral questions about the so-called cloud-piercing future visionary towers.

So the megastructures of the future may yet dwarf the tallest buildings of today. People have different views about ultra-tall buildings, but I find them remarkable and fascinating. I look forward to the mile-high buildings of the future.

14.2 How long do we have left? ★★

A good friend of mine used to test prospective chemistry students with this estimation question back in the 1970s and 1980s. In those days he was a professor of physical chemistry and was researching the respirative absorption of oxygen and other gases, as applied to anaesthetics and diving. He must have been doing something right because one of his students went on to win the Nobel Prize for chemistry.

The question relies on a basic knowledge of physiology, chemistry and physics, at lower than pre-university standard. Anyone interested in the world around us should be able to make at least a good *guesstimate*[4] of most of the quantities required. The question went as follows.

There are three of us in this room (whatever room you happen to be in). Assuming it to be perfectly sealed, how long will we live for?

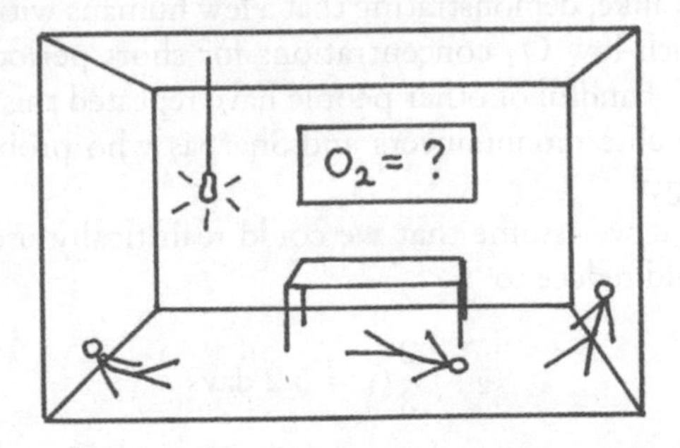

[4]Better than a guess, but worse than an estimate.

Answer

We need to start by estimating the size of the room (the particular room in which we are asked the question, that is). The volume of a normal office might be, say, $V = 5\text{ m} \times 4\text{ m} \times 3\text{ m} = 60\text{ m}^3$. Now for a bit of basic physiology. The *tidal volume* of a healthy adult is about 500 ml, or $v = 0.5 \times 10^{-3}\text{ m}^3$. It represents the volume of air we inhale and exhale in a typical breath. We should all have some experience of this from blowing up party balloons. We take a breath about once every four seconds, so the frequency is $f = 1/4$. There are $n = 3$ of us in the room. We can estimate the total volumetric rate, $\dot{Q}$, at which we are processing the air as

$$\dot{Q} = fvn = 3.75 \times 10^{-4}\text{ m}^3\text{s}^{-1}$$

Now for more physiology. Air is approximately 21% O_2 by volume (or, also, by *mole*). The air we breathe out under normal circumstances is approximately 16% O_2 and 5% CO_2. So with each breath we use up approximately 5/21 of the oxygen in that breath. Using this information we can work out how long it would take to use up all the oxygen in the room, at the nominal rate of O_2 consumption. The time taken is given by

$$t_1 = \frac{V}{\dot{Q}\,(5/21)} = 672,000\text{ s} \approx 7.8\text{ days}$$

Based on this estimate it looks like we are going to be fine for some time—certainly long enough for someone to raise the alarm. Unfortunately this does not represent an upper limit, because we are not able to use up 100% of the O_2 in the room.

At the summit of Everest (8,848 m above sea level), the amount of O_2 in the air is approximately one-third of that at sea level. For many years we thought it was impossible to survive at that altitude without supplementary O_2. For this reason, altitudes above 8,000 m are charmingly referred to as the *death zone.* In 1975, however, Reinhold Messner and Peter Habeler climbed Gasherbrum I (with an altitude of 8,080 m) without bottled oxygen, and went on to climb Everest in the same style in 1978. They astonished mountaineers and physiologists alike, demonstrating that a few humans with rare physiology can survive in such low O_2 concentrations for short periods of a few days. Since then a small handful of other people have repeated this feat. Those who have done so are elite mountaineers and Sherpas who probably have rather unique physiology.

In the room, if we assume that we could realistically use 2/3 of the O_2, our estimate would reduce to

$$t_2 = \frac{2}{3}t_1 = 5.2\text{ days}$$

We still have some time to ponder our fate. There is, however, one further physiological detail we've neglected to consider. CO_2 is, in fact, toxic. In small quantities it changes the blood pH and triggers increased breathing rates, but in large quantities it kills us. It is probably unreasonable to expect us to know what molar concentration is lethal—that, after all, is the job of doctors, not aspirant physicists, mathematicians and engineers, who have more important things to remember. We could guesstimate, however, that the toxic CO_2 level is similar to the level at which we exhale CO_2 under normal conditions: about 5%.[5] Our third estimate of the time we have remaining is

$$t_3 = \frac{V}{\dot{Q}} = 1.9 \text{ days}$$

14.3 Midas' storeroom ★

It is strange to think that the hugely significant part gold has played in human history, as an international currency for thousands of years, for example, is due primarily to its rarity. That is a shame, because if it didn't have such an entirely artificial price (rather than one based on some more fundamental value, such as its usefulness to industry, for example), it would be a very useful everyday engineering material. Among its interesting properties are extreme ductility, resistance to corrosion, high reflectivity and high electrical conductivity. Almost half the gold mined in human history is from the Witwatersrand basin in South Africa, where it is believed that 2 billion years ago a huge asteroid created what is known as the Vredefort crater, revealing even older rock that was rich in gold. It is thought that much of the gold that's in the Earth's crust today was, in fact, deposited by meteorites long after the formation of the Earth. Most of the gold that was present in the proto-Earth during its formation process (estimated at 4.5 billion years ago) is thought to have sunk to the Earth's core on account of its high density.

THE TOTAL AMOUNT of gold mined in recorded history is estimated at 170,000 tonnes. How many rooms the size of a typical office (or the size of the room you're sitting in) would you need to store that amount of gold?

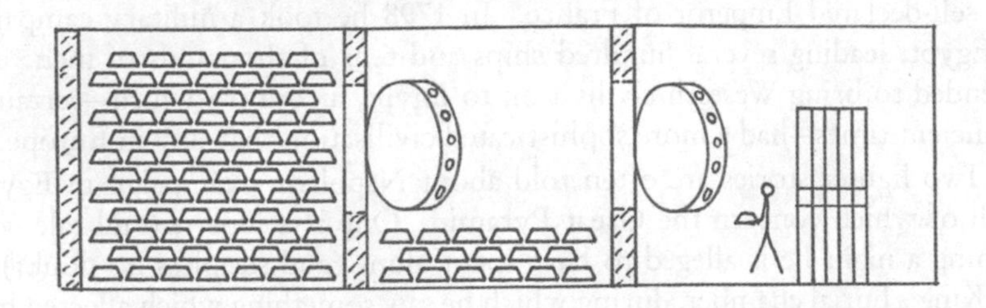

[5]The lethal level of CO_2 is between 7% and 10%, at which concentration it causes suffocation, typically leading to loss of consciousness within a few minutes. At concentrations anywhere above 1% it can start to cause headaches and drowsiness.

Answer

Previously we estimated the volume of a normal office to be approximately $V = 5\text{ m} \times 4\text{ m} \times 3\text{ m} = 60\text{ m}^3$. Anyone with a passing knowledge of the interesting elements in the periodic table will know that gold is very dense indeed. In round numbers, $\rho = 19 \times 10^3\text{ kg m}^{-3}$. To store all the gold ever mined would require a storage volume equal to

$$V_G = \frac{170 \times 10^6}{19 \times 10^3} = 8,950\text{ m}^3$$

How many offices is that? Using our educated guesses it is about

$$\frac{V_G}{V} = \frac{8950}{60} = 149\text{ offices}$$

Given how influential gold has been in human history, and how much effort has been expended in mining it, it's perhaps strange to think that to store all of it we'd only need about one office in each country in the world.

14.4 Napoleon Bonaparte and the Great Pyramid ★

According to standard chronology, the Great Pyramid of Giza was built in approximately 2560 BC to celebrate the Egyptian Pharaoh Khufu. It stands on the outskirts of modern day Cairo, just west of the river Nile, along which, it is believed, much of the stone used to build it was transported. The pyramid is really quite huge, standing almost 147 m high, with a square base of side 230 m. Amazingly, perhaps, for almost 3,900 years it was the tallest man-made structure in the world. It was eventually overtaken by the spire on the rather more humble Lincoln Cathedral in England, in 1311.

Over the years many great leaders have visited the Egyptian pyramids, among them Napoleon Bonaparte, then a French military leader, but later the self-declared Emperor of France. In 1798 he took a military campaign to Egypt, leading several hundred ships and tens of thousands of men. He intended to bring Western civilisation to Egypt, a country which—certainly in ancient times—had a more sophisticated civilisation than any in Europe.

Two lighter stories are often told about Napoleon's campaign in Egypt, both of which concern the Great Pyramid. One is a supernatural tale concerning a night he is alleged to have spent alone (and sleepless no doubt) in the King's burial chamber, during which he saw something which affected him deeply, but which remained a secret even to his deathbed. This tale is thought to be apocryphal. The second story is one of engineering and mathematics, and more historical writers seem to accept it. One nice account of it appears in Paul Strathern's Napoleon in Egypt.[6] On visiting Giza, Napoleon is said to have entertained himself by suggesting that his staff, who included a fifty-three year-old mathematician, race to the top of the Great Pyramid. When they returned exhausted, he apparently had lost interest in the race, and instead pronounced that the quantity of stone in the pyramid was sufficient to build a wall "three metres high and one metre wide around France". The mathematician supposedly concurred with Napoleon, though it would have taken a brave man to disagree.

TAKING THE DIMENSIONS of the Great Pyramid of Giza to be a square base of side 230 m and a height of 147 m, estimate whether the pyramid contains enough stone to build a 3 m high by 1 m wide wall around France. Were Napoleon Bonaparte and his mathematician, and Strathern and his mathematicians, correct?

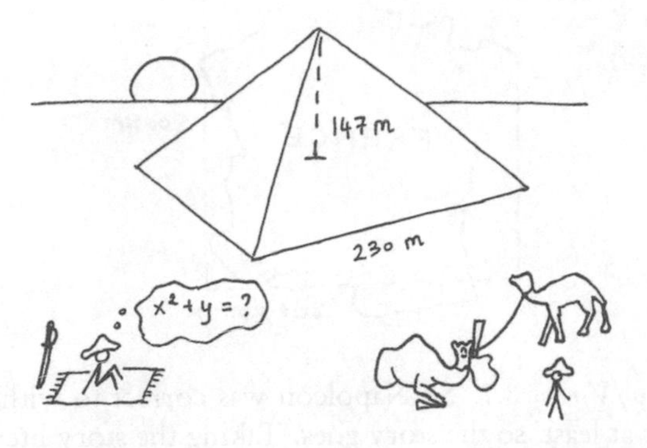

Answer

Let's first calculate the volume of stone in the Great Pyramid of Giza. For a regular pyramid of base side b and height h we can calculate the volume V_P by integration. The area of any horizontal section at height y is given by

$$\left(\frac{b(h-y)}{h}\right)^2$$

Integrating from 0 to h we have

$$V_P = \int_0^h \left(\frac{b(h-y)}{h}\right)^2 dy = \frac{b^2 h}{3}$$

This is the equation for the volume of a pyramid. Many people will remember it, but it is safer to be able to derive such formulae. For $b = 230$ m and

[6]Strathern, P., 2007, "Napoleon in Egypt: the greatest glory," Jonathan Cape Ltd., ISBN-10: 0224076817.

$h = 147$ m, we have $V_P = 2.6 \times 10^6$ m^3. This is really quite an accurate estimate because the dimensions of the Great Pyramid are easily determined.

Let us now estimate the perimeter of France. We might estimate France to be about 800 km across. The entire perimeter (which is what the story suggests Napoleon was working with) would therefore be about 3,200 km long. Here we ignore what is called variously the *coastline paradox* or the *Richardson Effect*. This is the now well-known explanation for reconciling different reported lengths of the same coastline, where measurement is complicated by the fractal geometry of the shape.[7] We note for now that 3,200 km is the smallest number we could take to be the length of France's border (approximating the country to be square).

What volume of stone V_W would be required to build a 3 m high by 1 m wide wall around France, according to Napoleon Bonaparte's calculations? The volume is trivially calculated to be $V_W = 9.6 \times 10^6$ m^3.

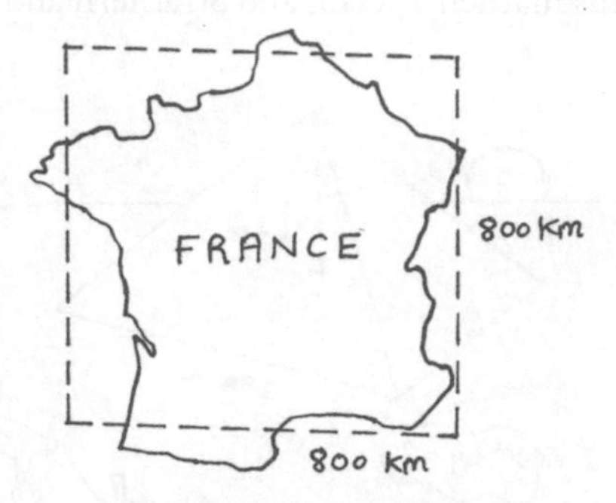

We see that $V_W/V_P \approx 3.7$. So Napoleon was correct to within an order of magnitude. Or at least, so the story goes. Taking the story literally, however, we see that Napoleon and his mathematician could have done a little better. A wall 0.5 m wide and 1.5 m high would be closer to the correct dimensions. It's rather odd to think that this story has been repeated so many times, and with so much speculation about whether the numbers are correct, when it's so easy to check. I think Napoleon should have hired a better mathematician.

14.5 Lawnchair Larry ★★

On 2 July 1982, Larry Walters, a truck driver and ex-army cook who had served in Vietnam, took flight from the back yard of his mother's house in a suburb of California. He rose gracefully above the trees in a home-made airship. Since he had been a boy, it had been his dream to fly. The craft, which

[7] The length of a fractal curve (a curve with scale-independent, self-similarity features) depends on the length scale at which it is being viewed. If we take account of this effect, we argue that France's perimeter is not well-defined. Instead, as we look at the border features with a closer and closer scale, the perimeter increases indefinitely. This effect is discussed in more detail in another question.

he dubbed *Inspiration I*, consisted of a Sears, Roebuck[8] lawn chair and four clusters of helium-filled weather balloons. It stood over 150 ft tall, arousing interest from the local police. He dismissed them on the pretence that he was preparing for an advertising stunt the following day. The plan had been to fly over the San Gabriel Mountains and into the Mojave Desert. His packing list for the maiden voyage included: a two-way radio; an altimeter; a compass; a flash-light; eight plastic bottles filled with water as ballast; beef jerky; a map of California; a camera; a bottle of Coca-Cola; and an air-gun, to allow him to descend by popping balloons.

The journey did not go quite as planned, and Larry rose rapidly to 15,000 ft, drifting into controlled airspace on one of the approach paths to Los Angeles International Airport. He was spotted by Delta Airlines and Trans World Airlines pilots, who radioed the peculiar sighting to the tower. In an interview with the New Yorker's George Plimpton,[9] Larry describes trying to lose altitude by shooting balloons with an air pistol. This worked perfectly until he dropped the gun over a residential area, watching in horror as it fell ten thousand feet to the ground.

Without his gun, Larry continued to gain altitude fast, and used his radio to make a distress call. Amazingly, the voice recording survives today, and can easily be found on the internet.[10] The operator asks again and again for the airport he took off from. The conversation goes like this:

> OPERATOR: "What airport did you take off from?" (Repeatedly. There is then a long pause.)
>
> LARRY: "My point of departure was 1633 West Seventh Street, San Pedro." (Larry's mother's home address.)
>
> OPERATOR: "Say again the name of the airport. Could you please repeat?"

The flight lasted only 45 minutes, and ended in the back garden of 432 45th Street in Long Beach. His arrival was announced by a blackout, when his balloons fell over power lines. It is one of the more peculiar chapters in the history of flight. Having survived the trip, in a New York Times interview the following day (3 July 1982), Larry said, "since I was 13 years old, I've dreamed of going up into the clear blue sky in a weather balloon. By the grace of God, I fulfilled my dream. But I wouldn't do this again for anything."

As with every bizarre act, there will be imitators. There have been many over the years. One of the latest, and perhaps most bizarre, was on 20 April 2008, when Adelir Antonio de Carli—a Brazilian Roman Catholic priest—set off from Paranaguá, Brazil, attached to no less than 1,000 party balloons. He

---

[8] A US department store.

[9] Plimpton, G., 1998, "The man in the flying lawn chair: why did Larry Walters decide to soar to the heavens in a piece of outdoor furniture?" The New Yorker, American Chronicles, 1 June, 1998.

[10] There is even a video of part of his flight, which again is easily found on the internet. It's worth looking up.

was caught in a storm, and his last radio transmission was from above the Atlantic Ocean. Three months later his body was recovered by the Brazilian Navy.

GIVE AN ACCURATE estimate of the minimum radius of a single large spherical helium balloon that would lift a man weighing 80 kg. Take the weight of the fabric into account.

Answer

Consider first the forces on the helium balloon. There are four forces.

- The weight of the man, which acts downwards: $F_1 = m_1 g$, where $m_1 =$ 80 kg.
- The weight of the skin of the helium balloon, $F_2 = m_2 g$, where m_2 is the mass of the skin. This acts in a distributed way over the entire surface of the sphere, but we represent it as acting from the centre of the sphere. The mass of the skin is equal to the surface area of the sphere, $4\pi r^2$, multiplied by the mass per unit *area* of the balloon material $\rho_{balloon}$. We write $F_2 = \rho_{balloon} 4\pi r^2 g$. We will assume we have balloon material available with a mass per unit area of 100 g m^{-2}.[11]
- The weight of the helium contained within the balloon. The mass of helium in the balloon is the density, ρ_{He}, multiplied by the volume, $(4/3)\,\pi r^3$. Thus the weight, which acts downwards, is given by $F_3 = \rho_{He}\,(4/3)\,\pi r^3 g$.
- The buoyancy force, F_B, which is equal to the weight of the displaced air, and acts upwards on the balloon. The force arises because of the pressure gradient in the atmosphere between the top and bottom of the balloon. We represent it as a single force. The magnitude of the force is $F_B = \rho_{air}\,(4/3)\,\pi r^3 g$.

[11] A friend suggested that it would be difficult to estimate the mass per unit area of a hypothetically suitable material to within even an order of magnitude. I disagree. If we are imaginative we could come up with any number of ways of making at least a *rough* guess. We have all handled bin bags, for example. We can estimate the weight of a roll of bin bags, the total surface area of the unrolled bags, and therefore the mass per unit area of this material. We probably need a material a bit thicker than that used for bin bags, though.

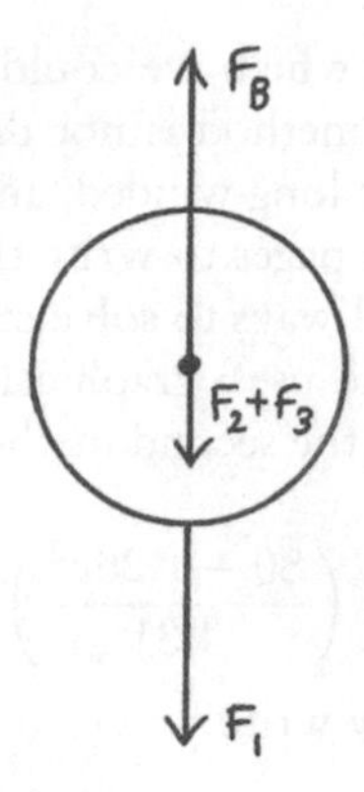

Summarising,

$$F_1 + F_2 + F_3 = F_B$$

Substituting using the equations for F_1, F_2, F_3 and F_B, we get

$$m_1 + \rho_{balloon} 4\pi r^2 + \rho_{He} \frac{4}{3}\pi r^3 = \rho_{air} \frac{4}{3}\pi r^3$$

We first reduce the equation to numbers by substituting known values into the equation. We are given $m_1 = 80$ kg. We estimated $\rho_{balloon} = 0.1$ kg m^{-2}. We should know that at STP[12] the density of air is approximately $\rho_{air} = 1.2$ kg m^{-3}. The relative molecular mass[13] of air is approximately 29, and the molecular mass of helium is 4. At atmospheric pressure the densities of many gases (where we assume the gases are *perfect*) scale approximately with the molecular weight.[14] Taking this assumption,[15] we get $\rho_{He} = (4/29)\,1.2 \approx 0.17$ kg m^{-3}. Substituting into the previous equation, we get

$$80 + 1.26r^2 \approx 4.31r^3$$

[12]Standard temperature and pressure (STP) conditions are agreed standard conditions at which comparisons can be made between the measurements of chemical properties. The International Union of Pure and Applied Chemistry (IUPAC) and the National Institute of Standards and Technology (NIST) have defined a standard temperature of 273.15 K (0 °C) and a standard absolute pressure of 100 kPa (0.987 atm, or 1 bar).

[13]The *relative molecular mass* is the mass of a molecule relative to a carbon 12 molecule, ^{12}C, which is used as a reference mass and given the value 12.

[14]Taking the *ideal gas law*, $pV = nRT$, and noting that $n = m/M$, we write $pV = (m/M)\,RT$. As $\rho = m/V$ we can write $p = \rho RT/M$. For two gases with the same p and T, we see that $\rho_1/\rho_2 = M_1/M_2$.

[15]This assumes that the balloon is made of a non-elastic material, in which case the pressure inside the balloon is approximately the same as that outside the balloon. This is a good approximation, at least as far as the calculation of density is concerned. Clearly for an elastic balloon with significant skin tension, such an assumption would not be reasonable. Large, high-altitude weather balloons are, in fact, often made of a non-elastic material, and are only partially filled. This is so that the volume of the contained gas can increase as it rises into surrounding air at lower pressure.

This is a *cubic* equation in r, which we could solve using the general solution method for cubics. The method is not dissimilar to that for quadratic equations, but it is much more long-winded, and involves the use of complex notation. It could take several pages to write the solution out in full. There are two quicker, more practical ways to solve cubics (other than using a computer, of course). The first is to use a graphical method, and the second is to solve by iteration. We will use the second method. Rearranging, we get

$$r = \left(\frac{80 + 1.26r^2}{4.31} \right)^{1/3}$$

To solve by iteration we simply write

$$r' = \left(\frac{80 + 1.26r^2}{4.31} \right)^{1/3}$$

We then guess a value r, and return the value of r'. We compare the values r' and r, and iterate the input r until these values agree.

Guess of r (m)	Returned r' (m)	Error (%)
2.00	2.70	35
2.50	2.73	9.2
2.70	2.75	1.9
2.75	2.75	0.1

In general, a cubic equation with real coefficients has three roots,[16] at least one of which is a real number. We expect one real root in this case, because we expect only one size r that satisfies our problem. We see from our iteration that a good estimate of the minimum radius of the required balloon is $r =$ 2.75 m. We should cross-check the answer, however. To do this we calculate the values of the forces we derived equations for. We get $F_2 = 93$ N, $F_3 =$ 145 N, and $F_B = 1,026$ N. We know that $F_1 = 785$ N. It seems that the equality $F_B = F_1 + F_2 + F_3$ holds to the degree of accuracy we are working to.

To see how good an estimate this is, I looked on the internet at the guidelines for meteorological balloons. A balloon of radius 2.75 m corresponds almost exactly to a diameter of 18 ft. According to *lift tables* produced by the University of Hawaii, a typical 18 ft diameter helium balloon has a measured lift capacity of 88.7 kg.[17] It seems we are very much in the right territory, but our estimate of the density of the balloon fabric (mass per unit area) was perhaps a little higher than that used in practice.

[16]The general cubic equation $ax^3 + bx^2 + cx + d = 0$, with $a \neq 0$, and with real coefficients a, b and c, has three roots (or solutions). It is possible to have three real distinct roots, three real roots of which two are the same, or one real and two complex roots. The methods of solving for the roots, and geometrically interpreting them, are interesting. They are no longer taught in most Further Mathematics courses, but you should look them up if you feel so inclined.

[17]The lift tables also include hydrogen and methane balloons, which have lift capacities of 96.4 kg and 42.3 kg respectively for the same size of balloon.

14.6 Do we get lighter by breathing? ★★

I have always thought that this would be an interesting question to ask, but it's more suited to subjects that involve some elements of higher-level biology. Breathing, however, is fairly important, and the basic chemistry is covered in almost all GCSE general science courses. So really it's just a matter of estimation. The question is as follows.

ESTIMATE THE RATE at which we get lighter by breathing alone.

Answer

First we need to consider the mechanisms that might cause us to get lighter. Under normal conditions we breathe in air and process some of the oxygen to CO_2. The net result is that some carbon leaves our body. The air we breathe out is also saturated with water vapour. So there are two mechanisms by which we lose mass. Now we need to estimate how fast this happens.

In a previous question we estimated the *tidal volume* (displaced volume of a typical adult breath) at about 500 ml, or $v = 0.5 \times 10^{-3}\text{m}^3$. We also estimated that we take a breath about once every four seconds, so the frequency is $f = 1/4$. We should know the density of air at room temperature and pressure. It is approximately 1.2 kg m^{-3}. The mass flow rate of air into our lungs is approximately

$$\dot{m}_{in} = \rho f v = 1.5 \times 10^{-4} \text{ kg s}^{-1}$$

Now for some chemistry. Air is approximately 21% O_2 by volume, with N_2 taking up the other 79%.[18] The air we breathe out is approximately 16% O_2 and 5% CO_2 by volume. The molecular weights are as follows: $O_2(32)$, $N_2(28)$ and $CO_2(44)$. Using this we see that the ratio of the mass we breathe out to the mass we breathe in is

$$\frac{\dot{m}_{out}}{\dot{m}_{in}} = \frac{(79 \times 28) + (16 \times 32) + (5 \times 44)}{(79 \times 28) + (21 \times 32)} = 1.021$$

We can now estimate the rate at which we lose carbon by breathing, $\dot{m}_C$, which is the difference between the mass flow rate in and out of our lungs.

$$\dot{m}_C = \dot{m}_{out} - \dot{m}_{in} = 3.2 \times 10^{-6} \text{ kg s}^{-1}$$

Now we should consider the rate at which we lose water vapour through breathing. We might, and I stress *might*, be able to make a very crude estimate of the density of water vapour in the air we breathe out. We have, after all, seen it condense on the inside of car windows in winter, or in party balloons on cold days. More likely, we would probably need to ask for help,

[18] We neglect the very small quantities of the inert gases, and CO_2, which together make up less than 1% by volume.

and a more accurate number. It's the kind of thing we can look up in reference tables. The density of water vapour in saturated air at body temperature (37°C) is very approximately $\rho_{H_2O} = 44 \text{ g m}^{-3}$. The net rate at which we exhale water vapour—taking the worst case in which we inhale completely dry air—is approximately

$$\dot{m}_{H_2O} = f v \rho_{H_2O} = 5.5 \times 10^{-6} \text{ kg s}^{-1}$$

The rate at which we lose carbon and water vapour is approximately the same. The total rate of mass loss is $\dot{m}_T = \dot{m}_C + \dot{m}_{H_2O} = 8.7 \times 10^{-6} \text{ kg s}^{-1}$. How much would this be in a 24-hour period?

$$\Delta m = \dot{m}_T \times 24 \times 60 \times 60 = 0.75 \text{ kg}$$

It's quite a significant number. According to this estimate, we exhale about 0.27 kg of carbon per day, and up to 0.48 kg of water vapour. So by breathing alone we get substantially lighter.

I had always assumed that one of the main ways the human body loses carbon is by breathing. That is, the primary way we lose non-water body mass is by breathing. An extension to the argument is that anything that enhances our normal respiratory rate will cause us to lose more weight—exercise, for example. If you go mountain climbing for two weeks, at high levels of exertion for over twelve hours a day, you lose a significant amount of weight. I imagine the CO_2 floating away and making us lighter, one carbon atom at a time. I've put this theory to more than one medical doctor over the years, and I've always been met with blank looks. I have yet to hear a convincing argument to the contrary, however. But in researching this question, I've discovered that I'm in reasonable company. In a recent paper entitled The Human Carbon Budget,[19] the authors write as follows:

> The average adult in the US contains about 21 kg C and consumes about 67 kg C year^{-1} which is balanced by the annual release of about 59 kg C as expired CO_2, 7 kg C as faeces and urine, and less than 1 kg C as flatus, sweat, and aromatic compounds.

I find this fascinating. We lose 88% of the carbon we take in by breathing! Our own estimate was a carbon loss of 0.27 kg per day due to breathing alone, or just over 98 kg per year. As estimates go, it's very similar to the number reached by the authors of the study I quote. If you want to diet, simply breathe faster.[20]

---

[19] West, T.O., Marland, G., Singh, N., Bhaduri, B.L., Roddy, A.B., 2009, "The human carbon budget: an estimate of the spatial distribution of metabolic carbon consumption and release in the United States," Journal of Biogeochemistry, Vol. 94, pp. 29–41, DOI 10.1007/s10533-009-9306-z

[20] By exercising, of course!

The Deadly Game of Puzzle Points

"Rébuffat points" must have been responsible for the deaths of quite a number of young mountaineers. In his 1973 book The Mont Blanc Massif: The Hundred Finest Routes, Gaston Rébuffat—French Alpinist extraordinaire—made the fatal error of organising his climbs roughly in order of difficulty, from route one to one hundred. Mountaineers are no less competitive creatures than mathematicians, and the deadly game of collecting Rébuffat points was quickly established. You got one point for a gentle traverse of the Clocher Clochetons on the Aiguilles Rouges, the first route in his book, and 100 points for dicing with death on the unrelentingly vertical walls of the Central Pillar of Freney, the last route in the book. If you got back alive, you could casually drop into conversation that you had "had a good season". If you were lucky, someone would ask how many Rébuffat points you had collected, and you could brag that you had collected x points in that particular year, giving you y points in total. It was difficult to know whether to go for broke on a few hard routes or tick off many easy ones.

So it is with Puzzle Points.

"The four-star puzzles are just so much harder than the one-star puzzles," I said to my friend Tet. "It would seem unfair to give just four times as many points."

"Then square the number of stars before adding," said Tet.[1]

And so the Deadly Game of Puzzle Point collecting was born.

The distribution of questions in this book (including optional questions for which solutions are given) is: 35 ★, 40 ★★, 28 ★★★, and 10 ★★★★. That makes 113 questions in total, and a potential 607 Puzzle Points if we sum the squares of the number of stars.[2]

"What's the point of points if you can't brag about them?" said Tet.

And so the Puzzle Hall of Fame was born. By keeping track of your

[1]In fact, a similar method of scoring was used for many years in Oxford maths exams. The raw marks for each question were squared before being added. Like any other arbitrary method of marking exams, it favours a certain kind of approach to the exam. In this case, it favours deep knowledge of particular topics rather than superficial knowledge of many topics.

[2]You calculate your points like this: $(35 \times 1^2) + (40 \times 2^2) + (28 \times 3^2) + (10 \times 4^2) = 607$.

progress with the ticklists and scoresheet on the pages that follow, you can calculate your Puzzle Points and your honorary title.

- < 35 Puzzle Points: Aspirant Puzzlist
- $35 \leq$ Puzzle Points < 195: Novice Puzzlist
- $195 \leq$ Puzzle Points < 447: Expert Puzzlist
- $447 \leq$ Puzzle Points < 607: Puzzle Master
- 607 Puzzle Points: Puzzle Overlord

You can enter and update your points in the Hall of Fame by going to the website created for this book:

www.PerplexingProblems.com

Here you can debate the questions, make the case for alternative or more elegant solutions, and propose questions of your own. You might think it strange, but I have avoided "clever" solutions in favour of more standard methods throughout this book. Of course, I appreciate clever solutions as much as the next person, but I thought it would be more instructive to focus on the basic techniques, especially as we're dealing with such unusual questions. There is ample scope, therefore, to "improve" upon many of my solutions in terms of simplicity and compactness. I look forward to seeing these "clever" solutions on the website.

Like all adventures, these puzzles are much more fun when they are shared. Hopefully you already have someone to share your puzzling with, but the Puzzle Forum on the website allows you to share your ideas with an even wider audience, and—more importantly—to learn from others. From time to time I will also post new questions on the website.

Finally, good luck. And I hope you enjoy the adventure, wherever it takes you.

★ Puzzle Ticklist

		No Hints	Hints
1.1	Shortest walk	○	○
1.2	Intercontinental telephone cable	○	○
1.6	Cube within sphere	○	○
1.7	Polygon inscribed within circle	○	○
1.9	Triangle inscribed within semicircle	○	○
1.12	Captain Fistfulls' treasure	○	○
1.16	An easyish fencing problem	○	○
2.1	Human calculator	○	○
2.2	Professor Fuddlethumbs' reports	○	○
2.3	More of Professor Fuddlethumbs' reports	○	○
2.11	The three envelope problem	○	○
3.1	Sewage worker's conundrum	○	○
4.1	Pulleys	○	○
4.5	Water-powered funicular	○	○
5.1	Friction at the superbike races	○	○
6.3	Dr Springlove's Oscillator	○	○
7.1	Stevin's clootcrans	○	○
7.5	Boyle's perpetual vase	○	○
8.2	The Unflinching Aviator	○	○
8.3	Target shooting	○	○
9.2	Resistor tetrahedron	○	○
9.5	Power transmission	○	○
9.6	RMS power	○	○
9.7	Boiling time	○	○
10.1	The hollow moon	○	○
11.1	Mote in a sphere	○	○
11.4	The Martian and the caveman	○	○
12.1	The heated plate	○	○
12.2	The heated cube	○	○
13.1	Archimedes' crown and Galileo's balance	○	○
13.3	Balanced scales	○	○
13.7	A quantitative piston puzzle	○	○
14.1	Mile-high tower	○	○
14.3	Midas' storeroom	○	○
14.4	Napoleon Bonaparte and the Great Pyramid	○	○

★★ Puzzle Ticklist

		No Hints	Hints
1.3	Chessboard and hoop	○	○
1.4	Hexagonal tiles and hoop	○	○
1.5	Intersecting circles	○	○
1.10	Big and small tree trunks	○	○
1.11	Professor Fuddlethumbs' stamp	○	○
1.13	Captain Fistfulls' treasure II	○	○
2.4	Ant on a cube I	○	○
2.7	A falling raindrop	○	○
2.8	The Three Door Problem	○	○
2.12	A card game	○	○
3.2	Sewage worker's escape	○	○
3.4	Aztec stone movers	○	○
3.6	The Wheel Wars II	○	○
3.7	Obelisk raiser	○	○
4.2	Dr Lightspeed's elastotennis match	○	○
4.4	The last flight of Monsieur Canard	○	○
5.2	Pole position at the superbike races	○	○
5.4	Derailed roller coaster	○	○
5.5	The last ride of Professor Lazy	○	○
5.7	Wall of Death: motorcycle	○	○
6.1	Oscillating sphere	○	○
6.4	Dr Springlove's Infernal Oscillator	○	○
7.2	Power-producing speed humps	○	○
7.3	The overbalanced wheel	○	○
7.4	Professor Sinclair's syphon	○	○
8.1	Professor Lazy	○	○
9.1	Resistor pyramid	○	○
10.2	Lowest-energy circular orbit	○	○
10.3	Weightless in space	○	○
10.6	Newton's cannonball	○	○
10.7	De la Terre à la Lune	○	○
10.9	Jet aircraft diet	○	○
12.3	Fridge in a room	○	○
12.5	The cold end of the Earth	○	○
13.2	Another Galileo's balance puzzle	○	○
13.4	The floating ball and the sinking ball	○	○
13.6	The hydrostatic paradox	○	○
14.2	How long do we have left?	○	○
14.5	Lawnchair Larry	○	○
14.6	Do we get lighter by breathing?	○	○

★★★ Puzzle Ticklist

		No Hints	Hints
1.8	Circle inscribed within polygon	○	○
1.14	Captain Fistfulls' treasure III	○	○
1.17	A hardish fencing problem	○	○
2.5	Ant on a cube II	○	○
2.9	Dr Bletchley's PIN	○	○
2.10	Mr Smith's coins	○	○
3.3	Sewage worker's resolution	○	○
3.5	The Wheel Wars I	○	○
3.9	The Ravine of (Not Quite) Certain Death	○	○
4.3	Accelerating matchbox	○	○
4.6	Sherlock Holmes and the Bella Fiore emerald	○	○
4.7	Equivalent statements for linear collisions	○	○
5.3	Roller coaster	○	○
5.6	Wall of Death: car	○	○
6.2	Professor Stopclock's time-manipulator	○	○
6.5	Dr Springlove's Improved Infernal Oscillator	○	○
7.6	The curious wheel	○	○
9.4	Resistor cube	○	○
10.4	Jump into space	○	○
10.5	Space graveyard	○	○
10.8	Professor Plumb's Astrolabe-Plumb	○	○
10.10	Escape velocity from the Solar System	○	○
10.11	Mr Megalopolis' expanding Moon	○	○
11.2	Diminishing rings of light	○	○
11.3	Floating pigs	○	○
11.5	Strange fish	○	○
12.4	Ice in the desert	○	○
13.5	Floating cylinders	○	○

★★★★ Puzzle Ticklist

		No Hints	Hints
1.10	Big and small tree trunks (★★★★ option)	○	○
1.15	The geometry of Koch Island	○	○
2.6	Ant on a cube III	○	○
3.8	Obelisk razer	○	○
5.7	Wall of Death: motorcycle (★★★★ option)	○	○
9.3	Resistor square	○	○
10.12	Asteroid games	○	○
11.4	The Martian and the caveman (★★★★ option)	○	○
11.5	Strange fish (★★★★ option)	○	○
13.8	The floating bar	○	○

Scoresheet

Date	Number of puzzles solved				Puzzle Points
	★	★★	★★★	★★★★	

Endnote

My good friend Tet Amaya, a physicist turned mathematician with a brain the size of a planet, was kind enough to independently attempt every single one of these questions. He did this in a matter of just a few days. It seemed appropriate to include a couple of annotations from the manuscript he gave back. They were designed to be helpful indicators of where more explanation might be required.

So unclear!!
Students will
cry!!

Or where he thought I had got the answer to my own questions wrong.

All wrong!!
Totally wrong...

If you found some of these questions hard, you can now see you're not the only one. If any of the answers are still lacking, or if you have a genuinely unusual question you think might be interesting to include in a later edition, please do get in touch.